Handbook of Marine Mammals

Volume 3 The Sirenians and Baleen Whales

Handbook of Marine Mammals

Volume 3 The Sirenians and Baleen Whales

Edited by

SAM H. RIDGWAY
*Naval Ocean Systems Center
San Diego, California, USA*

and

SIR RICHARD HARRISON, F.R.S.
*Downing College
University of Cambridge, UK*

1985

ACADEMIC PRESS
(Harcourt Brace Jovanovich, Publishers)
London   Orlando   San Diego   New York
Toronto   Montreal   Sydney   Tokyo

COPYRIGHT © 1985 BY ACADEMIC PRESS INC. (LONDON) LTD.
ALL RIGHTS RESERVED.
NO PART OF THIS PUBLICATION MAY BE REPRODUCED OR
TRANSMITTED IN ANY FORM OR BY ANY MEANS, ELECTRONIC
OR MECHANICAL, INCLUDING PHOTOCOPY, RECORDING, OR
ANY INFORMATION STORAGE AND RETRIEVAL SYSTEM, WITHOUT
PERMISSION IN WRITING FROM THE PUBLISHER.

ACADEMIC PRESS INC. (LONDON) LTD.
24-28 Oval Road
LONDON NW1 7DX

United States Edition published by
ACADEMIC PRESS, INC.
Orlando, Florida 32887

BRITISH LIBRARY CATALOGUING IN PUBLICATION DATA
Handbook of marine mammals.
    Vol. 3 : The Sirenians and Baleen whales
    1. Marine mammals
    I. Ridgway, Sam H.   II. Harrison, R.J. (Richard
John)
    599.5   QL713.2

LIBRARY OF CONGRESS CATALOGING IN PUBLICATION DATA
Main entry under title:

Handbook of marine mammals.

    Includes bibliographies and indexes.
    Contents: v. 1. The walrus, sea lions, fur seals,
and sea otter — v. 2. Seals — v.3. The Sirenians and
baleen whales.
    1. Marine mammals—Collected works.   I. Ridgway,
Sam H.   II. Harrison, Richard John, Date
QL713.2.H34   1985   599.5   80-42010
ISBN 0-12-588503-2 (v.3)

PRINTED IN THE UNITED STATES OF AMERICA

85 86 87 88   9 8 7 6 5 4 3 2 1

Contents

Contributors		vii
Preface to the Series		ix
Preface to Volume 3		xiii
Contents of Other Volumes		xv
1	Dugong—*Dugong dugon* MASAHARU NISHIWAKI and HELENE MARSH	1
2	Manatees—*Trichechus manatus, Trichechus senegalensis,* and *Trichechus inunguis* DAVID K. CALDWELL and MELBA C. CALDWELL	33
3	Gray Whale—*Eschrichtius robustus* ALLEN A. WOLMAN	67
4	Minke Whale—*Balaenoptera acutorostrata* BRENT S. STEWART and STEPHEN LEATHERWOOD	91

5 Bryde's Whale—*Balaenoptera edeni* 137
WILLIAM C. CUMMINGS

6 Sei Whale—*Balaenoptera borealis* 155
RAY GAMBELL

7 Fin Whale—*Balaenoptera physalus* 171
RAY GAMBELL

8 Blue Whale—*Balaenoptera musculus* 193
PAMELA K. YOCHEM and STEPHEN LEATHERWOOD

9 Humpback Whale—*Megaptera novaeangliae* 241
HOWARD E. WINN and NANCY E. REICHLEY

10 Right Whales—*Eubalaena glacialis* and
Eubalaena australis 275
WILLIAM C. CUMMINGS

11 Bowhead Whale—*Balaena mysticetus* 305
RANDALL R. REEVES and STEPHEN LEATHERWOOD

12 Pygmy Right Whale—*Caperea marginata* 345
ALAN N. BAKER

Index 355

Contributors

Numbers in parentheses indicate the pages on which the authors' contributions begin.

ALAN N. BAKER (345), National Museum of New Zealand, Wellington, New Zealand

DAVID K. CALDWELL (33), Marineland of Florida, St. Augustine, Florida 32086, USA

MELBA C. CALDWELL (33), Marineland of Florida, St. Augustine, Florida 32086, USA

WILLIAM C. CUMMINGS (137, 275), Oceanographic Consultants, San Diego, California 92122, USA

RAY GAMBELL (155, 171), International Whaling Commission, The Red House, Histon, Cambridge CB4 4NP, UK

STEPHEN LEATHERWOOD (91, 193, 305), Hubbs Marine Research Institute, San Diego, California 92109, USA

HELENE MARSH (1), School of Biological Sciences, Department of Zoology, James Cook University, Townsville, Queensland 4811, Australia

MASAHARU NISHIWAKI[†] (1), 17 Kawadacho, Shinjuku, Tokyo 162, Japan

RANDALL R. REEVES (305), Arctic Biological Station, Ste-Anne-de-Bellevue, Quebec H9X 3R4, Canada

NANCY E. REICHLEY (241), University of Rhode Island, Narragansett, Rhode Island 02882, USA

[†]Deceased.

BRENT S. STEWART (91), Hubbs Marine Research Institute, San Diego, California 92109, USA

HOWARD E. WINN (241), Graduate School of Oceanography, University of Rhode Island, Narragansett, Rhode Island 02882, USA

ALLEN A. WOLMAN (67), National Marine Mammal Laboratory, Seattle, Washington 98115, USA

PAMELA K. YOCHEM (193), Hubbs Marine Research Institute, San Diego, California 92109, USA

Preface to the Series

The idea of producing a Handbook of Marine Mammals of this type was the result of many discussions between the authors and the late Mr John Cruise of Academic Press during his visits to Cambridge over ten years ago. It would be, it was hoped, a comprehensive account of all marine mammal species with each chapter written by someone who had actually worked on a particular form. We felt that this would give a more personal flavour to each chapter, provide opportunity to include original observations, and increase accuracy. Some species of marine mammal are worldwide in their distribution and known to many investigators, others are restricted to certain rivers, lakes, seas and even to relatively short ranges along a coast. We know of no one person who claims to have seen every extant species alive in its environment. We hoped that such a series of chapters written by experts on each species would naturally draw attention to subtle differences in form, coloration, behaviour and many other characteristics, many of which might not be known to an ordinary reviewer.

There are obvious difficulties and problems to vex the editors of multi-author works and they all affected this one. We knew it would take time; active marine mammalogists spend long periods away from home on expeditions and at conferences. The majority of contributors has remained loyal to the project but a few have had to withdraw with inevitable difficulties over replacement and with continual postponement of completion dates. New information has become available about many species over the last ten years, due mainly to efforts by countries to find out more about their local species but also to enterprising attempts to display little-known species in captivity. These last ten years have also seen a much increased interest in conservation of marine mammals and an awareness by the public of their importance as interesting animals from many points of view. There has been much debate about whether a moratorium should be introduced by international agreement over the taking of large or even any cetaceans. The arguments have affected many of the classical views held on harvesting, culling, controlling or managing stocks of all marine mammal species. They have exposed our ignorance on many basic facts concerning reproduction, school competition and behaviour, and even what is really meant by many of the terms and concepts used so glibly in marine mammalogy in the past.

The last decade has seen much burgeoning in scientific investigation of marine mammals, though not always in the direction leading to important results. For example, the delightfully satisfying construction of mathematical models has occupied many folk. The results give joy to the constructor, the administrator, the politician and this whole game is not only fun but really quite cheap. What the computer tells us is only as good as the data put into it which are often incomplete or inadequate for detailed analysis. At the very best the results are always based on past events. What is happening now in the seas is still as mysterious as it always was.

Many advances have been made in our knowledge of how to keep marine mammals in captivity, about their diseases and the associated pathological changes, and on treatment. The life styles, life history, expectation of life, growth rates, vulnerability and responses to adverse effects of disease and pollution are becoming better known and understood. Formulae have been constructed to provide the optimum size of pool for holding marine mammals of varying number and size. Codes of practice have been issued to ensure the welfare of marine mammals during transport and in captivity. This concern has increased life expectation of captive animals, and has improved their lot from being mere objects of curiosity and participants in circus acts. The list of different species displayed to the public has steadily lengthened; to an extent that many countries demand permits to be issued for taking of rare and even not so rare animals from local stocks, and records must be kept about captive animals.

Film and television companies have realized the increasing popularity of productions involving marine mammals, and this has spread knowledge more than any other factor, and unfortunately in some cases misinformation, about the various species. Conservationists have also promoted public awareness of which nations still hunt whales and the purposes to which the carcasses are put. These groups have put pressure on national and international bodies to provide more accurate information about endangered species and the state of stocks generally. Again and again it has to be confessed that many species are still little known, their habits obscure, and assessment of their numbers mere guesswork.

Problems of nomenclature are bound to arise in presenting any group as diverse and widely dispersed as marine animals. We have avoided strict rules and have left decisions as to the use of particular names to each expert author. Authors were asked to summarize the scientific and common nomenclature of the genera and species about which they have written. As a guide to the various genera and species Dale W. Rice's "A List of Marine Mammals of the World" (NOAA Technical Report NMFS SSRF-711, US Department of Commerce, Washington, D.C., 1977, 15p) is especially useful.

This Handbook is intended as a guide to marine mammal types and is all about them as animal forms. It is meant for use in the field and laboratory as a practical aid to identification and to provide useful basic information. It is not concerned with management, husbandry or treatment of disease, stock levels or whether there should be a moratorium on the taking of certain marine mammals. We have both worked for many years on various aspects of the biology of what we have found an entirely fascinating group of mammals. We know that all contributors have an interest equal to ours and far more expertise, so we have tried to keep editorial intervention to a minimum. We dare to hope that only marine mammals will know about themselves what is not in this Handbook, or that it is the editors' fault that useful information has been overlooked.

We thank Academic Press for their patience and guidance.

Sam H. Ridgway
Richard J. Harrison

Preface to Volume 3

The first two volumes of this Handbook were devoted to amphibious marine mammals—the pinnipeds and the sea otter. In this third volume, we turn to the fully aquatic groups, beginning with the Sirenia or sea cows and continuing with the Mysticeti or baleen whales.

Regrettably, four years have passed since the first volume finished and we wrote the preface to the treatise. In that space of time some notable changes have occurred. First, biologists are becoming more knowledgeable in the use of computers for recording observations, analyzing data, and constructing models. Since computers in the last few years have become smaller, cheaper, more powerful, and can even be taken into the field, it now appears that the computer is becoming one of the more common biological research tools. Some of our colleagues were critical of our remarks in the series preface on mathematical models. We objected to mathematical modelling mainly because administrators and politicians often seem ready to support the production of computer models as an alternative to biological research. When sufficient biological data are available, models can be useful, not only as tools of wildlife management, but as means of refining biological theory and of focusing future data collection. Second, in the past year, we have lost two of our best cetologists: Masaharu Nishiwaki, for many years Director of the Ocean Research Institute, University of Tokyo, was recognized internationally for his work with cetaceans and sirenians; and Dr. Raymond Gilmore of the San Diego Natural History Museum was an astute observer of baleen whales and a key witness in the remarkable recovery of the California gray whale during the past few decades. The works of both of these dedicated scientists are frequently cited in this volume. Finally, the International Whaling Commission (IWC) has called for a moratorium on all commercial whaling, and we wait to see if the whaling nations comply.

Although receiving less attention, the Sirenia are probably more endangered than any of the whales discussed in this volume. In fact, a sirenian species has become extinct within the past 220 years. In 1741 Georg Wilhelm Steller, a member of the Bering expedition in the North Pacific, wrote of the sea cows found on Bering Island in the Commander Islands off Kamchatka. These animals probably grew to lengths of more than 8 m and to weights of nearly 3600 kg. The heart of a single animal weighed as much as 16 kg. Steller and the stranded explorers of the Bering expedition enjoyed the tender meat of the large, docile mammals later called the Great Northern sea cow or Steller's sea cow *Hydrodamalis gigas* (Zimmermann, 1780). The stranded explorers may have eaten the sea cows only to survive, but by 1768, hunters had apparently destroyed the last of the remaining great northern sea cows.

Because the sea cows provide a good visual transition between the seals and the baleen whales, we start this third volume of the series with the living relatives of the great northern sea cow, the dugongs and manatees. In Chapter 1, Masaharu Nishiwaki and Helene Marsh discuss the dugong, the only surviving member of the family Dugongidae, of which four subfamilies are extinct. We learn that, other than the manatee, the elephant is the closest living relative of the dugong. In Chapter 2, David and Melba Caldwell write about the small-brained, shy, slow manatees. Since these animals consume huge amounts of aquatic vegetation, manatees have even been put to work in Guyana consuming aquatic weeds that otherwise choke waterways. A canal in Georgetown has been kept weed-free by manatees for over 20 years.

With Chapter 3, we move from sirenians that graze on large aquatic plants to baleen whales that graze on small marine animals such as krill or zooplankton. Allen Wolman begins our examination of whales by introducing the California gray whale, the only baleen whale that feeds primarily on benthic invertebrates rather than on krill or zooplankton. About 15 000 of these large animals leave their northern feeding areas in the fall and migrate over 10 000 km to Baja California, where pregnant females calve in warm shallow lagoons. The gray whale is extinct in the North Atlantic, but there is a tiny surviving Korean stock consisting of animals that migrate along the western side of the Pacific from Korea and Japan to the Okhotsk Sea.

After Chapter 3, we move from our discussion of an animal that is the last survivor of its family, to the study of a family whose individual members make up over 90% of the baleen whales alive today. The next six chapters deal with the six species of whales that make up the family Balaenopteridae, commonly called rorquals. Brent Stewart and Stephen Leatherwood begin this section, writing in Chapter 4 about the smallest of the rorquals, the minke whale. W. C. Cummings continues in Chapter 5 with Bryde's whale, an animal sometimes passed over in accounts of whales. Ray Gambell of the IWC, a scientist with many years of first-hand experience with whales and whaling, has taken time from his busy schedule to write concise chapters on the sei whale (Chapter 6) and fin whale (Chapter 7). In Chapter 8, Pamela Yochem and Stephen Leatherwood provide a full account of the largest (weights of almost 200 000 kg have been recorded) animal on earth, the blue whale. All of these animals, from the minke whale to the blue whale, are quite closely related, belonging to the same genus, *Balaenoptera*. The remaining genus in the family Balaenopteridae is *Megaptera*, consisting only of one species.

In Chapter 9, Howard Winn and Nancy Reichley complete the discussion of rorquals with *Megaptera novaeangliae*, the humpback whale, an animal popularized in recent years by the discovery of its extensive song. Perhaps early cetologists knew more about the humpback's sounds than we sometimes give them credit for. Boenninghaus* (1904) wrote, "We know that the humpback whale has a voice. Rawitz observed a herd of 40 producing a loud howling, that began with deep tones and gradually rose to high tones, ending again with deep tones. They went through the tonal scale. These animals seemed to be in heat, because they swam in pairs." Recent research on the humpback's sounds seems to suggest that the song indeed has something to do with breeding.

The last three chapters concern the slower, more ponderous right whales. The whalers gave the animals the common name "right whale"; because of their slowness, fatness, and long baleen, these whales were the "right" whales for early whaling ships to catch. Today, these animals are the most endangered of the great whales. W. C. Cummings, a bioacoustics expert who has followed right whales around both hemispheres to record their sounds, presents some of his findings in Chapter 10. In Chapter 11, Randall Reeves and Stephen Leatherwood discuss the whale that has lately been of most concern to the United States Government. The lives of the two or three thousand remaining bowhead whales may be threatened by native Eskimo hunters and oil drillers on the coast of Alaska. Because of the recent flurry of research, stimulated largely by this issue, we know a good deal more about the bowhead than about the last animal discussed in this volume. Alan Baker from the National Museum of New Zealand presents in the final essay a discussion of the smallest baleen whale, the seldom seen pygmy right whale.

Naval Ocean Systems Center
San Diego, California, USA

Sam H. Ridgway

Downing College
Cambridge, UK

Sir Richard Harrison

---

*Boenninghaus, G. (1904). Das Ohr des Zahnwales. *Zool. Jahrb.* **19**, 189–360. (From an unpublished English translation by E. G. Wever.)

Contents of Other Volumes

Volume 1
THE WALRUS, SEA LIONS, FUR SEALS AND SEA OTTER

Walrus—*Odobenus rosmarus*
   Francis H. Fay
New Zealand Sea Lion—*Phocarctos hookeri*
   Gregory E. Walker and John K. Ling
South American Sea Lion—*Otaria flavescens*
   Raul Vaz-Ferreira
California Sea Lion—*Zalophus californianus*
   Daniel K. Odell
Australian Sea Lion—*Neophoca cinerea*
   Gregory E. Walker and John K. Ling
Steller Sea Lion—*Eumetopias jubatus*
   Ronald J. Schusterman
Northern Fur Seal—*Callorhinus ursinus*
   Roger L. Gentry
Southern Fur Seals—*Arctocephalus*
   W. Nigel Bonner
Sea Otter—*Enhydra lutris*
   Karl W. Kenyon
Index

Volume 2
SEALS

Harbour Seal—*Phoca vitulina* and *P. largha*
   Michael A. Bigg

Ringed, Baikal and Caspian Seals—*Phoca hispida, P. sibirica* and *P. caspica*
    Kathryn J. Frost and Lloyd F. Lowry
Harp Seal—*Phoca groenlandica*
    K. Ronald and P. J. Healey
Ribbon Seal—*Phoca fasciata*
    John J. Burns
Grey Seal—*Halichoerus grypus*
    W. Nigel Bonner
Bearded Seal—*Erignathus barbatus*
    John J. Burns
Hooded Seal—*Cystophora cristata*
    Randall R. Reeves and John K. Ling
Monk Seals—*Monachus*
    Karl W. Kenyon
Crabeater Seal—*Lobodon carcinophagus*
    Gerald L. Kooyman
Ross Seal—*Ommatophoca rossi*
    G. Carleton Ray
Leopard Seal—*Hydrurga leptonyx*
    Gerald L. Kooyman
Weddell Seal—*Leptonychotes weddelli*
    Gerald L. Kooyman
Southern Elephant Seal—*Mirounga leonina*
    John K. Ling and M. M. Bryden
Northern Elephant Seal—*Mirounga angustirostris*
    Samuel M. McGinnis and Ronald J. Schusterman
Index

1

Dugong

Dugong dugon (Müller, 1776)

Masaharu Nishiwaki and Helene Marsh

Genus and Species

Common names

Dugong is now the common name throughout much of the range. Although Malaysian in origin, it has been widely used throughout Asia, the southwest Pacific, and Japan. In addition, the Chinese have a character which can be pronounced as "dugong." The more general term "sea cow" is used to refer to any of the Recent sirenians.

Taxonomy

The dugong is the only extant member of the family Dugongidae (order Sirenia). The classification of this order is as follows:

Order Sirenia
    Family Prorastomidae—extinct. Eocene

Family Protosirenidae—extinct. Eocene
Family Dugongidae
   Four extinct subfamilies of which the most recent was the Hydrodamalinae [*Hydrodamalis gigas*, the huge (up to 7.5 m long) Steller's sea cow, was exterminated by man in the nineteenth century (Domning, 1978)].
   Subfamily Dugonginae
      Genus *Dugong*
Family Trichechidae
   Genus *Trichechus* [which includes the three extant species of manatees (see Chapter 2)].

Evolution

The dugong's closest nonsirenian relative is probably the elephant. Eye lens protein analyses have confirmed the monophyletic origin of the paenungulate orders Hyracoidea, Sirenia, and Proboscidea and place the paenungulates as the second offshoot from the main eutherian line after the edentates (de Jong, et al., 1981). Sirenians appear to have descended from terrestrial herbivores in the early Eocene or earlier. They have an extensive fossil history (Savage, 1976; Domning, 1978), but even at their zenith in the Miocene they numbered only about a dozen (mostly monotypic) genera worldwide (Domning, 1978).

External Characteristics

From a distance, a dugong looks rather like a rotund dolphin (Figs 1 and 2). Dugongs also resemble manatees in many ways but are usually regarded as being more specialized. Manatees are euryhaline or riverine, whereas the dugong is strictly marine. Compared with the manatees, the dugong appears to be a more efficient swimmer with a more streamlined body, whalelike flukes, and shorter and less mobile flippers, and to be more adapted to bottom feeding with its ventrally directed feeding apparatus.

Dorsally, the dugong's body is grey to bronze, becoming somewhat lighter ventrally. Many older animals have large areas of unpigmented skin and are extensively scarred, particularly dorsally. The skin is extremely thick and smooth (Fig. 2), unlike the wrinkled epidermis of some of the manatees. Although the dugong neck is very compressed, it can be rotated and moved in the horizontal and vertical planes to a limited extent. The muzzle is a most versatile and complex structure. The lateral and posterior margins of the

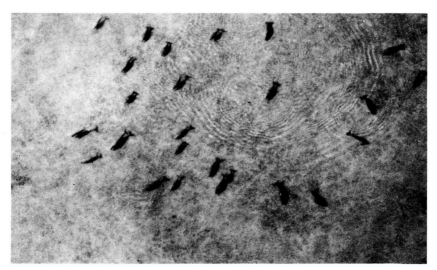

FIG. 1 Part of a herd of more than 100 dugongs in Moreton Bay, Queensland, in 1979.

rostral disc can be controlled separately to grasp plant material (including rhizomes) with precision and to convey it to the mouth, which opens ventrally (Anderson and Birtles, 1978).

The eyes are small and not prominent (Fig. 3). The ears lack pinnae and externally consist only of small openings on the sides of the head (Fig. 3). The nostrils lie close together and are situated anterodorsally (Fig. 3), thereby allowing the dugong to breathe with most of its body submerged. The nostrils are closed during diving (Fig. 2) by anteriorly hinged valves.

The flippers (Fig. 4a) are short [$\sim 15\%$ of total adult body length (Spain and Heinsohn, 1975)], rounded at the ends, and, unlike those of the Caribbean and West African manatees, lack nails (Harrison and King, 1965). There are two pectoral mammae, one on each side with the teat located in the axilla behind the flipper. The tail is horizontally flattened and variably notched (Fig. 4b) and quite different from its paddlelike manatee counterpart.

The testes are abdominal, and, unless the penis is extruded, it is the proximity of the genital aperture to the umbilicus which identifies the animal as a male. In females the genital aperture is located farther back, very close to the anus.

Detailed descriptions of the external morphology are given in Dexler and Freund (1906), Petit (1955), and Kingdon (1971).

FIG. 2 Dugong in captivity at Toba, Japan, with leaves of the seagrass *Zostera marina* trailing from its mouth. The nostrils are closed in contrast to those of the animal in Fig. 3.

FIG. 3 Anterior lateral aspect of a dugong at Jaya Ancol Oceanarium, Jakarta, Indonesia, showing the valvular nostrils, the eye, auditory meatus (arrowed), and sinus hairs.

FIG. 4a Two dugongs in a drained pool at the Jaya Ancol Oceanarium, Jakarta, Indonesia.

FIG. 4b Dugong cow and calf in Shark Bay, Western Australia. (Photograph courtesy of Ben Cropp.)

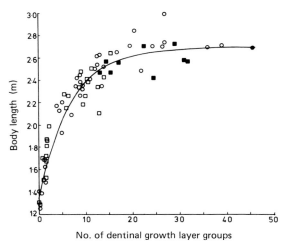

FIG. 5 Relationship between body length (L_A) in metres and age (A) in years (expressed by number of dentinal growth layer groups) for dugongs from Townsville, Queensland. The regression line fitted for data for animals with unworn tusks only is $L_A = 2.69 - 1.35 (0.86)^A$ (□) Male unworn tusks; (■) male worn tusks; (○) female unworn tusks; (●) female worn tusks. (Reprinted from Marsh, 1980, with permission.)

Size

The maximum weight recorded for a dugong is 1016 kg for a specimen of 4.06-m body length (Mani, 1960). This appears improbably high (Spain and Heinsohn, 1975). Petit (1955) recorded maximum lengths of 3.05 m for females and 3.15 m for males. B. E. T. Hudson (personal communication) measured a female 3.31 m long. Such large animals are unusual; individuals more than 3 m long constituted only ∼2% of the 310 adult-sized animals measured by Hudson. There is some evidence (Marsh, 1980) that the asymptotic length of females tends to be slightly greater than that of males. A growth curve (Fig. 5) describing the age–length relationship (based on number of dentinal growth layer groups) was developed by Marsh (1980). An empirical weight for length curve based on 44 animals (Fig. 6) was computed by Spain and Heinsohn (1975). The estimated weight at an approximately pubertal length of 2.4 m is 248 kg. A large animal of 3 m is estimated to weigh nearly 420 kg.

Distribution and Abundance

Dugongs occur in the tropical and subtropical shallow coastal and island waters of the Indo-Pacific as summarised in Fig. 7. The precise extent to

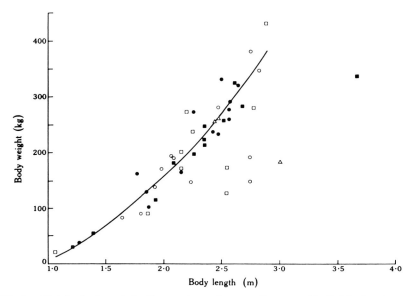

FIG. 6 Weight for length in *Dugong dugon*. Scatter diagram of individual points and fitted regression line $y = (-34.251) - 14.976x + 55.218 x^2$, where y is the body weight in kilograms and x the body length in metres from the tip of the snout to the notch in the tail fluke. (●) North Queensland males; (■) north Queensland females; (○) males from other locations; (□) females from other locations; (△) unknown sex from other locations. (Reprinted from Spain and Heinsohn, 1975, with permission.)

which the dugong's range has contracted is unknown, but over much of its present range it is now represented by relict populations separated by large areas where it is close to extinction or extinct.

Australia

Extensive aerial surveys for dugongs (e.g., Heinsohn et al., 1976a, 1978; Ligon, 1976a; Elliott, 1981; Marsh et al., 1981; Prince et al., 1981; Anderson, 1982a; Marsh et al., 1984a; Prince, 1984; Bayliss, in press) carried out along the coasts of northern Australia since 1974 have demonstrated that relatively large populations still occur in these waters (Fig. 1). In Queensland and Western Australia, a number of areas have been identified where more than 100 dugongs have been counted in a single survey (Table 1). It is impossible, at present, to convert these aerial survey counts into population estimates because of sighting difficulties in the variably muddy water.

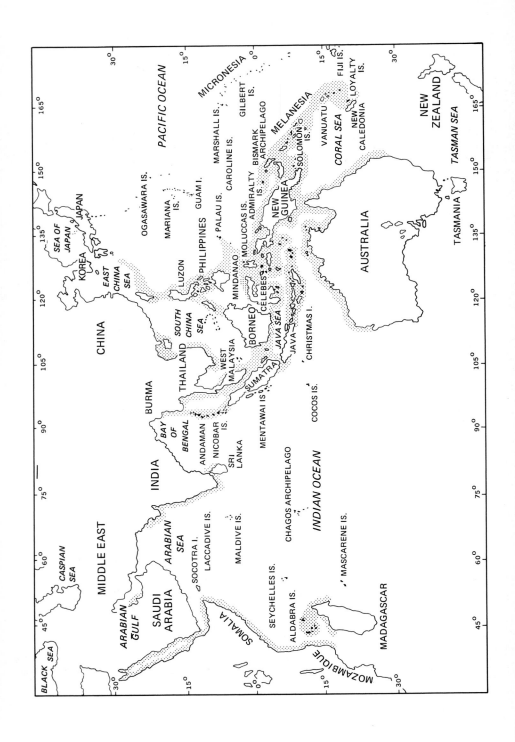

TABLE 1 Locations in northern Australia where more than 100 dugongs have been counted during a single aerial survey

Location	Number of dugongs counted
Queensland[a]	
Moreton Bay	343
Great Sandy Strait–Hervey Bay–Tin Can Inlet	156
Shoalwater Bay/Pt. Clinton	213
Cape Flattery–Cape Melville	600
Torres Strait	223
Weipa–Staaten River	247
Wellesley Islands	374
Western Australia[b]	
Exmouth Gulf	138
Shark Bay	496

[a]From Heinsohn and Marsh (1981) and Marsh et al. (1984a).
[b]From Anderson (1982a) and Prince et al. (1981).

Melanesia

Dugongs are reportedly not uncommon along some parts of the Papua New Guinea coast. In 1976, during an aerial survey covering 1207 km of coastline and coral reefs, 186 dugongs were counted. These included two concentrations of 28 and 39 dugongs in the Warrior Reef area of Torres Strait (Ligon and Hudson, 1977). A maximum of 38 dugongs has been sighted in aerial surveys of West New Britain (Hudson, 1980a). Fifty dugongs were sighted in similar surveys of Manus Island (Hudson, 1980b). After an aerial survey of parts of northern Irian Jaya, Salm et al. (1982) reported that two dugongs were sighted in the Auri Archipelago; one was sighted from the air at Maransabadi, and one was sighted from a boat at Anggremeos. Fishermen also reported one at Nubari. A further 13 dugongs were seen from the air along the west coast of Teluk Cenderawasih, Roon, and Mioswaar, and one was seen off Sorong.

Small numbers occur in the waters off New Caledonia and Vanuatu (Bertram and Bertram, 1973; R. A. Birtles, personal communication, 1975, Nishiwaki et al., 1979; Dickenson, 1981).

FIG. 7 The known distribution of the dugong today. It is possible that the dugong's range is, in fact, continuous around the southern coast of the Middle East and Asia, but the information from some areas is very scanty. (From the Cartographic Centre, James Cook University of North Queensland.)

Although significant numbers are reputed to occur in the Western Province of the Solomon Islands, only small numbers were seen during aerial surveys of the adjacent Bouganville Island area (B. E. T. Hudson, personal communication, 1982).

Micronesia

The Palau population was estimated to be about 50 after aerial surveys in 1977 and 1978 (Brownell *et al.*, 1981). Dugongs are very rare on Yap and Guam islands (Nishiwaki *et al.*, 1979).

Asia

Dugongs are very rare in the Ryukyus (Bertram and Bertram, 1973; Nishiwaki *et al.*, 1979) with about 10 dugongs captured over the last 20 years, mostly in the Okinawa area, but also as far north as Anami Oshima. One young female dugong was caught in 1979 (Uchida, 1979; Yamaguchi, 1979) but there is no evidence of their occurrence in Japan proper or Korea. Zhang (1979) reported that some occur off southern China and Taiwan and in the Mekong Delta area. Dong Jin hai (unpublished, 1983) cited incidental catch records from Qin Zou, Hepu County, Hainan Island in the Beibu Gulf, near Hong Kong, and off Taiwan. (However, the records from near Hong Kong and Taiwan date from the 1940s). Dong Jin hai observed groups of up to twenty dugongs along the coast of Hepu County in April 1979.

R. V. Salm (personal communication, 1984) reports that dugongs are scattered throughout Indonesia, usually in very low numbers. Little is known of the smaller populations other than that they occur around Kupang Bay (Timor), Arakan Reef (North Sulawesi), Togian Island–Teluk Tominy (Central Sulawesi) and other small bays and straits around Sulawesi including the Spermonde Island, South Kalimantan, and Bangka-Belitung Islands (Karimata Strait) (see also Sukiman Hendrokusomo *et al.*, 1981). Toward the end of 1979 dugongs were apparently still fairly numerous around the Aru Islands, their last known area of abundance.

At least small numbers are distributed throughout the Philippines including parts of Luzon, Palawan, Panay, the Sulu Archipelago, Masbate Island, Cataduanes Island, and in the provinces of Camarines Norte (Nishiwaki *et al.*, 1979; G. Hodgson, personal communication, 1984) and in East Malaysia (northern Borneo). Dugongs are believed to be rare in West Malaysia and Singapore (Bertram and Bertram, 1973) and in Thailand. Some occur along the coast of Burma (Nishiwaki *et al.*, 1979; Jones, 1981). There seems to be no resident population in Bangladesh, although there have been reports of dugongs straying into the region from the north coast of Burma (Jones, 1981).

Dugongs are present in the Andaman and Nicobar islands; they are present in apparently decreasing numbers in the Gulf of Mannar and Palk Bay between India and Sri Lanka, and the Gulf of Cutch in northwest India has a small population (Jones, 1981). They are probably rare along the west coast of India, and none were recorded along the east coast of India (Nishiwaki *et al.*, 1979). Although the coast of Pakistan does not provide suitable habitat, individual dugongs may stray into these waters from the Gulf of Cutch (Jones, 1981). It is thought that dugongs are extinct in the Mascarene, Laccadive, and Maldive islands (Husar, 1975, 1978; Jones, 1981).

Bertram and Bertram (1973) reported that dugongs were very rare in the Arabian (Persian) Gulf, where 38 dugongs died coincident with the major oil spills and leakages in late 1983. They are known to be present at Bahrain (Gallagher, 1976) and the Gulf of Salwa (Dean and Dean, 1981). They seem to be absent from Muscat and the south coast of the Arabian Peninsula (Bertram and Bertram, 1973) and present, but rare, in the gulfs of Suez and Aqaba and the rest of the Red Sea (Bertram and Bertram, 1973; Lipkin, 1975; Gohar, 1979). Aerial surveys conducted at the end of 1980 located a group of some 30 individuals in the inshore waters of the Republic of Djibouti (Robineau and Rose, 1982). The authors commented that this was the largest herd ever recorded in the Red Sea and the Gulf of Aden.

East Africa

Dugongs are generally scarce in East Africa and Madagascar, although there is one report of a large herd numbering in the hundreds on the south coast of Somalia (Bertram and Bertram, 1973; Nishiwaki *et al.*, 1979). Twenty-seven dugongs were observed during an aerial survey over the large estuary and mangrove-lined channels at Antonio Enes, northern Mozambique, in 1970 (Hughes and Oxley-Oxland, 1971). In 1975, an aerial survey was flown in a grid pattern giving 25% coverage of the entire shallow water area along the 482-km-long Kenya coast. Only eight dugongs were seen (Ligon, 1976b).

Population

The total population size is unknown. There are undoubtedly some thousands of dugongs in the waters of northern Australia. However, it has not been confirmed whether numbers are increasing, decreasing or stable at any location, and data from Torres Strait indicate a serious recent decline in numbers (Marsh, *et al.*, 1984a; Marsh, in press a). Substantially less is known about dugong numbers in other areas. It seems likely that the populations ranging from India and Sri Lanka westward along the west coast of India, Arabian Gulf, Arabian Peninsula, Red Sea, East Africa, and Madagascar are the most seriously endangered.

Internal Anatomical Characteristics

Skull

The skull has a remarkably enlarged and sharply down-turned premaxilla and a correspondingly elongated, symphysial region of the mandible (Figs 8a, 9). On the top of the skull, an expanded cavity containing the narial openings extends posteriorly behind the anterior margin of the orbits. Nasal bones are absent. Skull growth is dominated by the positively allometric development of the premaxillae and other anterior parts of the skull (Spain and Heinsohn, 1974). The premaxillae of the newborn dugong and the opposing parts of the mandibles are also directed relatively more anteriorly in the juvenile than in older animals. This reorientation is possibly associated

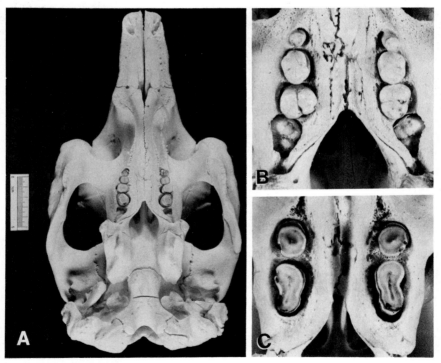

FIG. 8 (A) Ventral view of the skull of a 3-year-old dugong with premolars 3 and 4 and molars 1 and 2 erupted and in wear and molar 3 unerupted. The upper jaw cheek tooth dentition of (B) a presumed neonate (premolars 2–4 erupted, molar 1 unerupted), and (C) an animal more than 29 years old (only molars 2 and 3 in wear). (Photograph courtesy of R. Yeldham.)

FIG. 9 Lateral aspect of the skull and mandible of the same dugong as shown in Fig. 8. (Photograph courtesy of R. Yeldham.)

with the cessation of suckling. Although several features of the adult skull can be shown statistically to be sexually dimorphic (Spain et al., 1976; Spain and Marsh, 1981), the most obvious difference is in the thickness of the premaxillae. These are much more robust in the male, presumably because of the different pattern of tusk growth discussed in the next section.

Dentition

Dugong dentition is distinctly different from that of manatees, which lack incisors and have a different pattern of cheek tooth replacement. The following description is based on Marsh (1980). Other useful references include Heuvelmans (1941), Fernand (1953), and Mitchell (1973, 1976, 1978).

Two pairs of upper incisors are present in the juvenile dugong. The deciduous (first) incisors are small, do not erupt, and are resorbed. In the males, the deciduous incisors are lost around the time of tusk eruption, and their sockets disappear as the tusks expand. In the female, small partially resorbed incisors may persist until the animal is about 30 years old.

The posterior growth of the second incisors (tusks) through the premaxillae is similar in both sexes for about 10 years by which time the tusks are each about 10 cm long. The female tusks continue to grow in this manner throughout life and may reach a length of about 18 cm. In some old females, the tusks erupt and wear, presumably because they have reached the bases of the premaxillae and cannot grow posteriorly any farther. Tusk eruption alone cannot therefore be considered diagnostic of male dugongs.

The tusks of male dugongs erupt when they are about 12 to 15 years old. The lengths of the tusk alveoli remain relatively constant thereafter and never reach the base of the premaxillae. Adult male tusks may be up to 15 cm long but because they protrude only a couple of centimeters (Fig. 10), they do not seem very formidable weapons even though the erupted end is worn into a chisel shape, the cutting edge of which is reinforced with an enamel layer below the cementum.

Up to three pairs of vestigial incisors and one pair of canines may occur occasionally in the rudimentary sockets under the horny plate that covers the downturned symphysial portion of the lower jaw. However, apart from the erupted tusks, the only functional teeth are the cheek teeth.

In situ replacement of the milk dentition does not occur. During the life of the animal, there is a total of six cheek teeth in each jaw quadrant (premolars 2, 3, and 4, and molars 1, 2, and 3), but these are never simultaneously all erupted and in wear. Three premolars are erupted at birth (Fig. 8b), but the roots of the anterior teeth are progressively resorbed, causing them to fall out. Their sockets then become occluded with bone. The molars progressively erupt during growth, the whole process continuing until usually only molars 2 and 3 (Fig. 8c) (occasionally molar 1) are present in each quadrant in old animals. Molars 2 and 3 have persistent pulps and continue to grow axially and radially throughout life, so that the total occlusal area of the cheek teeth is maintained and increased even after the anterior cheek teeth have been lost.

This pattern of cheek-tooth development can be used to age dugongs up to about 9 years old and equations have been developed for the prediction of age from the size of molars 2 and 3 in animals from 10 to 30 years of age. However, much more precise age estimates can be obtained by counting dentinal growth layer groups in longitudinally bisected half teeth that have been etched in acid to produce a ridge–groove pattern, one ridge plus one groove representing 1 year's growth.

Skeleton

The dugong skeleton is of extremely dense bone that presumably acts like a built-in weight belt, to help overcome buoyancy problems while bottom-feeding in salt water. The ribs may also provide protection for the viscera during attacks by sharks and other large marine carnivores. There are 57 to

60 vertebrae. The sternum is reduced. The scapula has a short acromion and well-developed coracoid. The humerus has prominent tuberosities and the carpals show a tendency for fusion (Harrison and King, 1965). The pelvic girdle is vestigial; pubic bones are absent and the ischium and ilium, which are both rodlike, are fused in adults.

Myology

Domning (1977) concluded that most of the muscle systems closely resembled those of manatees. The shoulder muscles, however, are quite different in the two families, and Domning suggested that this may reflect the requirements of manoeuvering in different habitats, the marine habitat of the dugong being much more open than that of the manatee.

Nervous system

The brain is relatively small [250–300 g, i.e., about 0.1% of adult body weight (Dexler, 1913; Kamiya *et al.*, 1979)] and looks almost foetus-like because of the few and shallow sulci and the thin covering of leptomeninges (Hill, 1945). The corpora quadrigemina, especially the inferior, are prominent, suggesting auditory sensitivity. The cerebellum is well developed and possesses large flocular lobes composed mainly of the paraflocculus, probably associated with pronounced swimming and equilibratory activities (Harrison and King, 1965).

Native dugong hunters consider that the hearing of dugongs is acute but that their visual powers are less well developed (Jarman, 1966; Roughsey, 1971). Anderson (1982b) observed that dugongs could detect, locate, and make some preliminary and generalized identification of a stimulus at distances of up to 150 m when underwater visibility (perceived by a human diver) was 5 m or less. He suspected that sinus hairs on the body (Figs. 2 and 3) may assist in the detection of objects moving nearby. Kamiya and Yamasaki (1981) claimed that dugongs have the most developed sinus hairs of all animals. The vibrissae around the mouth (Fig. 10) are presumably very important for food detection.

Respiratory system

The lungs are long and extend posteriorly almost as far as the kidneys. They are separated from the abdominal viscera by a large obliquely sloped diaphragm (Hill, 1945). Dugong lungs are considered unusual both in terms of the nature of the bronchial tree and that of the respiratory tissue *sensu stricto*

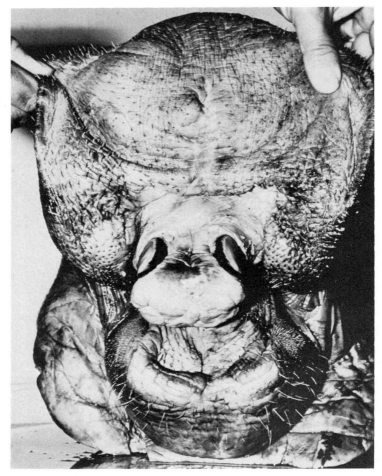

FIG. 10 The head of a 2.7-m-long adult dugong. There were 22.5 growth layer groups in each of the erupted and worn tusks. (Photograph courtesy of M. Lamont.)

(Engel, 1959a,b, 1962). The main bronchus runs almost the entire length of the lung with only a few side branches. The respiratory units (vesicles) arise laterally along the length of the bronchioli. Cartilage occurs throughout the length of the air passages and even the most peripheral bronchioli possess cartilagenous platelets within their walls. Engel considers the dugong lung to be extremely primitive. We do not know whether the lungs collapse on diving.

Circulatory system

The heart is considered primitive in structure (Harrison and King, 1965), with a deep interventricular cleft extending almost the full length of the ventricles (see Rowlatt and Marsh, in press). The most striking feature in the vascular system is the retia, which are in the form of vascular bundles rather than networks. Retia are also found in cetaceans and in some terrestrial mammals such as sloths and some primates. There has been much discussion about their function (see Harrison and King, 1965), but no entirely satisfactory hypotheses have been advanced.

Integument

The skin of the dugong is modified for life in the water to a lesser extent than in some marine mammals, particularly, as would be expected, in those characteristics that contribute to the regulation of body temperature. The blubber is thinner than in most other marine animals, and there are no arteriovenous anastomoses in the dermis. Arteriovenous anastomoses are very numerous in the papillary layer of the dermis in seals and are associated with thermoregulation (Bryden *et al.*, 1978). Because the dugong lives in subtropical and tropical waters, keeping warm is probably less of a problem than it is for many other marine mammals, except perhaps at the limits of the dugong's range (see Anderson, in press a).

The pelage consists of both bristles and fine hairs, all of the sinus type (Fig. 3). The hairs are lightly scattered over the body surface but become denser and very robust on the muzzle and around the mouth (Fig. 10).

Urogenital system

Dugong kidneys have an elongated external form quite unlike the lobulated kidneys of cetaceans and are dipelvic with a characteristic bilaminar central transverse plate. The extensive renal sinus is curious in outline and boundaries (Batrawi, 1953, 1957). Batrawi considered that the dugong kidney combines features similar to those found partly in the camel and partly in the horse. Unrinsed seagrasses, as consumed by the dugong, contain about 15 times more chloride and 30 times more sodium than most terrestrial pasture plants (Birch, 1975) and, unlike manatees, the dugong appears to be physiologically independent of fresh water. Its salt metabolism is thus potentially very interesting and should be investigated.

Each ovoid testis lies caudal and slightly lateral to the corresponding kidney. A distinct epididymis runs along the lateral edge of each testis

(Marsh *et al.*, 1984b). Large seminal vesicles are present but there is no discrete prostate gland. Bulbourethral type glandular tissue is dispersed among the muscles close to the root of the penis (see also Marsh and Glover, 1981).

The ovaries are very large (up to 15 cm long), flattened, ellipsoidal organs (Marsh, 1981; Marsh *et al.*, 1984c). Each lies in a peritoneal pouch hidden in the dorsal abdominal wall posterior to the kidney and lateral to the ureter. The hilum is almost as extensive as the ovary itself. Graafian follicles and corpora lutea tend to be approximately 1 cm in diameter and project enough to be readily visible. The corpora albicantia persist for at least several years as brownish scars. The uterus is bicornuate. Purple placental scars mark the sites of previous well-developed placental attachments. The vagina is a long straight tube with a very capacious lumen, possessing a raised shieldlike region of keratinized material in its vault through which the cervix projects. The clitoris possesses a large conical glans.

Endocrine system and lymphatic tissues

The thyroid gland is exceptional in the relatively great abundance and density of its intervesicular stroma. Its histological appearance would seem to agree with other evidence favouring the occurrence of a functional hypothyroidism, which may explain the osteosclerotic condition of the bones (Cave and Aumonier, 1967). Cave and Aumonier (1967) described the histological structure of certain lymph nodes, the thymus, and thyroid glands but failed to confirm the presence of parathyroids. The histology of the pituitary and adrenal glands was described by Fernand (1951).

Digestive system

The dugong has a simple stomach remarkable only in that all the chief cells and most of the parietal cells occur in a discrete pouch which the digesta do not enter and which communicates with the main sac by a single aperture (Marsh *et al.*, 1977).

Two diverticula enter the duodenum just posterior to the pylorus. The diverticula enlarge the surface area of the proximal duodenum and may allow a large volume of digesta to pass from the stomach to the proximal duodenum at one time (Marsh *et al.*, 1977), although why this should be necessary is unknown. An analysis of the gas taken from the diverticula indicates that it contains a significant quantity of carbon dioxide, suggesting that they may secrete bicarbonate to neutralize the digesta (R. A. Hungate, personal communication, 1977).

A small caecum occurs at the junction of the small and large intestines.

This caecum and the large intestine, the contents of which weigh twice that of the stomach, have been shown to be the principal areas of disappearance of the fibre fraction of the diet (Murray et al., 1977). In an adult dugong, the large intestine is up to 25 m long, about twice as long as the small intestine. Thus, the dugong belongs to that group of nonruminant herbivores that have a greatly enlarged hindgut with a rich microflora which enables them to digest cellulose and other fibrous carbohydrates.

Behaviour

Social organization

Because of the practical difficulties associated with their observation and their now usually low population densities, little is known of dugong behaviour. Dugongs, however, are social animals as evidenced by their propensity for forming herds (Fig. 1), even though solitary individuals are often seen. Nineteenth century accounts are available, especially from eastern Australia (see Welsby, 1967), of discrete herds of several hundred dugongs. Hunting pressure has reduced herd sizes in most areas, and it was only with the introduction of aerial surveys that large herds were again located on the Australian coast. Groups of dugongs swimming in rank formation have also been frequently observed during aerial surveys.

This tendency to form groups suggests some degree of communication between individuals. Australian aborigines report the presence of "whistlers" whom they regard as dominant animals which are thought to control herds; these have been named for the sounds they emit (Roughsey, 1971). The vocalizations of captive dugongs have been recorded by Nair and Mohan (1977) and Mercer et al. (reported by Marsh et al., 1978). Nietschmann (1984) made in 1976 the first recording of underwater dugong vocalisations in the wild. All sounds were in the 1 to 8 kHz range.

Feeding

Aerial surveys have shown that dugongs tend to occur in warm ($\sim$18–33°C) shallow, sheltered inshore and reef areas where extensive beds of seagrasses occur. It has been frequently stated that dugongs remain offshore during daylight hours and come inshore to feed only at night (e.g., Jonklaas, 1961; Jarman, 1966). However, aerial surveys (Heinsohn et al., 1976a; Heinsohn, 1981b) and shore-based observations (Anderson and Birtles, 1978) demonstrated that diurnal inshore feeding is normal in north Queensland. However, dugong feeding is restricted to the hours of darkness in Palau (Anderson, 1981).

Analysis of stomach (Lipkin, 1975; Marsh *et al.*, 1982; Nietschmann, 1984) and mouth (Johnstone and Hudson, 1981) contents, indicate that dugongs consume a wide variety of tropical and subtropical seagrasses, including *Enhalus, Halophila, Halodule, Cymodocea, Thalassia, Thalasodendron, Syringodium,* and *Zostera.* Algae are also eaten but usually in only very small amounts if seagrasses are abundant. When dugongs feed on small, delicate seagrasses such as *Halodule* and *Halophila,* they dig up the whole plant including the rhizomes, making a distinctive feeding trail (Anderson and Birtles, 1978; Heinsohn *et al.*, 1977). However, in localities where taller growing species, such as *Amphibolis antarctica,* predominate, dugongs do not dig up the substrate (Anderson, 1981, 1982b). Captive dugongs have often been observed using their flippers to guide seagrass into their mouths. Calluses are usually present on the anterior ventral part of the flippers on which they support themselves to feed.

Locomotion

Although a dugong can swim at ~10–12 knots while being chased by a speedboat, it tires in a few minutes (Marsh *et al.*, 1981). Most movement is much more leisurely, and Anderson (1981) estimated the normal crusing speed to match that of a fin-equipped diver. The tail flukes are the principal organ of locomotion and work vertically with slow, powerful beats [one every $0.45 \pm SE\ 0.48$ sec at normal swimming speeds (Anderson, 1982b)]. When turning, a dugong stops stroking and twists its flukes to effect the change in direction. During acceleration the flippers are raised and pressed against the body. At normal swimming speeds they are sometimes raised, but often hang down loosely (Anderson, 1982b). However, the flippers do have considerable range of movement [but rather less than that of manatees (Domning, 1977)] and are used in guiding turns and in stopping forward movement (Anderson, 1981).

Dugongs undergo daily (Anderson and Birtles, 1978) and seasonal movements (e.g., Marsh *et al.*, 1981; Anderson, in press a), but there is no evidence of large-scale migrations comparable to those of the baleen whales. Individual dugongs can certainly travel many hundreds of miles, as evidenced by the occasional sighting of a dugong as far south as Sydney, 700 km south of Moreton Bay, the nearest known dugong habitat area. However, colonization of remote island groups is probably a rare event.

Diving

There have been few observations of diving times. Kenny (1967) recorded the submergence times of a captive animal in a swimming pool and found that it varied up to a maximum of 506 sec. Anderson and Birtles (1978) recorded the times of 370 dives during field observations off the coast of central Queensland where dugongs feed by grubbing seagrass rhizomes. The mean time for each dive was 73.3 sec, with a maximum dive time of 400 sec. Submerged times of dugongs presumed to be feeding on *Amphibolis antarctica* in Shark Bay were significantly shorter than these (Anderson, 1982b). Anderson (in press b) concluded that the interval between appearances at the surface varied with locality, foraging mode and forage species, activity and reproductive status. His data, obtained from waters of varying depths, suggest a trend for dugongs to remain submerged longer in deeper water. However, given their essentially coastal distribution and dependence on seagrass for food, it is doubtful that dugongs dive to any considerable depth.

Captivity

Dugongs have not bred in captivity and are very difficult and expensive to maintain in zoos or oceanaria, mainly due to the difficulties of meeting their specialized food requirements. Since 1959, attempts have been made to maintain at least 30 dugongs in captivity at 13 institutions in Australia, Burma, India, Indonesia, Japan, New Caledonia, Thailand, and the USA. Nine of these animals survived less than one month; only 10 definitely survived longer than 6 months. However, two dugongs were kept at the Central Marine Fisheries Institute at Cochin, India, for 11 years (Jones, 1976; S. Jones, personal communication). In early 1985, we know of two dugongs in captivity at each of two institutions: Toba Aquarium, Toba, Japan, and Jaya Ancol Oceanarium, Jakarta, Indonesia. The high mortality rate can probably be attributed partially to the young age at which most of the animals have been captured. We judge by their size that at least 22 of the dugongs probably would have still been suckling their mothers when captured. A further problem is that many of the animals were transported considerable distances, for example, from the Philippines and Australia to Japan and from Palau to California. In addition, there is the possible problem of capture stress (Marsh and Anderson, 1983). We recommend that permits to capture and maintain this endangered marine mammal should be granted only to rigorously selected institutions.

Life History and Reproduction

Almost all our knowledge is based on incidental observations and analyses of data and specimens obtained from dugongs accidentally drowned in shark nets in northern Australia or killed for food by native hunters in northern Australia and Papua New Guinea. Carcass analyses (Marsh, 1980; Marsh *et al.*, 1984b,c,d; H. Marsh and B. E. T. Hudson, unpublished) suggest that dugongs have a maximum longevity of about 70 years and a minimum prereproductive period of 9 to 10 years for both sexes. Estimates of calving interval based on pregnancy rates or accumulation rates of placental scars or calf counts from aerial surveys, range from 3 to 7 years for various Australian/Torres Strait populations (Marsh et al., 1984d, Marsh, in press a). Population simulations indicate that even with the most optimistic combination of life history parameters, a dugong population is unlikely to increase at more than about 5 percent per year (Marsh, in press a). Females are polyovular and polyoestrous and may undergo a number of sterile cycles before becoming pregnant. Multiple corpora lutea (up to 90, H. Marsh, unpublished) are present per pregnancy, persisting until term. Males bred asynchronously and discontinuously in the tropical populations studied (Marsh et al., 1984d). Although there are some reproductively active males present in a population throughout the year, the proportion of active animals is much greater during the months when the ovaries are also active (H. Marsh and B. E. T. Hudson, unpublished).

There are descriptions of the courtship and/or mating behaviour of dugongs in Queensland (Roughsey, 1971; Anderson and Birtles, 1978). In contrast to the manatee (Hartman, 1979), there is a preliminary display and mating seems to involve rotating the female until her ventral surface is uppermost.

A single calf is born after a gestation period estimated to be about 13–14 months (H. Marsh and B. E. T. Hudson, unpublished). Neonates are about 1 to 1.2 m in length and weigh 20 to 35 kg (Marsh et al., 1984d). Calving in the Townsville–Cairns area of north Queensland is diffusely seasonal; most calves are born from September through December. Dugong births have been described by Macmillan (1955) and Marsh et al. (1984d). Although there are some differences in these accounts, all births occurred in shallow water. On two occasions, the mother was effectively aground; mother and baby were soon refloated on an incoming tide. During an aerial survey, G. E. Heinsohn and A. V. Spain (personal communication) also observed a dugong which appeared to be giving birth under substantially similar circumstances. Although a calf begins to eat seagrasses soon after birth, lactation can last for at least 18 months (Marsh et al., 1984d). Calves suckle beneath the surface, lying beside the cow and behind her axilla (Anderson, 1982b, 1984). The cow–calf bond seems to be extremely well developed, and fishermen often exploit it to net a cow after taking her calf (Welsby, 1967; R. V. Salm, personal communication, 1984).

Diseases

Allen et al. (1976) performed postmortem examinations on three dugongs that had died at the Jaya Ancol Oceanarium and attributed all deaths primarily to problems of the gastrointestinal system. Campbell and Ladds (1981) described pathological changes in 14 wild and 1 captive dugong from northeastern Australia. Diseases of the skin, gastrointestinal tract, and pancreas predominated. Specific parasitism was associated with trematode species *Lankatrematoides gardneri* in the pancreatic ducts and *Lankatrema* sp. in the intestine (see also Blair, 1981a). Fatal gastrointestinal salmonellosis (*Salmonella lohbruegge*) occurred in a single dugong from an oceanarium as detailed by Elliot *et al.* (1981). Blair (1981b) provided a checklist of dugong parasites. Marsh and Anderson (1983) have warned that dugongs may be particularly susceptible to the capture stress syndrome.

Exploitation and Protection

Man is the most important predator of dugongs and has been netting them for thousands of years. Specialized cultures based upon dugong hunting developed in places as far apart as Cape York (northern Australia) (Thomson, 1934) and the Persian (Arabian) Gulf (Bibby, 1969). Today the dugong is still of great significance in the ceremony, religion, economy, and culture of certain Australian aboriginal and Torres Strait islander societies, in which it has an important coordinating role (Anonymous, 1981a).

Dugong meat, which tastes like veal or pork, has long been prized. An average adult will yield 100 to 150 kg of fresh meat and from 5 to 8 gallons of oil. This oil, sought for its medicinal properties, formed the basis of a cottage industry in Queensland from the middle nineteenth century until dugongs were protected in the 1960s. The thick hide makes a good grade of leather, well suited for sandal-making. Dense bones and tusks can provide ivory for carving. The tusk ends are still sold in the Aru Islands as cigarette holders (Compost, 1978). The atlas vertebrae were used as wristlets in Palau during the eighteenth century, a practice which persists, although with modified cultural significance (Brownell *et al.*, 1981).

Many different cultures believed that various dugong products possessed medicinal and aphrodisiac properties, and fishermen in the Aru Islands still sell "dugong tears" as an aphrodisiac (Compost, 1978). Several superstitions exist concerning dugongs. Muslim fishermen in southern India kill them because they resemble pigs, while Hindu fishermen in the same area kill them because they fear that they will tear nets and eat their fish (Lincoln Young, personal communication, 1981).

Dugongs are still hunted for food over most of their range although much of this hunting is now illegal. Very limited data are available on the numbers taken but in some areas the take is still substantial. Compost (1978) estimated that 1000 animals are killed each year in the Aru Islands. Nietschmann (1984) conjectured that 750 dugongs were killed each year in

Torres Strait in the late 1970s. Hudson (in press) reported that 463 passed through the Daru market in Torres Strait between July 1978 and January 1983.

Netting is a major threat to dugongs (Heinsohn et al., 1976b). Accidental deaths in nets used in fisheries (e.g., in Sri Lanka, India, Kenya, Papua New Guinea and northern Australia) and to catch sharks near Queensland swimming beaches have caused significant local reductions in numbers (Bertram and Bertram, 1973; Heinsohn, 1972; Husar, 1975; Nair et al., 1975; Paterson, 1979).

Thirty-eight dugongs were found dead coincident with the major oil spills and leakages in the Arabian Gulf in late 1983 (Heinsohn, 1985).

The dugong is listed as vulnerable to extinction in the IUCN Red Data Book (Thornback and Jenkins, 1982). Trade in dugong products is regulated or banned (depending on the dugong population involved) by the Convention on International Trade in Endangered Species of Wild Fauna and Flora (CITES). The dugong is included in class A of the African Convention (1969), which means that it may be hunted or collected only on the authorization of the highest competent authority. Dugongs are totally protected in many countries but this protection is usually enforced inadequately (Anonymous, 1981b). In some countries, for example Australia and Papua New Guinea, limited traditional hunting by native peoples is allowed (Heinsohn, 1981; Hudson, 1981, in press; Marsh, in press b)

Dugongs are protected in the Paradise Islands National Park in Mozambique (Husar, 1975). In Papua New Guinea, two Wildlife Management Areas have been established to protect dugongs (Hudson, 1981). Marine reserves have also been proposed for nine areas in Indonesia, including western Teluk Cendera Wasih and the southeast Aru Islands (R. V. Salm, personal communication, 1984). In Australia, the Great Barrier Reef Marine Park will soon protect dugongs along a significant proportion of the Queensland Coast. In the Cairns section of this Park, the zoning plan specifically provides for the protection of dugongs in the Starcke River area. Net fishing has been banned from the dugong habitat below low water mark. Indigenous peoples' hunting in this area is regulated by both seasonal closure and permits (Marsh, in press b). Marine parks are also being developed in some dugong habitat areas in Western Australia and the Northern Territory.

Many important dugong areas especially in Australia and Papua New Guinea have been identified through aerial surveys, and the results of these surveys are now being used as the basis for planning and management. Aerial surveys need to be extended through the rest of the dugong's range to identify key areas for the establishment of an extensive series of marine reserves.

Without the rapid implementation of such protection, the dugong, aptly described by Anderson (1981) as a "graceful and gentle outlier in the spectrum of mammalian evolution," faces extinction.

Acknowledgements

We wish to thank George Heinsohn, Brydget Hudson, Paul Anderson, and Darryl Domning for their comments on this manuscript.

References

Allen J. F., Lepes, M. M., Budiarso, I. T., Sumitro, D., and Hammond, D. (1976). Some observations on the biology of the dugong (*Dugong dugon*) from the waters of South Sulawesi. *Aquat. Mamm.* **4,** 33–48.

Anderson, P. K. (1981). Dugong behaviour: Observations, extrapolations, and speculations. *In* "The Dugong: Proceedings of a Seminar/Workshop held at James Cook University, 8–13 May 1979" (Ed. H. Marsh), pp. 91–111. James Cook Univ., Queensland.

Anderson, P. K. (1982a). Studies of dugongs at Shark Bay, Western Australia. I. Analysis of population size, composition, dispersion, and habitat use on the basis of aerial survey. *Aust. Wildl. Res.* **9,** 69–84.

Anderson, P. K. (1982b). Studies of dugongs at Shark Bay, Western Australia. II. Surface and subsurface observations. *Aust. Wildl. Res.* **9,** 85–100.

Anderson, P. K. (1984). Suckling in *Dugong dugon*. *J. Mammal.* **65,** 510–511.

Anderson, P. K. (in press a). The Shark Bay dugong herd. *Proceedings of Symposium on Endangered Marine Animals and Marine Parks*, held at Cochin, India, 12–16 January, 1985.

Anderson, P. K. (in press b). Submerged times of dugongs. *Proceedings of International Reunion of the Mexican Society for the Study of Marine Mammals, 9th*, held in La Paz, Baja California Sur, Mexico, 29–31 March 1984.

Anderson, P. K., and Birtles, A. (1978). Behaviour and ecology of the dugong, *Dugong dugon* (Sirenia): observations in Shoalwater and Cleveland Bays, Queensland. *Aust. Wildl. Res.* **5,** 1–23.

Anonymous. (1981a). Additional recommendations relating to the involvement of Aboriginal and Torres Strait Islander communities in information gathering and conservation programs. *In* "The Dugong: Proceedings of a Seminar/Workshop held at James Cook University, 8–13 May 1979" (Ed. H. Marsh), pp. 215–216. James Cook Univ., Queensland.

Anonymous. (1981b). Recommendations of the Seminar/Workshop on the Biology and Conservation of the Dugong. *In* "The Dugong: Proceedings of a Seminar/Workshop held at James Cook University, 8–13 May 1979" (Ed. H. Marsh), pp. 205–214. James Cook Univ., Queensland.

Batrawi, A. (1953). The external features of the dugong kidney. *Bull. Zool. Soc. Egypt* **11,** 12–13.

Batrawi, A. (1957). The structure of the dugong kidney. *Publ. Mar. Biol. Stn. Ghardaqa, Red Sea,* **9,** 51–68.
Bayliss, P. (in press). Factors affecting aerial counts of marine fauna and their relationship to a census of dugongs in the coastal waters of the Northern Territory. *Aust. Wildl. Res.*
Bertram, G. C. L., and Bertram, C. K. R. (1973). The modern Sirenia: their distribution and status. *Biol. J. Linn. Soc.* **5,** 297–338.
Bibby, G. (1969). "Looking for Dilmun". Penguin, London.
Birch, W. R. (1975). Some chemical and calorific properties of tropical marine angiosperms compared with those of other plants. *J. Appl. Ecol.* **12,** 201–212.
Blair, D. (1981a). The monostome flukes (Digenea: families Opisthotrematidae Poche and Rhabdiopoeidae Poche) parasitic in sirenians (Mammalia: Sirenia). *Aust. J. Zool., Suppl. Ser.* **81,** 1–54.
Blair, D. (1981b). Helminth parasites of the dugong, their collection and preservation. *In* "The Dugong: Proceedings of a Seminar/Workshop held at James Cook University, 8–13 May 1979" (Ed. H. Marsh), pp. 275–285. James Cook Univ., Queensland.
Brownell, R. L., Jr., Anderson, P. K., Owen, R. P., and Ralls, K. (1981). The status of dugongs at Palau, an isolated island group. *In* "The Dugong: Proceedings of a Seminar/Workshop held at James Cook University, 8–13 May 1979" (Ed. H. Marsh), pp. 19–42. James Cook Univ., Queensland.
Bryden, M. M., Marsh, H., and Macdonald, B. W. (1978). The skin and hair of the dugong, *Dugong dugon. J. Anat.* **126,** 637–638.
Campbell, R. S. F., and Ladds, P. W. (1981). Diseases of the dugong in north-eastern Australia: a preliminary report. *In* "The Dugong: Proceedings of a Seminar/Workshop held at James Cook University, 8–13 May 1979" (Ed. H. Marsh), pp. 176–181. James Cook Univ., Queensland.
Cave, A. J. E., and Aumonier, F. J. (1967). Observations on dugong histology. *J. R. Microsc. Soc.* **87,** 113–121.
Compost, A. (1978). Pilot survey of exploitation of dugong and sea turtle in the Aru Islands. Unpublished report to Yayasan Indonesia Hijau, Bogor, Indonesia.
Dean, C. G., and Dean, L. T. (1981). The occurrence of the sea cow, *Dugong dugon* Müller (Sirenia: Dugongidae) in the Arabian Gulf. Abstract of paper presented at the Saudi Biological Society 5th Symposium on the Biological Aspects of Saudi Arabia, 13–16 April 1981. Univ. of Riyadh, Abha.
de Jong, W. W., Zweers, A., and Goodman, M. (1981). Relationship of aardvark to elephants, hyraxes and sea cows from α-crystallin sequences. *Nature (London)* **292,** 538–540.
Dexler, H. (1913). Das Hirn von *Halicore dugong* Erxl. *Morph. J.* **45,** 97–189.
Dexler, H., and Freund, L. (1906). External morphology of the dugong. *Am. Nat.* **40,** 567–581.
Dickenson, D. (1981). Have you seen a dugong recently? *Naika* **2,** 8–9.
Domning, D. P. (1977). Observations on the myology of *Dugong dugon* (Müller). *Smithson. Contrib. Zool.* **226,** 1–57.
Domning, D. P. (1978). Sirenia. *In* "Evolution of African Mammals" (Eds. V. J. Maglio and H. B. S. Cooke), pp. 573–581. Harvard Univ. Press, Cambridge.

Irian Jaya (1982). Unpublished report prepared for the Directorate of Nature Conservation, Directorate General of Forestry, Republic of Indonesia.

Savage, R. J. G. (1976). Review of early Sirenia. *Syst. Zool.* **25,** 344–351

Spain, A. V., and Heinsohn, G. E. (1974). A biometric analysis of measurement data from a collection of North Queensland dugong skulls. *Dugong dugon* (Müller). *Aust. J. Zool.* **22,** 249–257.

Spain, A. V., and Heinsohn, G. E. (1975). Size and weight allometry in a north Queensland population of *Dugong dugon* (Müller) (Mammalia: Sirenia). *Aust. J. Zool.* **23,** 159–168.

Spain, A. V., and Marsh, H. (1981). Geographic variation and sexual dimorphism in the skulls of two Australian populations of *Dugong dugon* (Müller) (Mammalia: Sirenia). *In* "The Dugong: Proceedings of a Seminar/Workshop held at James Cook University, 8–13 May 1979" (Ed. H. Marsh), pp. 143–161. James Cook Univ., Queensland.

Spain, A. V., Heinsohn, G. E., Marsh, H., and Correll, R. L. (1976). Sexual dimorphism and other sources of variation in a sample of dugong skulls from North Queensland. *Aust. J. Zool.* **24,** 491–497.

Thomson, D. F. (1934). The dugong hunters of Cape York. *J. R. Anthrop. Inst.* **64,** 237–264.

Thornback, J., and Jenkins, M. (1982). *IUCN Mammal Red Data Book*, Part 1. IUCN, Gland, Switzerland.

Uchida, S. (1979). Dugongs kept in Okinawa. *Abstr. Symp. Biol. Dugong (Dugong dugon)*, Abstract 2-1, 6–7 December 1979, pp. 8–11. Ocean Research Institute, Univ. of Tokyo, Japan.

Welsby, T. (1967). "The Collected Works" (Ed. A. K. Thomson), Vol. 1, pp. 102–110, Vol. 2, pp. 233–257. Jacaranda Press, Brisbane.

Yamaguchi, M. (1979). Distribution of sea-grass meadows in the Ryukyu Islands. *Abstr. Symp. Biol. Dugong (Dugong dugon)*. Abstract 3-2, 6–7 December 1979, pp. 25–26. Ocean Research Institute, University of Tokyo, Japan.

Zhang, Z. M. (1979). About dugong. The secret of mermaids. *Nat. Hist.* (Shanghai Natural History Museum) **1,** 33–36. (In Chinese).

2

Manatees

Trichechus manatus Linnaeus, 1758; *Trichechus senegalensis* Link, 1795 and *Trichechus inunguis* (Natterer, 1883)

David K. Caldwell and Melba C. Caldwell

Genus and Species

Taxonomy

All three manatee species belong to the family Trichechidae. The single genus *Trichechus* as presently recognised was named by Linnaeus in 1758. Manatees have also been referred to the genera *Manatus*, *Oxystomus* and *Halipaedisca*, but these genera have all been reduced to the one genus *Trichechus* as summarised by Husar (1977a, 1978a,b). That writer also listed the numerous other specific names that have been applied to two of the three species that are presently recognised. There are 11 such synonyms for *T. manatus* and five for *T. senegalensis*. *T. inunguis* is the only specific name that has been applied to that species.

Trichechus manatus has at times been considered divisible into two subspecies: *T. m. manatus* Linnaeus, said to range from Mexico south through Central and into South America, including the West Indies, and *T. m. latirostris* (Harlan) from the southeastern United States. There is considerable doubt as to the validity of such a division, and we follow earlier writers (e.g., Moore, 1951a; Lowery, 1974) in not using the subspecific designations.

Evolution

According to Anderson and Jones (1967), manatees have been distinct from the dugongids (the other members of the order Sirenia) since the Eocene. The fossil record is incomplete, however, and the relationships between manatees and their ancestors are still poorly known. Those same writers list the geological record for the family Trichechidae as being from the Miocene and Recent in South America, the Pleistocene to Recent in North America, and Recent in Africa. Harrison and King (1965) stated that the closest living evolutionary relatives of this group may be the Proboscidea (elephants) and the Hyracoidea (hyraxes).

Common names

Trichechus manatus, the West Indian manatee, is also widely known simply as manatee, sea cow, or Florida manatee. The species is also known less frequently as the American manatee.

Trichechus senegalensis, the African manatee, has but the one common name in English. There undoubtedly are local common names in the particular local languages where the species occurs.

Trichechus inunguis, the Amazonian manatee, is known in Brazil, at least in educated circles, as *peixe-boi*, but as with the African manatee there surely are local common names in local languages.

For the most part, the following discussion is based on data available for *T. manatus*, a species more widely studied than the other two. Where special data are available for the other two species, they are indicated. Where data for more than one species are available in the same category, they are so similar that most of the material on the life history, behavior, and anatomy of the West Indian manatee can be extrapolated to the other two species as well, but in this discussion when one of these two species is not specifically mentioned we are referring to *T. manatus*. A useful bibliography of the Sirenia, an order that includes the manatees, was provided, with annotations, by Whitfield and Farrington (1975).

External Characteristics and Morphology

The general appearance of West Indian manatees (Fig. 1) and of the other two species is that of a massive, almost blimplike, creature with a horizontally flattened and rounded tail that is well adapted for the animal's aquatic habitat. There are no hind limbs, but manatees possess paddlelike and flexible forelimbs, or flippers, that are used for aiding motion over the bottom, scratching, touching, and even embracing other manatees, and for moving

FIG. 1 Adult West Indian manatee. (Drawing by John Quinn, courtesy of Biological Systems, Inc.)

food into and cleaning the mouth. The head is relatively small, often surrounded by heavy folds, and there is no evident neck (Fig. 2). The snout is squarish, with soft, fleshy, movable lips. The snout bears a number of stiff bristles, but otherwise the body is nearly devoid of hair and that present is short and fine. The paired, semicircular nostrils can be closed when the animal dives but may be opened wide when it is on the surface to breathe. The colour is generally slate grey to brown in adults but darker in infants. Manatees, and especially adults, are often heavily scarred and covered with algae, barnacles, and other encrustations that obscure the basic pigmentation. The scars, many rather extensive, are usually caused by collisions with boats and running boat propellers. While unfortunate for the manatees, these scars, along with cuts and gashes on the trailing edge of the fluke, are often used by scientists to the benefit of the manatees as natural tags in population and behavioral studies.

FIG. 2 Upper torso of a live captive juvenile female West Indian manatee underwater. Note the fine hairs on the body and the short stiff bristles on the snout. One of the paired crescent-shaped nostrils that tightly close when the animal is submerged is shown. The somewhat peculiar lid structure around the eyes can be seen as well as the fleshy lips and lack of a distinct neck. (Courtesy of Marineland of Florida.)

There is no marked sexual dimorphism other than in the position of the genital apertures (Fig. 3). On the other hand, the females in general are said to be heavier bodied than males of the same length (Hartman, 1971). Lowery (1974) felt that males average somewhat larger than females. Husar (1977b), who summarised in detail data on manatee external morphology, also summarised the work of others in stating that adults measure from 2.5 to 4.5 m in total length, with corresponding weights of 200 to 600 kg. Lowery (1974) noted that newborns weigh about 18 to 27 kg and measure about 1 m in length. The skin of adults may be as much as 2 inches thick, and in the past was sometimes used to make a tough and durable leather.

There is a single teat in the axilla behind each flipper (Fig. 4).

Rudimentary nails normally are found at the tips of the second, third, and fourth digits (Harrison and King, 1965), but they often may be worn down to the point where they are no longer visible externally (Husar, 1977b).

Externally the African manatee is not readily distinguishable from the West Indian manatee, and the cranial characters also are so similar that

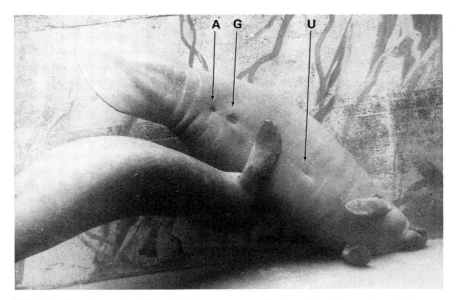

FIG. 3 Captive male West Indian manatee (left) embracing a female in precopulatory activity. Note the positions of the anus (A), the genital opening (G) and the umbilicus (U) for the female. In a male, the positions of the anus and umbilicus would be the same, while the penile opening would lie near the umbilicus in a position in this photograph just anterior to the embracing flipper of the male. (Courtesy of Marineland of Florida.)

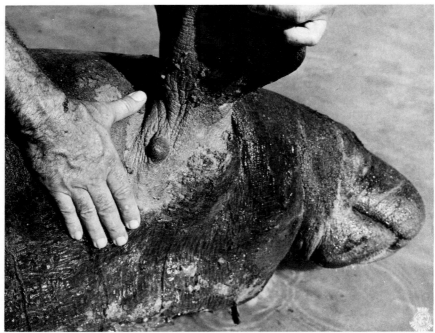

FIG. 4 Teat of an adult female West Indian manatee shown where it lies in the axilla just behind one of the flippers. (Courtesy of Miami Seaquarium.)

there is little basis for separating the species in that regard. The separation of the two species appears to be based primarily on their wide geographical separation. Husar (1978a) has given the best general account and summary of literature and data for this species that we have seen to this time.

The Amazonian manatee is similar in overall body shape to the other two members of the genus, but in general the adults are said to be smaller [a 2.8-m male is the largest on record while the other species may reach 4.5 m (see Husar, 1977a)] and more slender. The chest and abdomen of the Amazonian manatee bear distinct white markings (Fig. 5), and there are no nails on the flippers (these may be worn away in the other species and superficially appear to be absent).

Distribution

As clearly stated and detailed by Husar (1977b), the range of the West Indian manatee includes rivers, estuaries, and coastal areas of the tropical

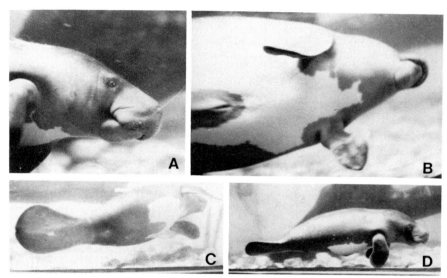

FIG. 5 Young male Amazonian manatee, originally from the Amazon River near Leticia, Colombia, live at the Fort Worth, Texas, Zoo on 26 June 1965. (A) Lateral view of head; (B) ventral view of anterior portion of body showing genital opening with extension of dark pigment into the white area; (C) overall ventral view showing extent of dark and white areas; (D) overall lateral view. (All photographs by David K. Caldwell.)

and subtropical regions of the New World Atlantic (Fig. 6). Manatees often enter man-made bodies of water such as canals. They live both in fresh and saline, shallow, well-vegetated waters, usually moving into the warm freshwater springs and around power plant warm-water outfalls in the colder months of the year and then dispersing into the warm saline waters during the warmer months. By like token, manatees reach into the more northern parts of their range in summer.

In the United States, the range of the West Indian manatee reportedly extends as far north as Virginia (Moore, 1951a), but records north of Georgia are rare though not unknown (see Browne and Lee, 1977, for a discussion of manatees in North Carolina). Records for the northern coast of the Gulf of Mexico are also sparse but manatees have been documented from Pensacola, Florida (Collard, et al., 1976), and from Louisiana (Lowery, 1974). Manatees occasionally seen in Texas are thought to have strayed northward from Mexico. The West Indian manatee's range then extends southward through Central and South America to about 19° S in Brazil. In the West Indies, manatees are known from the Greater Antilles (including Cuba, Hispaniola, Jamaica, and Puerto Rico). There is evidence from Indian middens that

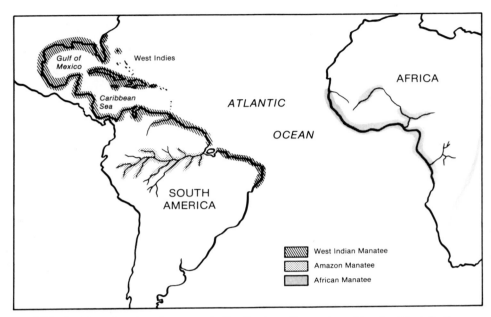

FIG. 6 Map showing the general distribution of manatees. For the West Indian manatee, the lighter shaded areas indicate the approximate normal range, while the darker shaded areas at the extreme north and south indicate the recorded extremes and area where West Indian manatees are considered unusual and perhaps accidental. The ranges are probably not as continuous as that suggested by this map and reference should be made to the text for details. There are a few fossil records from the Lesser Antilles not indicated here as there are no Recent records of living manatees from there.

manatees once inhabited the Virgin Islands, Antigua, Martinque, and St. Lucia (Ray, 1960; Wing et al., 1968). They once were reported from Trinidad, but their present status there is unknown (Husar, 1977b). There is a recent record from Grand Bahama Island (Anonymous, 1976).

It should be noted that the recorded coastal range of manatees, although extensive, is not necessarily continuous. Manatees tend to congregate in certain areas within their overall geographic range where ecological conditions best suit their requirements. Such local distribution may be extensive, as in inland waterway areas on the east coast of Florida, or very spotty, as in certain springs and rivers along the west coast of that state. In Puerto Rico the manatees are mostly marine in their distribution but have been found to congregate within 5 km of natural or artificial freshwater sources (Powell et al., 1981).

According to Husar (1978a), the range of the African manatee includes both shallow coastal and freshwater rivers, estuaries, and coastal regions of West Africa from Senegal to Angola (Fig. 6). They seem to prefer large, shallow estuaries and their swampy borders where weeds are abundant. Details of specific bodies of water from which African manatees have been reported, with appropriate literature citations, were listed in some detail by Husar (1978a). More recently, Nishiwaki et al. (1982) have surveyed much of the African manatee range. The area of greatest abundance was found to be the Niger River and its tributaries.

Amazonian manatees may be found in the Amazon River Basin of South America and possibly in the nearby Orinoco system as well (see map and summary in Husar, 1977a, and Fig. 6, this chapter). There are unverified claims that these manatees may also range along the Brazilian coast from the mouth of the Amazon to Espirito Santo (19°S), but it is likely that such supposed coastal records refer to *T. manatus* if valid for a manatee at all. Little appears to be known of the habitat preferences for this species, although it is said to prefer blackwater lakes, where acidity is relatively high, as well as lagoons and oxbows—all of which suggest a very placid water habitat preference similar to the other members of the genus.

Abundance and Life History

The population of the West Indian manatee is estimated as having been several thousand when the Europeans arrived in what is now the United States. Unfortunately for the species, the meat was considered highly palatable. Adding to this problem, particularly in those days when no refrigeration was available, was the fact that manatees could be maintained alive in pens or crawls until wanted for food or sale. Compounding this, the dried meat remained good for about 1 month unsalted. If salted, it was good for about 3 months (True, 1884). In addition to the use of manatees for food, Campbell and Powell (1976) noted that manatees have also been deliberately taken for oil and leather. It was man's predation for all of these reasons that has nearly exterminated the manatee in the West Indies and is still seriously depleting its populations elsewhere.

The manatee then filled a real need in the lives of the early Florida settlers, so much so that by 1893 the population was so reduced that the state of Florida passed a law safeguarding it. The law was reinforced with another in 1907 which imposed a $500.00 fine on anyone killing or molesting a manatee (Moore, 1956). These laws remained in effect until the passage of the U.S. Endangered Species Acts of 1969 and 1973 and the U.S. Marine Mammal Protection Act of 1972 when these animals were placed under even more stringent protection, with greatly increased penalties for violations of the laws. Despite the restrictions, man has decimated manatee populations, especially for meat, and this has continued into recent decades when meat was in short supply (as during times of war rationing) or expensive (as during times of financial depression).

More indirectly, man has also contributed to manatee deaths by altering the habitat in the disposal of sewage and other wastes, and by dredge and fill projects. Oil spills are a potential problem as offshore drilling and oil shipping increase in the Gulf of Mexico. Many of these factors either destroy manatee food (aquatic vegetation) directly or increase the turbidity of the water so that the vegetation cannot get enough sunlight to grow. It should be noted in all fairness to man, however, that the warm water effluents from an increasing number of power and other industrial plants do provide warm refuges for manatees and that new canals may open up new areas to manatees or within themselves provide new protected habitats.

Probably the single most frequent cause of manatee death in these modern times, at least in the United States, is major injury from collisions with boats and barges and especially with boat propellers. Moore (1956), who made special note of propeller injuries as early as 1949, utilised the scars on those animals that survived as natural tags in local population and behavioral studies. Poaching and vandalism have long been problems, along with habitat alteration as noted above. Flood control gates also present a major problem, especially in south Florida where so much of this work is done in manatee habitat (Reynolds and Odell, 1977). Manatees are swept into the gates during times of water-control interchange where they subsequently drown. Accidental entanglement in fishing lines and nets may also cause death through drowning (Brownell *et al.*, 1978). Some deaths can be attributed to cold and disease (see section on Disease below).

Hartman (1974) suggested that reduced boat speed and supervision of SCUBA divers (who harass the animals and often drive them into less suitable habitat) in known manatee refuges are solutions that should not arouse too much public resistance, as they threaten no one's economic survival. Poaching is a problem that comes and goes with the price and availability of other meat, and an increased incidence might be anticipated with high levels of inflation and should be guarded against. For whatever reason, manatees arouse actual hostility in a not-insignificant portion of the human populace, and a program of education has been proposed to help combat this (Brownell *et al.*, 1978).

Latest surveys of the manatee population in Florida indicate the presence of between 800 and 1 000 individuals (Brownell *et al.*, 1978). Estimates of a slight decline or a slight increase in numbers are in conflict, but a continuing program of strict conservation is deemed advisable. Suggestions for manatee breeding colonies are presently held to be unrealistic because of the slow reproductive rate and spatial requirements of the species.

Manatees apparently interact only marginally with animal species other than man. There is no evidence that they are preyed upon by alligators, crocodiles, sharks, or killer whales (Hartman, 1971). There is no evidence of interplay even with porpoises, those most playful of mammalian species, during their chance encounters. Future studies could show that feeding competition might conceivably exist between manatees and other aquatic animals such as turtles, fish, or crustacea, but no data are available on this subject (Campbell and Irvine, 1977).

It has been suggested that manatees could be used to control the weed overgrowth that often blocks waterways in the southeastern United States. Such practical use of these animals is presently unfeasible, as is the consideration that they might be raised for food (Campbell and Powell, 1976).

Direct depredations by man for food and his modification of its habitat have caused concern for the African manatee, and it has been placed on the threatened list by the United States Fish and Wildlife Service. There are few data to make direct comparisons between present stocks and those of the past, but the numbers are believed to be reduced. African manatees are also protected locally and any direct hunting that continues is now done in secret. On the other hand, nets set for sharks often kill a number of manatees incidental to the fishery and at least local impact on manatee populations may be severe. Predators other than man have not been reported, but sharks and crocodiles do occur in African manatee habitat and might be considered a danger.

The Amazonian manatee has long been hunted for food by Indians and in more recent years for hides. All of these activities are believed to have reduced the stocks of these animals; and even though they are now protected by law, there is still at least some poaching. As the Amazonian basin is developed for human activities, it is likely that these manatees will suffer from habitat destruction and consequent injury to their stocks. It has been claimed that Amazonian manatees may be preyed upon by crocodiles, sharks, and even jaguars.

Hartman (1971, 1974, 1979) has done a masterful job in dealing with the behavior, ecology, and conservation of the West Indian manatee. We have drawn heavily upon his work in the following sections as well as others. It will be noted that only a little experimental work has been published. We hope that research can be initiated towards eliminating some of the gaps in our knowledge of this species.

Manatees live only in tropical and subtropical waters, retreating into natural or man-made refuges when the air temperature drops below 10 to 15°C

(The African manatee seems limited to waters with temperatures of 18°C or higher. Irvine's (1983) studies of manatee metabolism supported behavioural observations of captive and free-living animals which suggested that 20°C is the minimum suitable water temperature for Florida manatees. West Indian manatees prefer shallow bays, estuarine waters, or slow-moving rivers. In the area between Cape Canaveral and Jacksonville, Florida, manatees may occur in the open ocean, but much more rarely than in more protected, less saline, environments. About half of the experienced fishermen that we interviewed had not seen a manatee in the open ocean beyond the immediate environs of inlets. William C. Raulerson of the University of Florida Whitney Marine Laboratory told us that he had once participated in the accidental capture off St. Augustine Beach of two adult-sized manatees in a beach seine being fished in water no deeper than 1.5 m and extending no more than 100 m off the beach. The fact that the two animals were not seen when the net was set indicated that they were fully submerged, probably in the deeper (more offshore) part of the area encompassed by the net. B. C. Townsend of Marineland of Florida reported that he had never seen manatees in the St. Augustine area more than 1 km out at sea, and then only very rarely. This agrees with observations on the Gulf of Mexico coast of Florida where manatees are found within 100 m of the shoreline along high-energy beaches (Hartman, 1979). Jerry Foreman, formerly of Marineland of Florida, told us that he had occasionally seen manatees cruise out from the St. Augustine Inlet (from the saline Matanzas River) to the ocean end of the rock jetties but turn back as soon as they reached the protective end of the breakwater. This suggests that the species avoids turbulent waters unless strongly motivated to migrate to another location. Our personal records on manatee sightings in the open ocean include none for the winter. In the years 1968 through 1973, all six such sightings off northeastern Florida were from June through October, indicating that open-ocean excursions in this area are even more limited in cold weather than warm. We should note here that manatees are found on occasion in the inland waterway just behind the open beach in winter, and we have a positive record of an adult-sized animal just north of St. Augustine on 17 February 1979. Studies based heavily on Florida–Gulf of Mexico coast observations indicate that manatees migrate into the open ocean there in the summer (Hartman, 1971).

Manatees are so flexible in their food requirements that they can sustain themselves in most underwater habitats that they can reach. There is considerable behavioral evidence that they have a need for drinking fresh water. Hartman (1979) cited instances of freshwater intake by several individuals at different sites. At one site in the North River, in the inland waterway about 5 km north of the St. Augustine Inlet, one or more animals occasionally drank from a freshwater outlet (sulfur water) at Usina's Fish Camp. On the other hand, A. Blair Irvine (in Brownell *et al.*, 1978) indicated that his studies on blood and urine suggest that a freshwater source may not be physiologically required by manatees.

Although movement is evident into warm waters at colder times of the year, Hartman (1979) concluded that no specific migration routes exist for the species. Individuals are believed to move into and out of areas virtually at random and for variable lengths of time. Their nomadic wanderings probably cover several hundreds of kilometers of coastline.

No evidence of territoriality has even been seen, either in defense of an underwater site or even a specific bit of vegetation upon which an individual might be feeding.

Internal Anatomical Characteristics

For this discussion we have drawn heavily on the summaries of Harrison and King (1965) and Husar (1977b).

Skeleton and teeth

The skeleton is of extremely dense (pachyostotic) bone that may have contributed to the downfall of manatee populations for the heavy ribs especially were once highly treasured for such things as pistol grips and for other uses to which ivory would have been put if available. Pachyostosis increases specific gravity, and it has been suggested that this condition may contribute to maintenance of neutral bouyancy in these slow-moving and rather sedentary animals. The West Indian manatee skull (Fig. 7 A–D) is broad with a relatively short snout and an expanded nasal basin. Husar (1978a) included photographs of the skull of the African manatee. She also (1977a) included photographs of an Amazonian manatee skull, and these suggest a structure narrower than that of the other two species. It is not clear as to whether or not this might be a function of age. As with all three species, the lower jaw of the West Indian manatee is massive. The molariform cheek teeth are replaced from the rear and, as the anterior teeth wear from the excessive amount of sand and grit in the manatee's normally vegetable diet, they drop out. Domning and Magor (1977) discussed this phenomenon in *T. inunguis*, and their data confirmed earlier observations that the increase in solid food intake after weaning, and hence more chewing, acts as a mechanical stimulus for the initiation and continuation of tooth row movement. It was once thought that this replacement continued throughout life, but there is recent evidence that this may not be the case (Husar, 1977b). There are usually 5 to 7 functional teeth in each upper and lower jaw, and it has been estimated

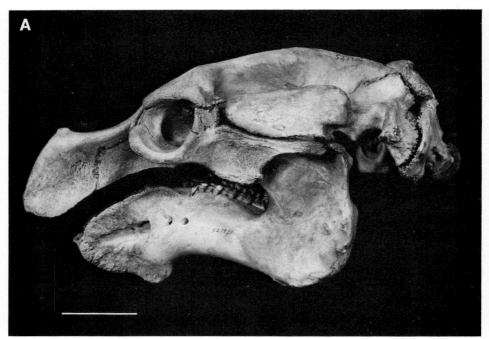

FIG. 7 Skull of a male West Indian manatee (USNM 527920). (A) Lateral view of cranium with mandible in place; (B) dorsal view of cranium with mandible in place; (C) ventral view of cranium; (D) ventral view of cranium and mandible. Scale (A–D) 5 cm (2 inches). (Courtesy of the U.S. National Museum of Natural History, Smithsonian Institution.)

that from 20 to 30 teeth per jaw are possible within the life span of a given individual. The cheek teeth are brachyodont, enameled, and without cement. According to Harrison and King (1965), there are two vestigial incisor teeth in each jaw at birth, but these are later resorbed.

Vertebrae are said to number 48 to 54 (6 cervical, 17 to 19 thoracic, and 27 to 29 lumbocaudal) (Husar, 1977b).

Musculature

Although giving special emphasis to the description of the musculature of *T. inunguis*, Domning (1978) also made comparisons between the muscles of this species and those of *T. manatus*. That writer speculated that *T. inunguis* is more specialised for surface feeding and for swimming than is *T. manatus*.

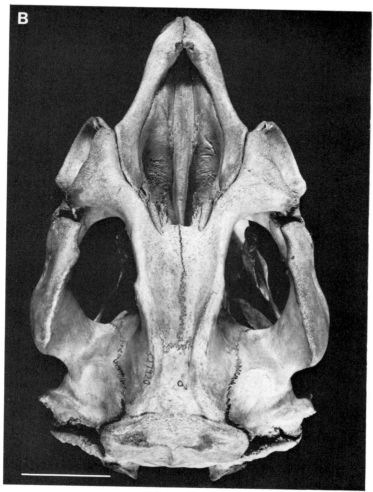

FIG. 7 *(continued)*

Digestive system

The fleshy lips are highly movable and well adapted to gathering in the vegetable matter that constitutes the primary diet of manatees. The animal often uses its flippers in this. Yamasaki *et al.* (1980) studied the gross anatomy and histology of the tongues of *T. manatus* and *T. senegalensis* and found, along with mucous glands, a few taste buds and papillae which may serve as

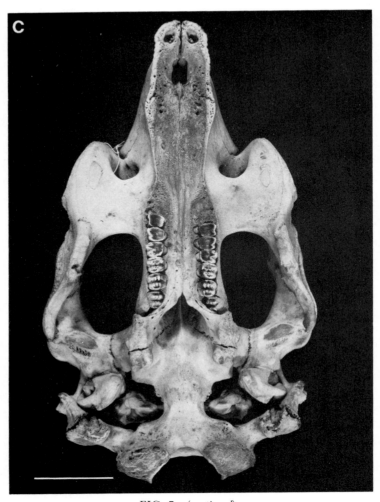

FIG. 7 (*continued*)

tactile organs. Once ingested (there are roughened mandibular and premaxillary pads which apparently are also used to assist in ingestion), the food is chewed and passed into the muscular oesophagus over a tongue that is not especially mobile. There is a strong cardiac sphincter at the lower end of the oesophagus before the main stomach, which has thick muscular walls with two appendages (gastric caeca) at its pyloric end. There are numerous digestive glands along this course. The second chamber of the stomach is not as thick walled as the first and leads directly into the very muscular small

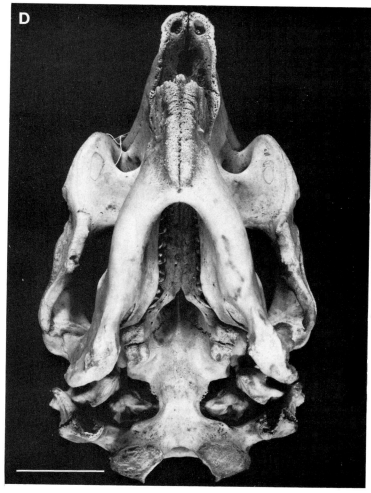

FIG. 7 (*continued*)

intestine. There are two conical caecal appendages close to a powerful ileocaecal valve. The large intestine is about twice the diameter of the small, and nearly as long. According to Harrison and King (1965), the small intestine of a 2.4-m male measured over 7.6 m, while the large intestine was over 5.5 m in length.

A more recent study by Reynolds (1980) of the structural and functional anatomy of the gastrointestinal tract of *T. manatus* agrees for the most part with the descriptions provided by Harrison and King (1965). Reynolds indicated that this system is basically similar to those of other nonruminant herbivores (e.g., horses) but that it has special adaptations not seen in most other mammals. He felt that the adaptations may relate to osmoregulation as well as to herbivory and that some structural features may provide glandular protection from the masses of rough vegetation typically ingested by manatees.

There is a large gall bladder. The liver is trilobed, and bile and pancreatic ducts open separately into the duodenum. More detailed descriptions of all of these organs and their associated glands are given by Harrison and King (1965) and summarised by Husar (1977b).

Respiratory system

There is a thin diaphragm separating the lungs from the abdominal viscera.

The epiglottis is rudimentary and there are no apparent vocal cords, these being replaced by fleshy cushions of ligamentous and fibroelastic tissue (Harrison and King, 1965; Husar, 1977b). Eight to 12 tracheal rings (several bifurcate) are present. The trachea divides into two bronchi which run parallel for several centimeters before entering the rather small, flattened, and elongate paired lungs near their summits. According to Harrison and King (1965), lungs are not lobed and have few notches on their borders. The air sacs are large, with unusually large amounts of fibroelastic and smooth muscle tissue, which may function to compress the air in the lungs in order to aid the manatee in sinking without propelling itself downwards or exhaling air (Husar, 1977b).

At each normal breath the manatee exchanges one-half its lung air, and compared with other aquatic mammals it has a very low resting consumption of oxygen, while utilizing the oxygen very highly (Scholander and Irving, 1941; Husar, 1977b). Bradycardia slowly develops during diving, and discussions of this and other aspects of manatee respiratory physiology are presented by Scholander and Irving (1941), Gullivan and Best (1980), and Irvine (1983).

Circulatory system

Husar (1977b) summarised earlier studies on and descriptions of the circulatory system, and White *et al.* (1976) provided data on the blood chemistry of manatees. The heart is globular, comparatively large, and has a large interventricular cleft. It normally beats at a rate of about 50 to 60 times per min (Harrison and King, 1965). There are two venae cavae, and two pulmonary veins enter the atrium at a single opening. The posterior vena cava system is double. Unusual vascular bundles are scattered throughout the body, including the body walls, face and jaw, tail, and spinal canal (Husar, 1977b), but they are most highly developed in the flippers. These bundles are not in the form of networks of vessels in the usual sense of rete mirabile, but have been described as "broomlike" where large arteries divide into large numbers of small vessels that run parallel for long distances with the further branching associated with a rete network.

Farmer *et al.* (1978) found that the hematocrit and oxygen binding capacity of the blood of *T. inunguis* are low compared with those values for other diving mammals but are similar to those for land mammals. Although not so stated, it is presumed that the blood of the other two manatee species would show similar values to those found in the Amazon manatee.

Urogenital system

As with much of our discussion of internal anatomy, we here draw on the summaries of Harrison and King (1965) and Husar (1977b).

The kidneys are only superficially lobulate and lie on the surface of the diaphragm. The testes are abdominal and their seminal vesicles are large. The nonglandular prostate is composed of erectile muscular tissue. The ovaries consist of large masses of beadlike spherules and lack a heavy capsular coat. The bicornate uterus narrows distally to a rounded cervix. The placenta is of the hemochorial, deciduate type.

Endocrine system

According to a summary by Husar (977b), the hypophysis is oval in shape with well-developed anterior and posterior lobes (but with reduced pars tuberalis and pars intermedia). Thyroid glands from three individuals ranged from 0.11 to 0.13 g/kg of body weight, and low thyroid activity is suggested by the presence of profuse colloid. Unusual thyroid structure (see summary in Husar, 1977b) is suggestively linked to the pachyostotic bone, the sluggish behavior typical of manatees, and the low rate of oxygen consumption. The cortex of the thymus is prominent, and the medulla is small and pale. The elongate adrenal glands, which also display atypical histology, are positioned retroperitoneally between the carotid artery and the bronchus. There is a small spleen.

Nervous system and senses

The brain of the adult manatee is small and weighs approximately 370 g. There is a deep longitudinal fissure separating the cerebral hemispheres and well-defined Sylvian fissure. Otherwise the cerebrum is relatively smooth with only a few shallow sulci on the frontal, temporal, and parietal lobes, and the rounded occipital lobe is concave and partially covers the somewhat convoluted cerebellum (Husar, 1977b). From above, the shape of the brain is quadrangular with rounded corners, and the height, breadth, and length are subequal (Harrison and King, 1965).

Behavioural observations suggest that manatees have good hearing. Bullock and associates (1980, 1982) have studied hearing in West Indian and Amazonian manatees. In the West Indian species the largest potentials were obtained at frequencies of 1.0 to 1.5 kHz, suggesting the most sensitive hearing in this range. Potentials were obtained to signals as high as 35 kHz. In the Amazonian species the greatest sensitivity was around 3.0 kHz. Manatees possess very large ear ossicles. There is no pinna to the ear, and the external auditory meatus is narrow and may be occluded for part of its course (Harrison and King, 1965). The meatus expands upon reaching the rather large tympanic membrane.

There is a well-developed nictitating membrane (or third eyelid) at the

FIG. 8 Captive West Indian manatees "kissing" or "nuzzling". The animal on the left is a female and the one on the right is a male. Note the mobility of the flippers. (Courtesy of Marineland of Florida.)

medial aspect of the eye, and the upper and lower eyelids cannot always be readily distinguished (Harrison and King, 1965). Those writers also noted that the eye is small compared with the size of the orbital cavity and that the optic nerve is slender. Hartman (1971) believed from behavioural observations in the wild that manatees can see for considerable distances in clear water, but that they may be farsighted inasmuch as they tend to bump into objects within close range. Their tapeta lucida shine pink in the dark, indicating the likelihood of good vision under low light intensities. Hartman never saw them raise their eyes above the surface to investigate objects in air but did note that underwater they usually approach new objects head-on, suggesting binocular vision.

The tactile senses of manatees apparently are good, and individuals frequently touch and even embrace one another. They are often seen to "mouth" or "kiss" in social encounters (Fig. 8). The heavy bristles covering the snout seem to be tactile–sensory in anatomy and function (D. K. Caldwell and M. C. Caldwell, 1972).

There is still conjecture as to the senses of smell and taste. Manatees have a large olfactory organ, and behaviourally they are often very selective in their choice of food; but whether the selection is through smell, taste, or touch (or a combination of these) is not known. Hartman (1971) did suggest that there may be a "smell–taste" function in which manatees may recognise chemical gradients in water.

Behaviour

Feeding

In nature, manatees are, for the most part, herbivorous, opportunistic feeders. They usually use their fleshy lips to grasp the food, and often use their flippers to help guide it into their mouths. Hartman (1979) listed in detail the wide variety of vegetation upon which West Indian manatees feed in nature in both saline and fresh waters. Husar (1978a) noted that African manatees consume large amounts of a wide variety of submerged aquatic vegetation and have been reported eating the leaves of mangroves hanging over the water. Like the West Indian manatee, the African manatee has been considered for organised use in aquatic weed control. According to a summary by Husar (1977a), Amazonian manatees feed primarily on vascular aquatic plants but have also been reported to eat floating palm fruits.

Hartman (1979) provided a list of food upon which West Indian manatees are maintained in captivity, where they consume about 1 kg food per 5 cm of body length. Captive manatees will eat a variety of unlikely vegetable foods such as most garden vegetables, cultivated legumes and pasture grasses, many commercial fruits, lawn grass, various weeds, palmetto fronds, and bread (Hartman, 1971). Lettuce forms the bulk of the food fed to captives. William C. Raulerson told us that as a boy he fed slices of common white bread to wild manatees in the St. Augustine area whereupon, over a period of several days, at first they only came up and touched it, then mouthed it, and finally ingested it. After that, the animals continued to return to the same spot each day until Raulerson tired of the routine.

Wild manatees are said to eat vegetation along the banks of the waterways in hard times (Hartman, 1971), but D. K. Odell (in Brownell *et al.*, 1978) reported seeing wild manatees pass up submerged vegetation that was readily available in favor of feeding on the emerged plants on the bank. On the other hand, Domning (1980) found that both *T. manatus* and *T. inunguis* prefer to feed as low in the water column as possible and to pull food below the surface if conditions permit.

Best (n.d.) provided a good summary of the foods and feeding habits of manatees and noted that primary differences in foods eaten result from differences in habitat (i.e., *T. inunguis* is restricted to fresh water while *T. manatus* and *T. senegalensis* live a large part of their lives in the inshore marine environment).

According to Hartman (1971) the species feeds for 6 to 8 hr per day in 1- to 2-hr sessions.

Apparently manatees ingest good numbers of lower-form animals that are attached to the vegetation that the manatees eat, but these are taken in only adventitiously (Hartman, 1979). In captivity, manatees will eat a variety of dead fishes, and similar behavior in the wild in which they eat fish still caught in gill nets (these fish probably also dead) has been reported in Jamaica (Powell, 1978). We know of no evidence that manatees catch live fish to eat. Whatever the source, it has been suggested that the animal material ingested may be an important source of protein for manatees, especially the large numbers of arthropods that occur in the vegetation that manatees eat (Hartman, 1979).

Swimming and diving

The manatee, while a wholly aquatic mammal, is not a deep-water species. Ten meters is the greatest depth to which an individual has been seen diving (Hartman, 1971). Normally individuals cruise between 1 and 3 m below the surface and feed from just below the surfact to a depth of 3 m (Hartman, 1971).

The length of time that an animal remains below the surface without breathing is dependent upon its size and activity level. Resting adults may go without breathing for 15 or 20 min (Reynolds, 1977). Four minutes is about average for resting individuals, with one breath being taken at each surfacing (Hartman, 1971). Active animals need to breathe more frequently.

Manatees are excellent swimmers, propelled by the up and down motion of their tail (Hartman, 1979). Their top speed in short bursts is about 25 km/hr, cruising speed is between 4 and 10 km/hr. The tail is also used in steering (Hartman, 1979).

The flippers are held in close to the chest while swimming, but they are the sole instruments used in "walking" on the bottom (Hartman, 1971). They are also used in turning when the animal is moving very slowly (Hartman, 1979). Neonates begin life by swimming, using their flippers exclusively, but quickly learn to use their tails instead (Moore, 1956, 1957).

Manatees show no constancy in the time of day when they feed, rest, socialise, or engage in other activities. Their nocturnal behaviour parallels that during the day, with similarly unpredictable bouts of activity (Hartman, 1971).

Sound

Schevill and Watkins (1965) were the first to record a few manatee sounds that they described as "squeaky chirps" and "ragged squeals." Those writers felt that the sounds were communicative in function and were not being used for echolocation. The calls were short (< 0.5 sec), and low in both frequency (< 5 kHz) and volume (about 10 to 12 dB above background at a distance of 3 to 4 m). Those writers also found that the sounds usually consisted of two or more frequencies that were not harmonically related, but that each component may exhibit harmonics.

The few vocalisations reported for the Amazonian manatee are short and with a fundamental frequency of some 6 to 8 kHz (Evans and Herald, 1970).

Hartman (1979) reported that the West Indian manatee sounds he heard at Crystal River, Florida, were "high-pitched squeals," "chirp–squeaks," and "screams," but that normally the manatees there were silent. He noted that the sounds were produced with the mouth and nostrils closed and that there was no accompanying escape of air. It was Hartman's finding that the sounds were made only under conditions of fear, aggravation, protest, internal conflict, male sexual arousal, and play. Like Schevill and Watkins, Hartman found no evidence that any of the manatee sounds were navigational and he felt that rather than being truly communicative they were more impulsive. For example, a frightened animal (usually a juvenile) might "scream" in alarm, a cow might "squeal" in apparent annoyance when embraced by a bull, or an animal might "chirp–squeak" in apparent pleasure while rubbing against a log. Hartman (1979) included a number of such associations of sounds produced and behaviour performed. White (1984a, 1984b) noted that a captive newborn produced loud, high-pitched cries as it swam to the surface for its first breath.

Hartman (1979) found that the only predictable vocal exchange was an "alarm duet" between a cow and her calf. Here the cow would call her endangered calf to her side with repeated screams, and each of these screams triggered an immediate reply from the calf, which responded even more vigourously than its mother. A calf in distress would also vocalise and this would elicit an immediate response from its mother. Calves would also vocalise when not disturbed and the mother would usually respond unless she was resting. Finally, Hartman (1979) found that mother–calf vocalisations appeared to increase when the two were in turbid water, indicating that the animals were maintaining contact under conditions of impaired vision. He felt that the two were never out of vocal communication range even in clear water, and that the distances involved might be fairly great.

Hartman (1979) believed that manatees might be able to recognise other individual manatees by subtle variations in each animal's sounds.

Reynolds (1977) recorded underwater sounds related to chewing. As in cetaceans, manatees might also be expected to produce sounds related to digestive functions (stomach rumblings).

Aggregation and movement

Hartman's (1971) experiments on tracking or monitoring the movements of this species all ended in failure. A radio transmitter was attached to three different individuals for a total of five trials. On each trial the animal had rid itself of the device within 30 min. Hartman (1979) expressed confidence that a better method of radio tracking was possible and that the results would be rewarding. In the meantime, various scars and personalised markings on individuals, first detailed by Moore (1956), can continue to serve us well as natural tags.

To a large degree, manatees are only moderately social animals. Except for basic mammalian activities of copulating and mother–infant interaction, each individual appears to pursue its own activities much of the time and is largely independent of the others. On the other hand, Reynolds (1979) noted that West Indian manatees are capable of coordinated behaviour and may swim synchronously. According to Husar (1978a), one group of 15 African manatees cruising together was believed to be unusual. Although large aggregations of Amazonian manatees have been reported in the lakes and rivers of the middle Amazon, such large groups seem to be rare today, with loose groups of perhaps 4 to 8 being reported in a feeding area (Husar, 1977a).

They do come together in *aggregations*, usually with a group (of either or both sexes) consisting mainly of animals of similar size (excluding infants). Such associations should probably not be termed *herds* as the latter term infers a stability not present in manatees. The aggregation often occurs in response to environmental stimuli acting simultaneously on various individuals.

Cold-weather-induced aggregations are common in manatees. Hartman (1979) provided excellent maps of the warmer-water sites in Florida where these animals can be expected to congregate. He made splendid use of these refugia for study sites and population estimates of total population. Individual manatees do not necessarily return to the same refuge in winter but appear to wander into whatever spring, estuary, or river is convenient. While there, some interplay occurs between individuals, but it consists largely of stroking and nuzzling.

Mother–calf and sexual behaviour

Long-term individual associative relationships have not been noted in the species beyond that of mother and calf. This association is of 1 to 2 years' duration (Hartman, 1971).

Cows in oestrus do attract temporary associations of both adult and juvenile males for periods of up to 1 month. As many as 17 males have been seen actively engaged in attempting to copulate with one female in oestrus.

Sexual behavior between two manatees stimulates other males in the area to engage in seuxal activity. Their attention may be directed towards the cow in oestrus or nonreceptive females or other males in the vicinity. Similarly, social facilitation is also seen in resting, eating, and breathing when the activities of one animal induce others to engage in the same behaviour (Hartman, 1979).

Herd cohesion

The absence of herd cohesion in this species is in stark contrast to that reported for a close relative, the Arctic sea cow. The latter species, now extinct, fortunately was reported on rather thoroughly by Steller, who spent 10 months observing it from a hut in the Bering Sea (True, 1884). Steller reported a family group as consisting of one adult male, one adult female, one large offspring, and one smaller calf. If a member of one such group was injured by a harpoon, Arctic sea cows from nearby groups sometimes made directed attacks on the line or the boat. Arctic sea cow males also tried vigourously to defend the females. If the female was killed the male remained in the area of the dead body for up to 3 days (see True, 1884). Also, when the herd was moving, the young were kept in the middle. None of the above cohesiveness is seen in the solitary manatee.

While no great affectional bonds are known to exist for manatees, neither is there evidence of aggression. They do not react with irritation even when their passage to the surface for air is blocked by another animal (Moore, 1956).

No data on their intelligence are available, but one captive was trained to perform a few simple acts on cue (Moore, 1951b).

Reproduction

Manatees in south Florida apparently breed year-round (Reynolds, 1977) and probably do so elsewhere as well. Also in Florida, and probably elsewhere, they are known to calve throughout the year, but Blair Irvine has suggested that, on the basis of relative instances of calf mortality, most are born during the spring and fall (see Brownell et al., 1978). This agrees with Steller's description of year-round breeding and calving in the Arctic sea cow, with peak breeding in spring and peak calving in autumn (translated in True, 1884). It is not known if manatees are seasonally polyoestrous or have one long oestrous period (Reynolds, 1977).

There is still some controversy regarding the age at sexual maturity. Hartman (1979) estimated that the manatees at Crystal River, Florida, matured at an age of from 3 to 5 years, although earlier (1971) he had suggested 4 to 6 years. On the other hand, growth curves led Daniel K. Odell to conclude that females may not become reproductively mature until 7 or 8 years of age, and males not until the age of 9 or 10 (see Brownell, et al., 1978). Two captive females produced their first young in their ninth year (White, 1984b).

In truth, it appears that it is the manatee males that are sexually active year-round. Females rarely appear to be receptive even when in apparent oestrus and surrounded by a group of males that have been attracted to her presence. Hartman (1971) observed female receptivity to copulation only once during his extensive field studies. In this one instance the cow permitted intromission by three males over a 45-min interval. During mating, both animals lie horizontally in the water, deep enough to support both of them, in a ventral to ventral position (Hartman, 1971).

Precopulatory activity is intensive (Hartman, 1971). Both juvenile and adult males pursue the presumably oestrous but nonreceptive female, pushing and shoving at her while she tries to keep her genital area away from them to the point of nearly stranding herself in shallow water (Hartman, 1971; M. C. Caldwell and D. K. Caldwell, 1972; Moore, 1956). We again observed similar behaviour in the inland waterway near Marineland, Florida, on 25 July 1978. In this case a large individual presumed to be a female was attended by at least seven animals thought to be males. The activity was first reported in midmorning and continued throughout the day in an area in the waterway extending over a distance of perhaps 5 km. From time to time the presumed female would steadily move away with the presumed males strung out behind. She would then stop and the group again gathered around her.

The prolonged, often frantic, precopulatory activity presumably serves some physiological purpose. Manatee females may have several sterile oestrous periods prior to successful conception, similar to elephants (Hartman, 1979).

The gestation period for wild West Indian manatees is not known, but it is believed now to be about 12 to 13 months (Hartman, 1971). White (1984a) gave a gestation period of about 14 months, based on captive animals. Earlier accounts had suggested a much shorter period. For the Amazonian manatee, Husar (1977a) also noted a gestation period of about 1 year with a single calf usually born.

Three captive manatee births have been reported. Tail-first delivery was seen in two of these and illustrated by photographs (White, 1984b); head first delivery occurred in the third was also shown in photographs (White, 1984a). Wild newborn calves have been found to measure from about 1.2 m (Reynolds, 1977) to about 1.3 m (Hartman, 1979). Moore (1957) gave the length of a 38-day-old calf as 1.3 m, with a weight of 31.9 kg. White (1984b) noted that calves born at the Miami (Florida) Seaquarium have ranged from 27.7 kg to 36.3 kg, with a male of 1.2 m in length weighing 31.8 kg.

Manatee calves nurse from one of the mother's two nipples (Fig. 9), and there is no apparent preference for one side over the other. The calf and mother float in a horizontal position and the calf comes in at an angle. The mother lends no particular assistance other than sometimes holding her flipper out of the way. The age at which the young is weaned is uncertain. Manatee calves begin to graze at least by the end of their third month, and probably much earlier (Hartman, 1979). A captive infant was eating lettuce leaves when 38 days old (Moore, 1957). The nursing period is prolonged and a calf that has been taking solid food will often continue to return to nurse for a year or more. Captive cows have been observed nursing both a newborn and a much larger calf, and in addition these captives have been seen to exchange calves for nursing. A lactating cow would, therefore, be a good candidate as a wet nurse for an orphaned calf as she will accept for nursing

FIG. 9 Captive West Indian manatee calf nursing underwater. The calf may lower its tail to varying degrees so that it is almost perpendicular to the plane of its mother as in this instance, or more often it may nurse at an angle in a plane more nearly the same as that of its mother with its tail only slightly drooped. Very early accounts incorrectly described the mother as cradling her calf even in air at the surface in a somewhat upright position, similar to that assumed by a human. (Courtesy of Miami Seaquarium.)

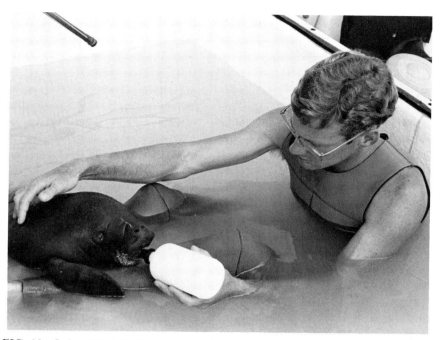

FIG. 10 Infant West Indian manatee during successful bottle feeding in captivity. The dark color is typical of such very young individuals. (Courtesy of Sea World of Florida.)

calves that are not her own (see also Hartman, 1971). Several oceanaria have had some success in maintaining wild orphaned calves on artificial formulae (Fig. 10).

It has been reported that a female will raise and lower her newborn calf on her back or tail for several hours, possibly to help the calf establish a breathing rhythm (Moore, 1951b; Reynolds, 1977).

Hartman (1971) estimated that manatees calve approximately every $2\frac{1}{2}$ to 3 years, but noted that a female losing her calf shortly after its birth might have another after as little as 2 years. On the other hand, other workers consider an interval of about 5 years as a more likely calving interval under completely normal conditions (see Brownell et al., 1978).

Disease

Few data are available on diseases in manatees. Husar (1978b) provided summary material that indicates that these animals appear to be highly susceptible to pneumonia and other bronchial disorders. It has been suggested that the winter activities of man may aggravate these conditions by boating and SCUBA activity in large warm side springs where manatees congregate, thereby driving the animals out into the colder main rivers where their resistance to such disorders is lowered. Bloated live manatees have been observed at the surface, and being unable to submerge they too may be more susceptible to cold if the condition, the cause of which is unknown, occurs in winter.

Infections from damage caused by ingesting such foreign objects as fish hooks have been listed as a cause of death (Forrester *et al.*, 1975). Manatees both in the wild and in captivity have developed pus-filled tumors which, while probably uncomfortable to the animal, have not been implicated positively in manatee deaths.

R. L. Jenkins (in Brownell *et al.*, 1978) reported that a nursing orphaned manatee died at Marineland of Florida from salmonella poisoning. A young orphaned manatee held at Sea World of Florida died soon after contracting a clostridial infection which manifested itself as a huge swelling on the side of the animal. This condition is called "fat neck" and is not infrequent in sea lions (E. Asper, in Brownell *et al.*, 1978).

At Sea World of Florida, E. Asper (in Brownell *et al.*, 1978) reported that a large wild female found breathing labouriously at the surface was captured and treated for septicaemia which was thought to have resulted from a retained foetus. The animal recovered. As with many cetaceans, wild deaths in manatees have been related to problems during the birth process in which the mother and/or calf may die.

Manatees appear to suffer a good deal from skin problems in captivity and these can usually be reversed by changing the salinity of the water in which the animals are maintained. It may be that some of the movements by this species between fresh and saline waters in the wild may be related to this.

Although not clearly demonstrated, there have been mortalities of manatees, in greater numbers than might otherwise be expected in a limited amount of time in a restricted area, following at least one outbreak of the infamous red tide in Florida (Layne, 1965).

Manatees have been recorded as having both internal and external parasites, and a number of external commensals. Husar (1977b), Hartman (1971), and Dailey and Brownell (1972) have listed a number of these, and the external commensals include algae as well as zoophytes. Captives may develop both bacterial and fungal infections.

As with many captive West Indian manatees, an Amazonian manatee suffered from curable skin problems and another developed osteomyelitis as a result of a harpoon wound inflicted in the wild. Yet another captive of this species was reported to have died of pleurisy. A trematode, presumably from a wild individual, is the only parasite reported for this species [see Husar's summary (1977a) for further details on these problems].

Captivity

Manatees have been maintained in captivity in both fresh and saline water for considerable lengths of time. For example, it was reported (in Brownell *et al.*, 1978) that one had been maintained at Bradenton, Florida, for 29 years and that a breeding pair had been at the Miami Seaquarium for at least 18 years. After a considerable length of time this pair produced a healthy calf. A female, captured as a juvenile, was held at Marineland of Florida for nearly 12 years before being transferred to another facility where at this writing she has continued to live for nearly 2 more years. Most institutions that maintain manatees do so most successfully if they occasionally alternate between fresh and salt water. This seems to help eliminate skin problems and also ectoparasites and commensals. Although slightly higher and lower water temperature can be tolerated, it has been found that captive manatees thrive best in waters with temperatures between 16°C (60°F) and 30°C (85°F) (see Brownell *et al.*, 1978). A wide variety of foods have been used in feeding captive manatees and these are discussed along with feeding habits elsewhere in this section. There has been some success with feeding and maintaining orphaned manatees brought in from the wild.

With a few exceptions (Bullock *et al.*, 1980, 1982; Gullivan and Best, 1980; Irvine, 1983), we have found little reference in the literature to formal, experimental, physiological studies, although some work has been accomplished on the still somewhat controversial matter of the need for manatees to consume fresh water (see Brownell *et al.*, 1978). Some work on diving physiology was mentioned in the section on internal anatomy above. There is still much to learn, but because of the low numbers of manatees being maintained in captivity, and those mostly on public display, such studies may be slow in coming.

Acknowledgements

The following persons were of special assistance in lending literature, unpublished reports, and general information: A. Blair Irvine, Howard W. Campbell, and Robert L. Brownell, Jr, all of the U.S. Fish and Wildlife Service; Sam H. Ridgway of the U.S. Naval Ocean Systems Center in San Diego, California; James N. Layne of the Archbold Biological Station, Lake Placid, Florida; and Robert L. Jenkins and Cecil M. Walker, Jr, of Marineland of Florida. Assistance with photographs was generously provided by the Miami Seaquarium (Warren Zeiller), Marineland of Florida (Charlene G. List), Sea World of Florida (Edward D. Asper), the U.S. National Museum of Natural History of the Smithsonian Institution (Charles W. Potter), and Biological Systems, Inc.

References

Anderson, S., and Jones, Jr., J. K. (1967). "Recent Mammals of the World; A synopsis of families". Ronald Press, New York.

Anonymous. (1976). The first sighting of a manatee in the Bahamas since 1904. *Mar. Fish. Rev.* **38**(1), 40.

Best, R. C. (n.d.). Foods and feeding habits of wild and captive Sirenia. Departamento de Biologia de Mamiferos Aquaticos, Instituto Nacional de Pesquisas da Amazonia (INPA), Projeto Peixe-Boi (Brazilian Manatee Project).

Browne, M. M., and Lee, D. S. (1977). The manatee in North Carolina. *ASB Bull.* **24**(2), 40.

Brownell, R. L., Jr., Ralls, K., and Reeves, R. R. (Eds). (1978). Report of the West Indian manatee workshop, Orlando, Florida, 27–29 March 1978. Cosponsored by the Florida Audubon Society, Florida Dept. Nat. Res., Natl. Fish and Wildlife Lab. of the U.S. Fish and Wildlife Serv., and Sea World of Florida.

Bullock, T. H., Domning, D. P., and Best, R. (1980). Evoked brain potentials demonstrate hearing in a manatee (Sirenia: *Trichechus inunguis*) *J. Mammal.* **61**, 130–133.

Bullock, T. H., O'Shea, T. J., and McClune, M. C. (1982). Auditory evoked potentials in the West Indian manatee (Sirenia: *Trichechus manatus*). *J. Comp. Physiol.* **148**, 547–554.

Caldwell, D. K., and Caldwell, M. C. (1972). Senses and communication. *In* "Mammals of the Sea: Biology and Medicine" (Ed. S. H. Ridgway), pp. 466–500. Thomas, Springfield, Illinois.

Caldwell, M. C., and Caldwell, D. K. (1972). Behavior of marine mammals. *In* "Mammals of the Sea: Biology and Medicine" (Ed. S. H. Ridgway), pp. 419–465. Thomas, Springfield, Illinois.

Campbell, H. W., and Irvine, A. Blair. (1977). Feeding ecology of the West Indian manatee *Trichechus manatus* Linnaeus. *Aquaculture* **12**, 249–251.

Campbell, H. W., and Powell, J. A. (1976). Endangered species: The manatee. *Fla. Nat.* April, 15–20.

Collard, S. B., Rubenstin, N. I., Wright, J. C., and Collard, S. B. III. (1976). Occurrence of a Florida manatee at Pensacola Bay. *Fla. Sci.* **39**(1), 48.

Dailey, M. D., and Brownell, R. L., Jr. (1972). A checklist of marine mammal parasites. *In* "Mammals of the Sea: Biology and Medicine" (Ed. S. H. Ridgway), pp. 528–589. Thomas, Springfield, Illinois.

Domning, D. P. (1978). The myology of the Amazonian manatee, *Trichechus inunguis* Natterer (Mammalia: Sirenia). *Acta Amazonica* **8**(2), Suppl. 1, 81 pp.

Domning, D. P. (1980). Feeding position preference in manatees (*Trichechus*). *J. Mammal.* **61**(3), 544–547.

Domning, D. P., and Magor, D. M. (1977). Taxa de substituicao horizontal de dentes no peixe-boi. *Acta Amazonica* **7**(3), 435–438.

Evans, W. E., and Herald, E. S. (1970). Underwater calls of a captive Amazon manatee, *Trichechus inunguis*. *J. Mammal.* **51**(4), 820–823.

Farmer, M., Weber, R. E., Bonaventura, J., Best, R. C., and Domning, D. P. (1978). Propriedades funcionais de hemoglobina e sangue completo em um mamifero aquatico, o peixe-boi (*Trichechus inunguis*). *Acta Amazonica* **8**(4), Sup., 311–321.

Forrester, D. J., White, F. H., Woodard, J. C., and Thompson, N. P. (1975). Intussusception in a Florida manatee. *J. Wildl. Dis.* **11**, 566–568.

Gullivan, G. J., and Best, R. C. (1980). Metabolism and respiration of the Amazonian manatee (*Trichechus inunguis*) *Physiol. Zool.* **53**, 245–253.

Harrison, R. J., and King, J. E. (1965). "Marine Mammals". Hutchinson Univ. Library, London.

Hartman, D. S. (1971). Behavior and ecology of the Florida manatee, *Trichechus manatus latirostris* (Harlan), at Crystal River, Citrus County. Unpublished Ph.D. Dissertation, Cornell Univ., Ithaca, New York.

Hartman, D. S. (1974). Distribution, status, and conservation of the manatee in the United States. Unpublished report to the U.S. Dept. of the Interior, Fish and Wildlife Serv., 247 pp.

Hartman, D. S. (1979). Ecology and Behavior of the Manatee (*Trichechus manatus*) in Florida. *Spec. Publ. Am. Soc. Mammal.* **5**, 153 pp.

Husar, S. L. (1977a). *Trichechus inunguis*. *Am. Soc. Mammal., Mammalian Species* **72**, 1–4.

Husar, S. L. (1977b). The West Indian manatee (*Trichechus manatus*). *Res. Rep. U.S. Fish Wildlife Serv.* **7**, 1–22.

Husar, S. L. (1978a). *Trichechus senegalensis*. *Am. Soc. Mammal., Mammalian Species* **89**, 1–3.

Husar, S. L. (1978b). *Trichechus manatus*. *Am. Soc. Mammal., Mammalian Species* **93**, 1–5.

Irvine, A. B. (1983). Manatee metabolism and its influence on distribution in Florida. *Biol. Conserv.* **25**, 315–334.

Layne, J. N. (1965). Observations on marine mammals in Florida waters. *Bull. Fla. St. Mus., Biol. Sci.* **9**(4), 131–181.

Lowery, G. H., Jr. (1974). "The Mammals of Louisiana and Its Adjacent Waters". Louisiana State Univ. Press, Baton Rouge.

Moore, J. C. (1951a). The range of the Florida manatee. *Q. J. Fla. Acad. Sci.* **14**(1), 1–19.

Moore, J. C. (1951b). The status of the manatee in the Everglades National Park, with notes on its natural history. *J. Mammal.* **32**(1), 22–36.

Moore, J. C. (1956). Observations of manatees in aggregations. *Am. Mus. Novit.* **1811**, 1–24.

Moore, J. C. (1957). Newborn young of a captive manatee. *J. Mammal.* **38**(1), 137–138.

Nishiwaki, M., Yamaguchi, M., Shokita, S., Uchida, S., and Kataoka, T. (1982). Recent survey on the distribution of the African manatee. *Sci. Rep. Whales Res. Inst.* **34**, 137–147.

Powell, J. A., Jr. (1978). Evidence of carnivory in manatees (*Trichechus manatus*). *J. Mammal.* **59**(2), 442.

Powell, J. A., Jr., Belitsky, D. W., and Rathbun, G. B. (1981). Status of the West Indian manatee (*Trichechus manatus*) in Puerto Rico. *J. Mammal.* **62**(3), 642–646.

Ray, C. E. (1960). The manatee in the Lesser Antilles. *J. Mammal.* **41**(3), 412–413.

Reynolds, J. E., III. (1977). Aspects of the social behavior and ecology of a semi-isolated colony of Florida manatees, *Trichechus manatus*. Unpublished Master's thesis, Univ. of Miami, Miami.

Reynolds, J. E., III. (1979). The semisocial manatee. *Nat. Hist.* **88**(2), 44–53.

Reynolds, J. E., III. (1980). Aspects of the structural and functional anatomy of the gastrointestinal tract of the West Indian manatee, *Trichechus manatus*. Unpublished Ph.D. dissertation, Univ. Miami, Miami.

Reynolds, J. E., and Odell, D. K. (1977). Observations on manatee mortality caused by flood control dams in south Florida. Unpublished mss. included as an appendix to Reynolds (1977), pp. 196–206.

Schevill, W. E., and Watkins, W. A. (1965). Underwater calls of *Trichechus* (manatee). *Nature (London)* **205**, 373–374.

Scholander, P. F., and Irving, L. (1941). Experimental investigations on the respiration and diving of the Florida manatee. *J. Cell. Comp. Physiol.* **17**(2), 169–191.

True, F. W. (1884). The sirenians or sea-cows. The Fisheries and Fishery Industries of the United States, section 1, Nat. Hist. of Useful Aquatic Animals, Part 1, article C, pp. 114–136.

White, J. R. (1984a). Man can save the manatee. *Natl. Geogr.* **166**(3), 414–418.

White, J. R. (1984b). Born captive, released in the wild. *Sea Frontiers* **30**(6), 369–375.

White, J. R., Harkness, D. R., Isaacks, R. E., and Duffield, D. A. (1976). Some studies on blood of the Florida manatee, *Trichechus manatus latirostris*. *Comp. Biochem. Physiol. A* **55**, 413–417.

Whitfield, W. K., and Farrington, S. L. (1975). An annotated bibliography of Sirenia. *Fla. Mar. Res. Pub.* **7**, 1–44.

Wing, E. S., Hoffman, C. A., Jr., and Ray, C. E. (1968). Vertebrate remains from Indian sites on Antigua, West Indies. *Caribb. J. Sci.* **8**(3–4), 123–139.

Yamasaki, F., Komatsu, S., and Kamiya, T. (1980). A comparative morphological study on the tongues of manatee and dugong (Sirenia). *Sci. Rep. Whales Res. Inst.* **32**, 127–144.

3

Gray Whale

Eschrichtius robustus (Lilljeborg, 1861)

Allen A. Wolman

Genus and Species

Nomenclature

The first specific name sometimes thought to apply to the gray whale, *Balaena gibbosa* Erxleben, 1777, was based on the New England scrag whale (Dudley, 1725). In 1861, Lilljeborg described the gray whale as a new species, *Balaenoptera robusta* (erroneously assigning it to the rorquals) on the basis of subfossil bones found in Gräsö, Sweden, in 1859. Because of irreconcilable difficulties in identifying the scrag whale (Schevill, 1954; Rice and Wolman, 1971), Lilljeborg's specific name is used for the Atlantic gray whale. In 1864, Gray established for the species the name *Eschrichtius* as a subgenus of *Megaptera;* he upgraded it to generic level the next year, assigning a single species, *Eschrichtius robustus,* to it. The Pacific gray whale was named *Agaphelus glaucus* by Cope (1868). Pacific and Atlantic gray whales showed no consistent differences when compared by van Diense and Junge (1937) and by Cederlund (1939).

The gray whale has a number of common names, such as *mussel digger* which refers to its habit of descending to the soft mud bottom; *hard head* and *devil fish* which arose from dangerous open boat encounters; and *grey back*, referring to the animal's color. Various names in other languages are *seryi kit* (Russian), *grähval* (Norwegian), *angtuchhaq* (Alaska Eskimo Yupik), *chikakhluk* (Aleut), and *koku kujira* (Japanese).

External Characteristics and Morphology

The body is more slender than that of right whales and more robust than that of most balaenopterids. Head size is small in proportion to total body length (Fig. 1), with two to four ventral grooves about 1.5 m long on the throat. There is no dorsal fin, but a series of 6 to 12 crenulations extends along the last one-third of the body. The body is grey with white mottling and is colonised by barnacles and three species of whale lice (cyamids). Frequently there are tooth scars, suggesting unsuccessful attacks by killer whales. The upper jaw has 140 to 180 coarse yellowish baleen plates on each side; the left and right baleen rows are separated anteriorly. There are more tactile hairs

FIG. 1 Gray whale surfacing on Laguna Ojo de Liebre, Baja California, Mexico. Note eye at gape of mouth. (Photo by David Withrow.)

on the jaws than on any other species of balaenopterid. The flipper has four phalanges, the first finger having disappeared.

Newborn calves are about 4.6 m long and weigh 500 kg. The whales are sexually mature at a mean age of 8 years and at a mean length of 11.1 m for males, 11.7 m for females. At physical maturity, about 40 years of age, the average male is 13.0 m long, the average female, 14.1 m. Two immature males, 9.25 m and 9.90 m, weighed 8808 kg and 8876 kg, respectively, when caught on their northbound migration. Similarly, two adult males of 11.72 m and 12.40 m weighed 15 686 kg and 16 594 kg (Rice and Wolman, 1971). A pregnant 13.35 m female (Zenkovich, 1934) on the feeding grounds weighed 31 466 kg. The largest male ever recorded was 14.6 m (Andrews, 1914) and the largest female, 15.0 m (Tomilin, 1957).

Anatomical Characteristics

As expected of a primitive species closely related to an extinct family, the gray whale exhibits characteristics intermediate between right whales and balaenopterids. The skull has a medially convex rostrum, less arched than in right whales, but more so than in rorquals (Nishiwaki, 1972) (Fig. 2). The frontal bone orbital process is narrower than in rorquals and wider than in right whales (Tomilin, 1957). The relatively long humerus resembles that of the fossil genus *Plesiocetus*. The vertebral formula is C7, D14, L12, Ca23, with the seven cervical vertebrae free throughout the animal's life. The proximal ends of the third to seventh ribs have well-developed tubercles, necks, and heads. The scapula is more massive than in rorquals and lighter than that of right whales (Andrews, 1914; Tomilin, 1957).

Organ weights taken from a 31 466-kg, 13.35-m female were (percent of total body weight): blubber, 29.0; tongue, 4.0; lungs, 1.0; heart, 0.6; intestine, 4.1; liver and kidneys, 2.0; brain, 0.0139 (4.376 kg) (Tomilin, 1957). Pilleri and Gihr (1969) weighed six gray whale brains which ranged from 3900 to 4800 g, averaging 4317 g.

Of 316 individuals examined by Rice and Wolman (1971), 314 had a thick-walled, fluid-filled, cystlike structure, 10 to 25 cm in diameter, on the ventral surface of the caudal peduncle between the blubber and the muscles. One animal had two in a line at the same location, and one animal had none. It seems unlikely that the structure is pathological, as Andrews believed (1914), although its function is not known. Based on histological evidence, Zimushko (1970a,b) believes it to be similar to sebaceous glands of land mammals. Durham and Beierle (1976) suggested that it is a "track laying" scent gland. It has not been found in any other whale species.

Based on gross and histological structure, Cowan and Brownell (1972) postulate a complex lymphoendothelial function to an organ in the elongated

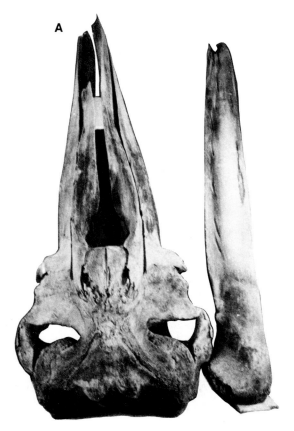

FIG. 2 Dorsal (A), ventral (B), and lateral (C) views of an 11.4 m female, stranded near Waldport, Oregon, on 1 May 1974. (Photo by M. L. Johnson.)

anal canal of the gray whale. The structure, an "anal tonsil," has not been noted in any other mysticete, although Simpson and Gardner (1972) reported gut-associated lymphatic tissue well developed in cetaceans. Uys and Best (1966) noted the presence of extrapharyngeal lymphoendothelial organs in sperm whales, presented as numerous small lumps, external to the anal opening.

Distribution

There are two stocks of gray whales in the North Pacific—the Korean stock on the western side and the California stock on the eastern side (Fig. 3). The

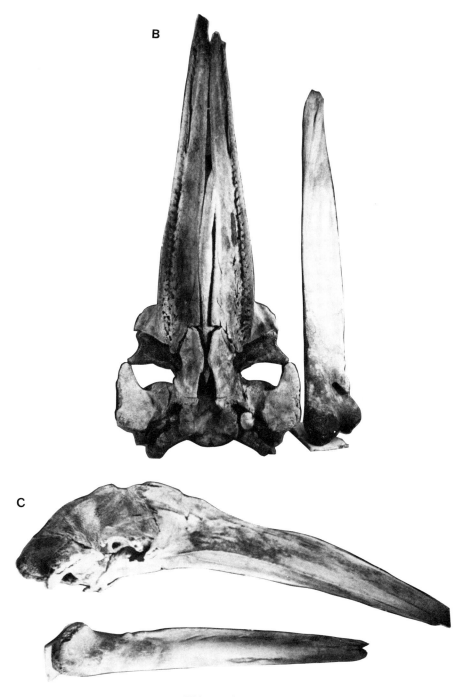

FIG. 2 (*continued*)

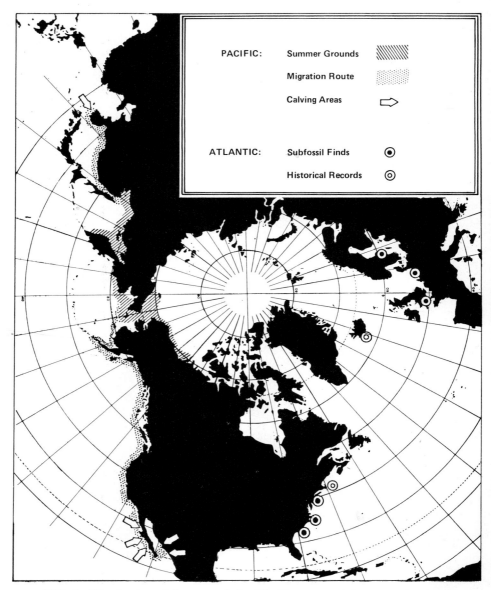

FIG. 3 Distribution of the gray whale. The Korean stock is near extinction.

Korean stock of the gray whale is probably below the critical population size and is nearly extinct (Bowen, 1974; Nishiwaki and Kasuya, 1970). At one time these animals fed in the northern Sea of Okhotsk and wintered off southern Korea and Japan (Andrews, 1914; Krasheninnikov, 1755; Omura, 1974). Nishiwaki and Kasuya (1970) and Bowen (1974) hypothesised that three recent sightings in Japan represented California stock migrants.

From the end of May through September, the California stock of gray whales feeds mostly in the northern and western Bering Sea, the Chukchi Sea, and the western Beaufort Sea, with a few individuals scattered along the west coast of North America as far south as Baja California (Patten and Samaras, 1977, Sprague *et al.*, 1978). Studies carried out by Darling since 1974 (Hatler and Darling, 1974; Darling, 1978) indicate that a summer population of about 50 to 70 animals utilises the west coast of Vancouver Island, Canada; many of the same animals return each year. On the Asian side, they are occasionally found from Kronotskiy Bay on the Kamchatka Peninsula and the Commander Islands northward to Cape Navarin and are common off the southwestern and northeastern shores of the Gulf of Anadyr, around the Chukotski Peninsula as far northwest as Tynkurginpil'gyn Lagoon and north to Wrangel Island (Berzin and Rovnin, 1966; Zimushko, 1970a). Gray whales are found on the North American side from Barter Island in the Beaufort Sea (Maher, 1960) to Cape Thompson, are common in Kotzebue Sound and through the Bering Strait to St. Matthew Island, and several are occasionally seen as far south as St. George Island. Small groups have been sighted as far east as 130° W (Rugh and Fraker, 1981) in the Beaufort Sea, and as far west as 174°08' E in the East Siberian Sea (Marquette *et al.*, 1982). They have not been found in deep waters of the southwestern Bering Sea; they are apparently restricted by feeding habits to shallow waters of the continental shelf, generally in areas of high benthic biomass.

From November through December, California gray whales move south through Unimak Pass in the Aleutian chain (Rugh and Braham, 1979), and then follow the coast around the Gulf of Alaska to the lagoons of Baja California (Pike, 1962). Near the central California coast, they hug the shoreline; 94% pass within 1.6 km between Monterey and Pt. Sur. In Mexico, the major calving areas are Laguna Guerrero Negro, Laguna Ojo de Liebre, Laguna San Ignacio, and Estero Soledad, with 85% of the calves (Rice *et al.*, 1981). A few continue around Cape San Lucas to the southeastern shore of the Gulf of California, as far north as Yavaros, Sonora (Gilmore *et al.*, 1967). Females generally migrate earlier than do males, and adults migrate earlier than sexually immature whales. Southbound pregnant females appear first, followed in sequence by recently ovulated females, by immature females and adult males, and by immature males. Northbound, newly pregnant females travel first, followed by anoestrous females, adult males, and immatures (Rice and Wolman, 1971). Females with calves migrate last, the peak passing Pt Piedras Blancas, California, about 1 May (Poole, 1981).

Abundance

Whaling was conducted by aboriginal groups from Washington state and the west coast of Vancouver Island (Swan, 1870; Swanson, 1956), around Kodiak Island, the eastern Aleutians, and along the Kamchatka Peninsula on the Asian side (Heizer, 1943), and by Eskimos of Arctic Alaska and eastern Siberian Chukchi waters for thousands of years (Tomilin, 1957). The American fishery for gray whales began in 1846 in Baja California and spread to the Bering Sea. About 11 000 were killed between 1846 and 1874 in California and Baja California (Scammon, 1874). Scammon estimated the stock of California gray whales prior to exploitation at "not more than" 30 000 (Rice, 1967) and about 8000 to 10 000 by 1874, but Henderson (1972) presents good evidence for an original stock size of 15 000. Shore whaling on the wintering grounds stopped about 1900 (Starks, 1922). Floating factory ships took gray whales from their wintering grounds and on the southern portion of the migration route intermittently in 1914 and from 1925 to 1929.

Although Japanese whaling began over a thousand years ago, commercial whaling of the Asian stock actually began during the seventeenth century (Fraser, 1937; Omura *et al.*, 1953). Shore whaling by modern methods began in 1899 in Korea. Almost 1500 whales were killed from 1910 to 1933, when Korean whaling stopped. The Korean stock probably numbered about 1000 to 1500 in 1900 and was virtually exterminated by 1933 (Rice and Wolman, 1971).

A Soviet floating factory took California grays in the Bering and Chukchi seas from 1933 to 1946, and a Japanese factory ship worked the Chukchi Sea in 1940. Between 1933 and 1946, 681 whales were taken. Since 1946, in accordance with the International Whaling Commission regulations, gray whales were not taken except by or for aborigines and for scientific research (Table 1). From 1948 to the present, a minimum of 4140 animals has been taken. In addition, 67 gray whales were reported taken in South Korean waters between 1948 and 1966, probably from the Korean stock (Brownell and Chun, 1977).

Recent estimates of population size were based on intermittent shore counts of migrating animals off California and on aerial censuses during calving in lagoons of Baja California. Between 1952–53 and 1959–60, four

TABLE 1 Recent gray whale harvests[a]

Year	Canada	United States California[b]	United States Alaska[c]	Soviet Union[d,e,f]	Korea[g]
1948				19	9
1949				26	4
1950				10	
1951	1[h]		1	12	7
1952				42	1
1953	10[b]		2	37	7
1954			2	36	
1955				59	
1956				121	
1957			1	95	
1958			2	145	7
1959		2	6	187	7
1960				156	8
1961			1	207	3
1962		4		147	
1963				179	2
1964		20	1	188	3
1965				175	4
1966		26		194	5
1967		125		125	
1968		66		135	
1969		73		139	
1970			5	146	
1971			3	150	
1972			1	181	
1973			1	173	
1974			1	181	
1975				171	
1976				163	
1977			1	186	
1978			2	182	
1979			4	178	
1980			3	178	

[a]Data from G. Y. Harry, Jr., and W. M. Marquette (personal communication), G. C. Pike (1962), and Rice and Wolman (1971).
[b]Special scientific permit, December to April.
[c]Not complete prior to 1970.
[d]For aborigines; no pre-1965 figures available.
[e]Eskimo whaling, spring to fall.
[f]Ivashin and Mineev (1978), Anonymous (1980), Ivashin (1981), and Ivashin (1982).
[g]Brownell and Chun (1977).
[h]Taken by error.

shore counts were made (Gilmore, 1960; Rice, 1961), with estimates of about 3000 to 6000 animals, respectively.

Hubbs and Hubbs (1967) made wintering gound aerial surveys in most years from 1952 to 1964 and stated that the population stood at 3000 thereafter, based on an "admittedly rather intuitive estimate that about half of the population was observed on the flights" (p. 25). A better correction of Hubbs' observations might be based on the fact that gray whales are visible about 1 of every 5 min between soundings, giving an estimate of 7500 animals, which is somewhat closer to Rice's figure. Adams (1968), on the basis of extremely limited shore counts of about 1 hr per day and only during days of optimum visibility, estimated 18 300 gray whales—seemingly too high at the time to be realistic. Gard (1974) made aerial counts of gray whales during 1970 and 1973 in the larger calving grounds of Baja California, finding that although the total number remained stable the use of individual lagoons and bays varied with human disturbance.

Rice and Wolman (1971; Wolman and Rice, 1979) have conducted shore counts annually south of Monterey, California, and intermittently at Point Loma, near San Diego, California, since 1967–1968. The whale watch was made from 0700 to 1700 hr by two observers working 5-hr shifts, using binoculars. A pod of whales may be in sight for 45 min as it passes. Vessel

TABLE 2 Counts of southward migrating gray whales off central California, population estimates, and standard deviations 1967–1968 to 1979–1980[a]

Year	Raw count	Population estimate	Standard deviation
1967–1968	3 077	13 095	1 276
1968–1969	3 265	11 954	1 545
1969–1970	3 399	12 408	1 619
1970–1971	3 264	11 177	1 625
1971–1972	2 667	10 414	918
1972–1973	3 684	14 534	1 348
1973–1974	3 889	14 676	1 558
1974–1975	3 836	13 110	1 366
1975–1976	4 295	15 919	1 803
1976–1977	4 720	16 621	1 798
1977–1978	3 717	14 811	2 272
1978–1979	3 927	13 676	1 127
1979–1980	4 924	17 577	2 364

[a]From Reilly (1981).

transects from Point Loma in 1968 by Rice and Wolman (1971) indicated that only 41% of the whales migrating south passed within sight of shore, some of the stock passing as far off as 195 km, although that may not have been true in earlier years. Sund and O'Connor (1974) made five flights in 1973 to test the observers and the migration path width. They found that at the northern site, 94% passed within 1.6 km offshore and 96% within 4.8 km and that "ground and aircraft observers apparently were equally adept at initially sighting whales" (p. 51). The estimates of gray whales migrating past the Granite Canyon counting station ranged from about 11 000 to 18 000 animals (Table 2). Studies carried out in 1978–1979 and 1979–1980 off Granite Canyon indicate a population size of 15 647, with an annual increase since 1965 of 2.51% (Reilly, 1981, Reilly et al., 1983).

In 1977, Rugh and Braham (1979) censused southbound gray whales at Unimak Pass, Alaska, where they leave the Bering Sea. Extrapolating from the count, they estimated the population at about 15 000. If we assume that not all gray whales migrate the entire distance to Mexico, the actual population size is probably on the order of 16 000 animals.

Behaviour

Swimming and diving

Most gray whales summer in the shallow coastal areas of the Chukchi, Bering, and Okhotsk seas. They migrate up to 18 000 km, the longest migration of any mammal; normal swimming speed is 7 to 9 km per hr. Individuals usually remain submerged for 3 to 5 min, then surface and blow five or six times. Wyrick (1954) followed four separate migrating whales; their average speed was 8.5 km/hr. Cummings et al. (1968) tracked nine whales, recording an average speed of 10.2 km/hr. Tomilin (1957) recorded a male swimming at almost 13 km/hr for 4 hr while being chased. According to Gilmore (1956), frightened gray whales develop a speed of 16 km/hr.

The blow is 3 to 4.5 m high and may be heard, under certain conditions, 0.8 km away.

Ordinary swimming, below 11 km/hr, is without turbulence in the form of eddies ("whale tracks"), but above that speed, swirls appear at the surface as the flukes move more powerfully (Gilmore, 1961). Shallow dives of 15 to 50 m are common (Tomilin, 1957), but on some feeding grounds, bottom-dwelling food organisms necessitate dives of about 120 m. During migration and in calving areas, gray whales may swim into the surf zone, in water so shallow that they are barely awash.

The respiratory pattern of a captive gray whale calf at rest (Wahrenbrock *et al.*, 1974) indicated a 1-min interrespiratory pause and about 2 sec for expiration and inspiration together. Tidal volume equalled about 50% of resting lung volume, smaller than that reported for mature diving mammals (Irving *et al.*, 1941; Olsen *et al.*, 1969; Scholander, 1940).

Before sounding, the animal raises its tail flukes into the air. Tomilin (1957), however, reported that sometimes gray whales turn on their body axis so that the flukes describe a semicircle before touching the water. When killer whales approach, gray whales often become motionless, with the blowhole barely protruding above the water. Others, when chased by boats or killer whales near shore, approach the land closely, even going into the surf zone. The habit of swimming rather close to shore has been observed on the migration route and on calving grounds by many individuals. Scammon (1874) reported whales in water as shallow as 4 m. Preference for shallow waters is one of the most interesting features of gray whale behaviour.

Breaching is a form of gray whale behaviour similar to that of humpback whales, as is the habit of "boltering," that is, lying on the side while waving a pectoral appendage in the air. The trait of thrusting the head vertically from the surface ("spyhopping") is characteristic of this species, somewhat similar to that of killer whales looking for prey near ice floes (Fig. 4).

Other behaviour

Unlike other baleen whales, gray whales are primarily, although not exclusively, bottom feeders. Infaunal benthic species, especially of gammaridean amphipods such as *Anonyx nugax, Pontoporeia femorata, P. affinis, Ampelisca macrocephala, A. eschrichti, Nototropis ekmani,* and *N. brueggeni* predominate among stomach contents from northern waters (Pike, 1962; Rice and Wolman, 1971; Zimushko and Lenskaya, 1970; Bogoslovskaya *et al.*, 1981). Polychaete worms and molluscs are poorly represented, suggesting that the whales are selective feeders, although small, densely schooling fish and crab juveniles have been reported in a few cases. (Mizue, 1951; Ray and Schevill, 1974; Rice and Wolman, 1971; Sund, 1975; Walker, 1949). Nerini (1984), however believes that benthic faunal representation is probably reflective of area community composition rather than true selection. It is possible that gray whales stir up bottom sediments with their snouts, then filter the turbid water immediately above the bottom from which the heavier molluscs have settled out. The occurrence of sand, silt, and gravel in the stomachs provides further evidence (Andrews, 1914; Tomilin, 1937; Zenkovich, 1934), although Ray and Schevill (1974) consider feeding to be a sucking action involving use of the strongly muscled tongue and flexible lips. Muddy snouts or trails have been observed several times in the Chukchi Sea (Pike, 1962; Scammon, 1874; Wilke and Fiscus, 1961; Rugh and Braham, 1979; Fig. 5), while the same

FIG. 4 (A) Gray whale spyhopping. Note that the eye is open. (B) Gray whale boltering. (Photos by David Withrow.)

FIG. 5 Gray whale feeding in Bering Sea. Note mud plume. (Photo by H. Braham.)

behavioural pattern has been reported on the Baja California grounds, in spite of there being little or no significant food quantities available. Oliver *et al.* (1983) surveyed six lagoons in Baja California and benthic invertebrate communities in the Bering Sea, comparing signs of gray whale feeding such as feeding excavations and faecal slicks. They concluded that gray whale feeding on benthic invertebrates is rare in the calving lagoons of Baja California and along the open coast near Scammon's Lagoon.

Greater wear of the baleen on the right side suggests that gray whales swim on their right sides while feeding (Kasuya and Rice, 1970). Ray and Schevill (1974) reported a left-sided feeding pattern from a captive calf, but this pattern may have been due to human training. Some pelagic feeding behaviour (Swartz and Jones, 1981; Wellington and Anderson, 1978) probably represents incidental opportunistic feeding. Zimushko and Lenskaya (1970) consider a filled gray whale stomach to hold about 300 kg. They estimated that while on the feeding grounds, one gray whale, feeding four times/day consumes 170 tons of food in 130 to 140 days. This figure is close to Rice and Wolman's (1971) estimate of 1 tonne/day per whale. Feeding gray whales often change course and sometimes break surface in the same place where they had dived (Tomilin, 1957). Nerini (1984) estimates northern Bering Sea food consumption as between 9 and 27% of the available benthic community.

Gray whale courtship behaviour has been observed by numerous individuals on feeding grounds, along the migration route, and on calving grounds (Fay, 1963; Gilmore, 1961; Houck, 1962; Sauer, 1963; Tomilin, 1957; Darling, 1977). Sometimes three whales engage in the event; presumably the third whale is an immature individual or a complemental male looking for a female, although Darling (1977) reported homosexual activity among groups of three males. Preliminary activity consists of arching out of the water, rolling around the body axis, and swimming in line. Mating involves swimming in line and in circles; swimming in lateral position, frequently with one flipper held above the water; exaggerated arching; touch display, copulation, including pumping movements; and an after-copulation dive. Copulation itself, with the male's penis visible, and pumping last 30 to 60 sec. The entire display may last up to 30 min.

Epimeletic (both succorant and nurturant) behaviour has been reported by Andrews (1914), Caldwell and Caldwell (1966), Limbaugh (in Norris and Prescott, 1961), Scammon (1874), and Zenkovich (1956). Limbaugh observed four whales supporting another one at the surface; his observation, made underwater, describes the positions of the individuals.

A case of cleaning symbiosis between gray whales and topsmelt, *Atherinops affinis*, was reported by Swartz (1981), in which the fish picked at sloughed epidermis and external parasites, with whale louse appendages found in topsmelt stomachs.

Gray whales interacting with other marine mammals (excluding killer whales) were observed by Leatherwood (1974). Upon release from captivity, a yearling was seen swimming with bottlenose dolphins in the surf zone off southern California. At other times, bottlenose dolphins, northern right whale dolphins, Pacific striped dolphins, Pacific common dolphins, and Dall's porpoises were seen "bow-riding" on pressure waves created by gray whales. Pilot whales have also been observed swimming about a gray whale. Surf riding by adults in a manner similar to that performed by humans was reported by Caldwell and Caldwell (1963).

Whale–human interactions cited as "friendly" behaviour (Gilmore, 1976; Larsen, 1978; Swartz and Jones, 1978) in Laguna San Ignacio and appearing in Laguna Ojo de Liebre by 1981 consist of seeming attraction towards skiffs, air bubble release underwater, lifting skiffs, and soliciting physical contact (petting, stroking, etc.) from passengers in skiffs. Swartz applies the term "curious whales" for animals engaged in this behaviour, which appears to be spreading and may be a learned behaviour.

Underwater sounds have been recorded from migrating gray whales off southern California (Cummings et al., 1968; Gales, 1966; Wenz, 1964); whales on the Mexican calving grounds (Eberhardt and Evans, 1962; Painter, 1963; Poulter, 1968); whales off Vancouver Island, Canada; and a gray whale in captivity at San Diego (Fish et al., 1974). Sounds have been described as low-frequency mumbles, croakerlike grunts, moans, bubble-type noise, knocks, bangs, and clicks. A captive whale produced some of the above sounds but the most common was a metallic-sounding pulsed signal of 8 to 14 pulses in bursts lasting up to 2 sec. Sounds recorded in Canada were click trains. The number of clicks per train varied from 1 to 96 with repetition rates of 8 to 40 per sec, occupying a band from about 2 to 6 kHz, centered at 3.5 to 4.0 kHz. The average click duration was just under 2 msec.

Reproduction

The complete reproductive cycle of the female gray whale occupies 2 years. Sexual maturity is attained at a mean age of 8 years (range, 5 to 11 years). The mean annual ovulation rate is 0.52 (1.20 for nulliparous; 0.96 for parous females), and the mean pregnancy rate is 0.46 (Rice and Wolman, 1971). Females come into oestrus during about a 3-week period in late November and early December. They usually conceive following their first ovulation, but if they fail to do so, they may undergo another oestrous cycle about 40 days later. Pregnancy lasts about 13.5 months. Parturition occurs within a period of 5 to 6 weeks from late December to early February. The calf is nursed for about 7 months. After weaning the calf about August, females are in anoestrus for 3 to 4 months until November or December, when most of them go into oestrus and commence a new pregnancy. A few either fail to ovulate, or ovulate but fail to conceive, and are in anoestrus for another year. There is no evidence for postpartum or postlactation ovulation. Evidence for the possible occurrence of ovulation following stillbirth or early loss of the calf is suggestive but inconclusive.

The reproductive cycle of the female gray whale is basically similar to that of the larger rorquals, one important difference being that the extremely long migration route and restricted calving grounds impose a much stricter annual schedule. The majority of gray whale calves are born during a period of 5 to 6 weeks, peaking about 27 January, as contrasted with about 5 months in rorquals.

The gray whale's ovulation rate is less than that reported by some authors for rorquals. Rorquals, apparently unlike gray whales, sometimes experience postpartum and postlactation or summer ovulations.

Population density may influence reproduction through lack of nutrition.

In the gray whale, nutrition could be of critical importance. The pregnant female must accumulate enough energy during the summer to support herself and her foetus through a long migration, to support herself and the newborn calf for a month or more on the wintering grounds, and to sustain herself and the rapidly growing calf during the long return migration to the summer feeding grounds. Under such conditions, selective pressure for suppression of ovulation at times when the female is not physiologically capable of carrying a new pregnancy to term might be expected. It might be that in the gray whale the potentiality for a postpartum oestrous cycle has been genetically eliminated from the population.

Testes weighing more than 5 kg are a reliable indication of maturity in gray whales. A penis length greater than 1.1 m also separates most mature males from those that are sexually immature. The much heavier testes and larger seminiferous tubules of males taken during southward migration compared with those of males collected during northward migration and on the summer grounds indicate that male gray whales have a marked seasonal sexual cycle (Rice and Wolman, 1971). There is a peak of spermatogenic activity in late autumn or early winter, correlating closely with the time females come into oestrus. The courtship behaviour during the summer in the Bering Sea, described earlier, certainly never results in successful conception.

Diseases

The gray whale is more heavily infested with a greater variety of ectoparasites and epizoites than any other species of cetacean. The host-specific barnacle *Cryptolepas rhachianecti* was present on all 316 gray whales examined by Rice and Wolman (1971). The whale louse *Cyamus scammoni* was found on 315 (99.7%) animals; *C. ceti* on 314 (99.4%); and *C. kessleri* on 310 (98.1%). These infestations may be partly due to the fact that the gray whale swims slowly and lives throughout the year in shallow coastal waters rich in nutrients. In contrast, the gray whale is infrequently infested with endoparasites, mostly in the liver (*Lecithodesmus goliath*), stomach (*Anisakis simplex*), intestine (*Orthosplanchnus pygmaeus*, *Ogmogaster pentalineatus*, *O. antarcticus*, *Priapocephalus eschrichtii*, *Priapocephalus* sp., *Diplogonoporus balaenopterae* (?), *Corynosoma* spp., *Bolbosoma* sp.), and rectum (*O. pentalineatus*), possibly due to its long period of fasting each year.

Besides parasites, little is known of gray whale diseases. Andrews (1914) described three apparent examples of disease, one of which was obviously the structure found by Rice and Wolman (1971) in 315 of 316 whales examined

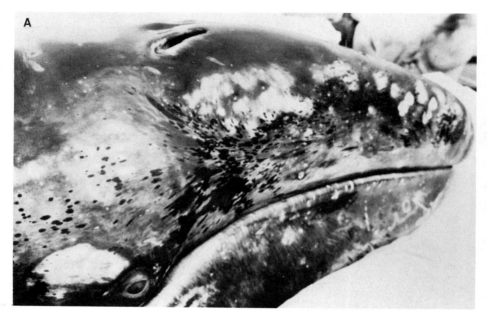

FIG. 6 (A) Live captive gray whale, March 1971; (B) keeper puts his hand in mouth of gray whale calf and mouth opens, showing baleen. (Photograph courtesy of Dr. J. P. Schroeder.)

(see description under Anatomical Characteristics). The second example may possibly have been a wound. The third example consisted of blubber exposed on the snout "drawn into small circular patches" (p. 241).

Of 16 gray whales tested, calicivirus antibodies were found in the sera of five animals (Akers *et al.*, 1974). This virus, also isolated from California sea lions and indistinguishable from vesicular exanthema of swine virus, has been transmitted to feral pigs on islands near the coast of California.

Captivity

In 1971 a young gray whale calf (Fig. 6.) was captured in Scammon's Lagoon, Baja California Sur, Mexico, and transported to Sea World at San Diego, California. This whale arrived in San Diego in March of 1971 and was measured at 5.84 m and weighed 1952 kg. A year later it was released in the Pacific just off San Diego with an attached radio-tracking device (Evans, 1974.)

References

Adams, L. (1968). Census of the gray whale, 1966–67. *Nor. Hvalfangst-Tid.* **57**(2), 41–43.
Akers, T. G., Smith, A. W., Latham, A. B., and Watkins, H. M. S. (1974). Calicivirus antibodies in California gray whales (*Eschrichtius robustus*) and Steller sea lions (*Eumetopias jubatus*). *Archiv Gesamte Virusforsch.* **46**, 175–177.
Andrews, R. C. (1914). Monographs of the Pacific Cetacea. 1. The California gray whale (*Rhachianectes glaucus* Cope). *Mem. Am. Mus. Nat. Hist. New Ser.* **1**(5), 227–286.
Anonymous. (1980). U.S.S.R. progress report on cetacean research, June 1978–May 1979. *Rep. int. Whal. Commn* **30**, 167–171.
Berzin, A. A., and Rovnin, A. A. (1966). Distribution and migrations of whales in the north-eastern section of the Pacific Ocean and in the Bering and the Chukotka seas. *Izv. TINRO* **58**.
Bogoslovskaya, L. S., Votrogov, L. M., and Semenova, T. M. (1981). Feeding habits of the gray whale off Chukotka. *Rep. int. Whal. Commn.* **31**, 507–510.
Bowen, S. L. (1974). Probable extinction of the Korean stock of the gray whale (*Eschrichtius robustus*). *J. Mammal.*, **55**(1), 208–209.
Brownell, R. L., Jr., and Chun, C. (1977). Probable existence of the Korean stock of the gray whale (*Eschrichtius robustus*). *J. Mammal.*, **58**(2), 237–239.
Caldwell, D. K., and Caldwell, M. C. (1963). Surf-riding by the California gray whale. *Bull. South. Calif. Acad. Sci.* **62**(2), 99.
Caldwell, M. C., and Caldwell, D. K. (1966). Epimeletic (care-giving) behavior in

cetacea. *In* "Whales, Dolphins, and Porpoises", (Ed. K. S. Norris), pp. 755–789. Univ. of California Press, Los Angeles.
Cederlund, B. A. (1939). A subfossil gray whale discovered in Sweden in 1859. *Zool. Bidr. Uppsala* **18**, 269–286.
Cope, E. D. (1868). On *Agaphelus*, a genus of toothless cetacea. *Proc. Acad. Nat. Sci. Philadelphia* **20**, 225.
Cowan, D. F., and Brownell, R. L., Jr. (1972). Gut-associated lymphoendothelial organ ("anal tonsil") in the gray whale. *In* "Functional Anatomy of Marine Mammals" (R. J. Harrison, Ed.), Vol. 2, pp. 321–327. Academic Press, New York.
Cummings, W. C., Thompson, P. O., and Cook, R. (1968). Underwater sounds of migrating gray whales, *Eschrichtius glaucus* (Cope). *J. Acoust. Soc. Am.* **44**(5), 1278–1281.
Darling, J. D. (1977). The Vancouver Island gray whales. *Waters* **2**(1), 5–19.
Darling, J. D. (1978). Summer abundance and distribution of gray whales along the Vancouver Island coast. Unpublished Report to the Natl. Mar. Fish. Serv. 10 pp.
Deinse, A. B. Van, and Junge, G. C. A. (1937). Recent and older finds of the California gray whale in the Atlantic. *Temminckia* **2**, 161–188.
Dudley, P. (1725). An essay upon the natural history of whales. *Philos. Trans. R. Soc. London* **33**(387), 256–269.
Durham, F. E., and Beierle, J. W. (1976). Investigations on the postanal sac of the gray whale (*Eschrichtius robustus*). *Bull. South. Calif. Acad. Sci.* **75**(1), 1–5.
Eberhart, R. L., and Evans, W. E. (1962). Sound activity of the California gray whale (*Eschrichtius glaucus*). *J. Audio Eng. Soc.* **10**, 324–328.
Erxleben, J. C. P. (1777). "Systema regni animalis . . . Classis I. Mammalia" Weygandianis, Leipzig.
Evans, W. E. (Ed.) (1974). The California gray whale. *Mar. Fish. Rev.* **36**(4), 64 pp.
Fay, F. H. (1963). Unusual behavior of gray whales in summer. *Psychol. Forsch.* **27**, 175–176.
Fish, J. S., Sumich, J. L., and Lingle, G. L. (1974). Sounds produced by the gray whale, *Eschrichtius robustus*. *In* "The California Gray Whale", (Ed. W. E. Evans), pp. 38–45. *Mar. Fish. Rev., Spec. Issue* **36**(4).
Fraser, F. C. (1937). Early Japanese whaling. *Proc. Linn. Soc. London* **150**(1), 19–20.
Gales, R. S. (1966). Pickup, analysis, and interpretation of underwater acoustic data. *In* "Whales, Dolphins and Porpoises", (Ed. K. S. Norris), pp. 435–444. Univ. of California Press, Berkeley.
Gard, R. (1974). Aerial census of gray whales in Baja California lagoons, 1970 and 1973, with notes on behavior, mortality, and conservation. *Calif. Fish Game* **60**(3), 132–143.
Gilmore, R. M. (1956). The California gray whale. *Zoonooz* **29**(2), 4–6.
Gilmore, R. M. (1960). Census and migration of the California gray whale. *Nor. Hvalfangst-Tid.* **49**(9), 409–431.
Gilmore, R. M. (1961). The story of the gray whale. San Diego, Priv. Publ. 2nd edition.
Gilmore, R. M. (1976). The friendly whales of Laguna San Ignacio. *Terra* 15, 24–28.
Gilmore, R. M., Brownell, R. L., Jr., Mills, J. G., and Harrison, A. (1967). Gray

whales near Yavaros, Southern Sonora, Golfo de California, Mexico. *Trans. San Diego Soc. Nat. Hist.* **14**(16), 197–204.

Gray, J. E. (1864). Notes on the whalebone whales; with a synopsis of the species. *Ann. Mag. Nat. Hist. (3)* **14**(83), 345–353.

Gray, J. E. (1865). Notice of a new whalebone whale from the coast of Devonshire. *Proc. Zool. Soc. London*, p. 40.

Hatler, D. F., and Darling, J. D. (1974). Recent observations of the gray whale in British Columbia. *Can. Field-Nat.* **88**, 449–459.

Heizer, R. F. (1943). Aconite poison whaling in Asia and America: An Aleutian transfer to the New World. *Bull. Bur. Am. Ethnol.* **113**, 415–468.

Henderson, D. A. (1972). "Men and Whales at Scammon's Lagoon." Dawson's Book Shop, Los Angeles.

Houck, W. J. (1962). Possible mating of gray whales on the northern California coast. *Murrelet* **43**(3), 54.

Hubbs, C. L., and Hubbs, L. C. (1967). Gray whale censuses by airplane in Mexico. *Calif. Fish Game* 53, 23–27.

Irving, L., Scholander, P. F., and Grinnell, S. W. (1941). The respiration of the porpoise, *Tursiops truncatus*. *J. Cell. Comp. Physiol.* **17**, 145–168.

Ivashin, M. V. (1981). U.S.S.R. report on cetacean research June 1979–May 1980. *Rep. int. Whal. Commn* **31**, 221–226.

Ivashin, M. V. (1982). U.S.S.R. progress report on cetacean research June 1980 to May 1981. *Rep. int. Whal. Commn* **32**, 221–224.

Ivashin, M. V., and Mineev, V. N. (1978). Osostayanie zapasov serykh kitov (The state of the gray whale stock). *In* "Rybnoe Khozyaistvo" Vol. 3, pp. 15–17. Izdatel'stzo Pishchevaya Promyshlennot', Moscow. (In Russian). (Transl. by S. Pearson, April 1979).

Kasuya, T., and Rice, D. W. (1970). Notes on baleen plates and on arrangement of parasitic barnacles of gray whales. *Sci. Rep. Whales Res. Inst.* **22**, 39–43.

Krasheninnikov, S. P. (1755). Opisanie zemli Kamchatki (Description of the land Kamchatka). 2 volumes. Imper. Akad. Nauk, St. Petersburg.

Larsen, K. (1978). Close encounters (of the whale kind). *Sea Frontiers* **23**(4), 195–202.

Leatherwood, J. S. (1974). A note on gray whale behavioral interactions with other marine mammals. *Mar. Fish. Rev.* **36**(4), 50–51.

Lilljeborg, W. (1861). Hvalben. Funna i Jorden par Grayson i Roslagen i Sverige. *Forhandlinger vid et Skandinaviska Naturforskaremotet* **1860**, 599–616.

Maher, W. J. (1960). Recent records of the California gray whale (*Eschrichtius glaucus*) along the north coast of Alaska. *Arctic* **13**(4), 257–265.

Marquette, W. M., Nerini, M. K., Braham, H. W., and Miller, R. V. (1982). Bowhead whale studies, autumn 1980–spring 1981: Harvest, biology and distribution. *Rep. int. Whal. Commn* **32**, 357–370.

Mizue, K. (1951). Gray whales in the East Sea area of Korea. *Sci. Rep. Whales Res. Inst.* **5**, 71–79.

Nerini, M. (1984). A review of gray whale feeding ecology. *In* "The Gray Whale" (Eds M. L. Jones, S. Leatherwood, and S. L. Swartz), pp. 423–450. Academic Press, Orlando.

Nishiwaki, M. (1972). General biology. *In* "Mammals of the Sea: Biology and Medicine" (Ed. S. H. Ridgway), pp. 3–204. Thomas, Springfield, Illinois.

Nishiwaki, M., and Kasuya, T. (1970). Recent record of gray whale in the adjacent waters of Japan and a consideration on its migration. *Sci. Rep. Whales Res. Inst.*, **22**, 29–38.

Norris, K. S., and Prescott, J. H. (1961). Observations on Pacific cetaceans of Californian and Mexican waters. *Univ. Calif. Publ. Zool.* **63**, 291–402.

Oliver, J. S., Slattery, P. N., Silberstein, M. A., and O'Connor, E. F. (1983). A comparison of gray whale, *Eschrichtius robustus*, feeding in the Bering Sea and Baja California. *Fish. Bull.* **81**, 513–522.

Olsen, C. R., Elsner, R., Hale, F. C., and Kenney, D. W. (1969). "Blow" of the pilot whale. *Science* **163**, 953–955.

Omura, H. (1974). Possible migration route of the gray whale on the coast of Japan. *Sci. Rep. Whales Res. Inst.* **26**, 1–14.

Omura, H., Maeda, K., and Miyazaki, I. (1953). Whaling in the adjacent waters of Japan. *Nor. Hvalfangst-Tid.* **42**, 199–212.

Painter, D. W., II. (1963). Ambient noise in a coastal lagoon. *J. Acoust. Soc. Am.* **35**, 1458–1459.

Patten, D. R., and Samaras, W. F. (1977). Unseasonable occurrences of gray whales. *Bull. South. Calif. Acad. Sci.* **76**(3), 205–208.

Pike, G. C. (1962). Migration and feeding of the gray whale (*Eschrichtius gibbosus*). *J. Fish. Res. Board Can.* **19**(5), 815–838.

Pilleri, G., and Gihr, M. (1969). Das Zentralnervensystem der Zahn- und Bartenwale. *Rev. Suisse Zool.* **76**(4), 995–1037.

Poole, M. (1981). The northward migration of the California gray whale, *Eschrichtius robustus*, off the central California coast. *In* "Proceedings of the Fourth Biennial Conference on the Biology Marine Mammals", Abstr., p. 96. San Francisco.

Poulter, T. C. (1968). Vocalization of the gray whales in Laguna Ojo de Liebre (Scammon's Lagoon), Baja California, Mexico. *Nor. Hvalfangst-Tid.* **57**(3), 53–62.

Ray, G. C., and Schevill, W. E. (1974). Feeding of a captive gray whale (*Eschrichtius robustus*). *Mar. Fish. Rev.* **36**(4), 31–38.

Reilly, S. B. (1981). Population assessment and dynamics of the California gray whale (*Eschrichtius robustus*). Ph.D. dissertation, Univ. of Washington, Seattle.

Reilly, S. B., Rice, D. W., and Wolman, A. A. (1983). Population assessment of the gray whale *Eschrichtius robustus*, from California shore census, 1967–80. *Fish. Bull.* **81**(2), 267–281.

Rice, D. W. (1961). Census of the California gray whale. *Nor. Hvalfangst-Tid.* **50**(6), 219–225.

Rice, D. W. (1967). Cetaceans. *In* "Recent Mammals of the World: A Synopsis of Families" (Eds S. Anderson and J. K. Jones, Jr.), pp. 291–324. Ronald Press, New York.

Rice, D. W., and Wolman, A. A. (1971). Life history and ecology of the gray whale (*Eschrichtius robustus*). *Am. Soc. Mammal., Spec. Publ.* **3**, 142 pp.

Rice, D. W., Wolman, A. A., Withrow, D. E., and Fleischer, L. A. (1981). Gray whales on the winter grounds in Baja California. *Rep. int. Whal. Commn.* **31**, 477–493.

Rugh, D. J., and Braham, H. W. (1979). California gray whale (*Eschrichtius robustus*) fall migration through Unimak Pass, Alaska, 1977: a preliminary report. *Rep. int. Whal. Commn* **29,** 315–320.

Rugh, D. J., and Fraker, M. A. (1981). Gray whale (*Eschrichtius robustus*) sightings in eastern Beaufort Sea. *Arctic* **34**(2), 186–187.

Sauer, E. G. F. (1963). Courtship and copulation of the gray whale in the Bering Sea at St. Laurence Island, Alaska. *Psychol. Forsch.* **27,** 175–176.

Scammon, C. M. (1874). "The marine mammals of the northwestern coast of North America." John H. Carmany & Co., San Francisco.

Schevill, W. E. (1954). On the nomenclature of the Pacific gray whale. *Breviora* **7,** 1–3.

Scholander, P. F. (1940). Experimental investigations on the respiratory function in diving mammals and birds. *Hvalradets Skr.* **22,** 1–131.

Simpson, J. G., and Gardner, M. B. (1972). Comparative microscopic anatomy of selected marine mammals. *In* "Mammals of the Sea" (Ed. S. H. Ridgway), pp. 298–418. Thomas, Springfield, Illinois.

Sprague, J. G., Miller, N. B., and Sumich, J. L. (1978). Observations of gray whales in Laguna de San Quintin, northwestern Baja California, Mexico. *J. Mammal.* **59**(2), 425–427.

Starks, E. C. (1922). A history of California shore whaling. *Calif. State Fish Game Comm., Fish. Bull.* **6,** 1–38.

Sund, P. N. (1975). Evidence of feeding during migration and of an early birth of the California gray whale (*Eschrichtius robustus*). *J. Mammal.* **56**(1), 265–266.

Sund, P. N., and O'Connor, J. L. (1974). Aerial observations of gray whales during 1973. *Mar. Fish. Rev.* **36**(4), 51–52.

Swan, J. G. (1870). The Indians of Cape Flattery at the entrance to Strait of Fuca, Washington Territory. *Smithson. Contrib. Knowl.* **16**(8), 1–108.

Swanson, E. A. (1956). Nootka and the California gray whale. *Pac. Northwest Quart.* **47,** 52–55.

Swartz, S. L. (1981). Cleaning symbiosis between topsmelt, *Atherinops affinis*, and gray whale, *Eschrichtius robustus*, in Laguna San Ignacio, Baja California Sur, Mexico. *Fish. Bull.* **79**(2), 360.

Swartz, S. L., and Jones, M. L., (1978). The evaluation of human activities on gray whales. *Eschrichtius robustus*, in Laguna San Ignacio, Baja California, Mexico. U.S. Dept. Commerce, NTIS PB-289 737.

Swartz, S. L., and Jones, M. L. (1981). Demographic studies and habitat assessment of gray whales, *Eschrichtius robustus*, in Laguna San Ignacio, Baja California Sur, Mexico. *Mar. Mamm. Commn Rep.* No. MMC-81/05, 56 pp.

Tomilin, A. G. (1937). Kity Dal'nego Vostoka (The whales of the Far East). Uchenye Zapiski Moskovskogo Gosudarstvennogo Universiteta, *Ser. Biol. Nauk* **12,** 119–167.

Tomilin, A. G. (1957). "Mammals of the U.S.S.R. and Adjacent Countries. Vol. IX: Cetacea" (Ed. V. G. Heptner). Nauk S.S.S.R., Moscow. (English Translation, 1967, Israel Program for Scientific Translations, Jerusalem).

Uys, C. J., and Best, P. B. (1966). Pathology of lesions observed in whales flensed at Saldanha Bay, South Africa. *J. Comp. Pathol.* **76,** 407–412.

Wahrenbrock, E. A., Maruschak, G. F., Elsner, R., and Kenney, D. W. (1974).

Respiration and metabolism in two baleen whale calves. *Mar. Fish. Rev.* **36**(4), 3–8.

Walker, L. W. (1949). Nursery of the gray whales. *Nat. Hist.* **58**(6), 248–256.

Wellington, G. M., and Anderson, S. (1978). Surface feeding by a juvenile gray whale, *Eschrichtius robustus. Fish. Bull.* **76**(1), 290–293.

Wenz, G. M. (1964). Curious noises and the sonic environment of the ocean. *In* "Marine Bioacoustics", (Ed. W. N. Tavolga), pp. 100–101. Macmillan, New York.

Wilke, F., and Fiscus, C. H. (1961). Gray whale observations. *J. Mammal.* **42**, 108–109.

Wolman, A. A., and Rice, D. W. (1979). Current status of the gray whale. *Rep. int. Whal. Commn* **29**, 275–280.

Wyrick, R. F. (1954). Observations on the movements of the Pacific gray whale, *Eschrichtius glaucus* Cope. *J. Mammal.* **35**, 506–598.

Zenkovich, B. A. (1934). Some data on whales of the Far East. *C. R. Acad. Sci. URSS*, **2**(6), 388–392.

Zenkovich, B. A. (1956). Whale observations during the third voyage of the Soviet Antarctic Expedition (1957–1958). *Sov. Antarct. Exped. Inf. Bull.* **3**, 143–144.

Zimushko, V. V. (1970a). Aerovisual calculation of abundance and observation for distribution of grey whales in inshore waters of Chukotka. *Izv. TINRO* **71**, 289–294.

Zimushko, V. V. (1970b). About anlage of grey whale's sebaceous gland. *Izv. TINRO* **74**, 341–343.

Zimushko, V. V., and Lenskaya, S. A. (1970). Feeding of the gray whale (*Eschrichtius gibbosus* Erx.) at foraging grounds. *Ekologiya* **3**, 205–212.

4

Minke Whale

Balaenoptera acutorostrata Lacépède, 1804

Brent S. Stewart and Stephen Leatherwood

Genus and Species

Common names

The smallest of the rorqual whales, *Balaenoptera acutorostrata*, is most commonly known as the minke[†] whale. The name reportedly derives from the propensity of a certain Norwegian whaler, named Minke, for overestimating lengths of whales taken by his ship. Based on this, others came to call all small or undersized whales "Minke's whales." This name was formally adopted to denote the populations of the northern and, later, the southern hemisphere (Mitchell, 1978). This species is also called the lesser rorqual;

[†]Nomenclature follows the list used by The International Whaling Commission (IWC) in that "minke," though a man's name, is not capitalized.

TABLE 1 Common names of *Balaenoptera acutorostrata*

English
  minke, little piked whale, lesser rorqual, little finner, sharpheaded finner, grampus (Newfoundland), gibord (Quebec), pike-headed whale, summer whale, sprat whale, lesser finback, pike whale, Davidson's whale, bag whale, bay whale, New Zealand piked whale
Aleutian
  agamakhchik
Chukchi dialect
  kauchikan uiiut
Czechoslovakian
  plejtvak stikovity
Danish
  sildeskiper
Dutch
  dwergvinvis
Eskimo
  chikaqulik, tikagulik
French
  baleine d'este a bec, rorqual a museau pointu, rorqual a rostre, petit rorqual
German
  Zwergwal
Japanese
  koiwashi kujira, minku kujira
Norwegian
  minke hval, minkies hval, rebbehual, vaaghval, vagehval
Russian
  karlikovyi polosatik, zalivov, ostromordyi, polosatik, ostrogolovyi polosatik, minke, malyi polosatika kit
Swedish
  vinkval, vikval, vikarehval, spetsnabbad finnfisk

---

rorqual, a common term applied to all the balaenopterid whales, has a Norse derivation meaning red whale, referring to the pinkish tint of the many throat grooves when distended. Other commonly used names are piked whale or little piked whale, referring to the very pointed, V-shaped rostrum of the species. In addition, there are at least 14 English names and 30 foreign names used by people who harvest or have frequent contact with this species (Table 1).

Taxonomy

The taxonomy of the minke whale is plagued by confusion and debate. Omura (1975) and Rice (1977) recognise three subspecies based on geographical distributions: *B. a. acutorostrata* in the North Atlantic, *B. a. davidsoni* in the North Pacific, and *B. a. bonaerensis* in the Southern Hemisphere. Specimens from Sri Lanka have even been tentatively described as a separate subspecies, *B. a. thalmaha* (Deraniyagala, 1963), but the validity of the subspecies has been questioned (Rice, 1977). Cowan (1939) examined specimens from British Columbia and concluded that they were correctly referred to, using Scammon's nomenclature, as *B. a. davidsoni*. Species originally identified as *Agaphelus gibbosus*, *B. bonaerensis*, *B. huttoni*, and *B. rostrata* are currently considered synonymous with *B. acutorostrata* (van Beneden and Gervais, 1880; Flower, 1885; Oliver, 1922; Jonsgård, 1951; Williamson, 1959). Williamson (1961), however, examined five specimens from the Antarctic and concluded that the Southern Hemisphere species differs from the Northern Hemisphere species and is correctly called *B. bonaerensis*. Wada and Numachi (1974) presented biochemical evidence that they interpreted to mean that there were genetically isolated stocks in the Southern Hemisphere. Van Beek and de la Mare (1981) and van Beek and van Biezen (1982) examined the same evidence and argued that it did not clearly discriminate stocks in the Southern Hemisphere. However, further analysis suggests that there is probably more than one biological population in the Southern Hemisphere, with a division likely between samples taken in areas IV and V (Wada, 1982; IWC, 1982). The questions of specific, subspecific, racial, and stock differences in minke whales worldwide remain confused despite the significant numbers taken in recent years by whaling countries. Until a comprehensive comparative morphological and biochemical study of all areas is undertaken, subspecific designation will probably remain little more than speculation.

External Characteristics and Morphology

The minke whale is the smallest of the rorquals and the second smallest of the baleen whales, seldom exceeding 10.1 m in length (the smallest is apparently the poorly known pygmy right whale, *Caperea marginata*, for which the largest specimen is a 6.4 m female; see Chapter 12). Body lengths of minke whales in Korean waters range from 5.8 to 6.5 m and weights between 2.0 and 2.7 metric tons (Gong, 1981). Newborn calves in California waters are reportedly 269–284 cm long (Norris and Prescott, 1961).

The rostrum is extremely narrow and pointed with a single head ridge similar to but much sharper than that of a fin whale. The dorsal fin is relatively tall, very sickle shaped, and located quite far forward on the posterior one-third of the body (Tomilin, 1957; Leatherwood *et al.*, 1976, 1982b). The pectoral fins are small, slightly greater than one-eighth the total body length (True, 1904; Nishiwaki, 1972). Omura and Sakiura (1956) state that the minke whale in Japanese waters has a shorter mouth, more posteriorly situated and higher dorsal fin, shorter flippers, broader tail flukes, and a more posterior genital aperture than does the Atlantic minke whale. Sergeant (1963) found morphological differences between minkes from the Norwegian seas and those from the Barents Sea. Pacific minke whales may have a more posterior dorsal fin position, narrower flukes and fewer baleen plates than do Atlantic minkes (Cowan, 1939; Scattergood, 1949). Compared with Northern Hemisphere minke whales, Antarctic minkes are apparently more slender, have longer rostrums, and have dorsal fins that insert closer to the tail flukes (Zemsky and Tormosov, 1964). Numbers of baleen plates and ventral grooves appear variable with region (Table 2).

Coloration of the body and baleen plates varies both individually and regionally (Tables 2 and 3). Minke whales are generally black, brown, or grey dorsally and light ventrally. In the Northern Hemisphere, the white flipper stripe is diagnostic of the species; in the Southern Hemisphere stocks, the presence and extent of the stripe are variable (Table 3, Fig. 1). Doroshenko (1979) reported marked differences between Southern Hemisphere populations: 55, 75.2, 26.1, and 48% of whales examined in the Indian, New Zealand, Brazilian, and Chilean–Peruvian populations, respectively, had white flipper stripes. Ohsumi et al. (1970) reported that all whales observed in area IV (see Fig. 2) lacked the white flipper patch. Best (1979) noted that 45% of the whales landed at Durban, South Africa, had a grey band on one or both flippers and another 75% had a white blaze on the shoulder that extended on to the flipper. Gaskin (1972) indicated that some whales in New Zealand waters possessed white flipper patches while some whales lacked them. Williamson (1959) and Zemsky and Tormosov (1964) reported that Antarctic specimens examined by them lacked white flipper stripes. Leatherwood et al. (1982) found no trace of white on most minkes in the Ross Sea.

Kato (1979a) reported a minke whale caught in area IV (see Fig. 2) of the Antarctic which was pink on the ventral surface; the blubber, connective tissue, and baleen plates were also flushed with a carotenoid colour.

Distribution and Movements

The minke whale is widespread and seasonally abundant in the North Atlantic Ocean. In the northwest Atlantic the species ranges from Davis Strait and Baffin Bay during the summer months south to the Florida Keys, Gulf of Mexico, and West Indies in the winter (Brown, 1868; True, 1904; Allen, 1916; Howell, 1934; Moore, 1953; Moore and Palmer, 1955; Sergeant, 1963; Jonsgård, 1966; Struhsaker, 1967; Erdman et al., 1973; Schmidly, 1981). Most records in the Gulf of Mexico and Florida Keys are of immature whales

TABLE 2 Characteristics of baleen plates and ventral (throat) grooves of the minke whale

Region	No. baleen plates (per side of jaw)	Baleen colour	No. ventral grooves	Source
North Pacific	270			Scammon (1873)
	231			Cowan (1939)
	285			Carl (1946)
	256–272	Yellowish white		Scattergood (1949)
	266–295	Yellowish white		Omura and Sakiura (1956)
	231–270		50–70	Tomilin (1957)
	280	Yellowish white, brown, or black on posterior portions		Haley (1978)
North Atlantic	287–290; 270–274; 311			Turner (1893)
	290; 300			Collett (1912)
	316			Allen (1916)
	270–348 ($\bar{x} = 304$)			Jonsgård (1951)
	300–325			Tomilin (1957)
		Creamy white		Sergeant (1963)
		White at front, grey and white at back		Williamson (1959)
Southern Hemisphere	261–359 ($\bar{x} = 299$)		52–60 ($\bar{x} = 56$)	Ohsumi et al. (1970)
		White		Best (1974)
	105–415		30–70	da Rocha (1980)
	308		57	Singarajah (1981)

TABLE 3 Characteristics of the colour pattern of the minke whale

Region	Colour pattern	Source
North Pacific	Area of blue-grey that extends up sides to dorsolateral surface from a region "touched by tip of flexed pectoral"	Cowan (1939)
	Characteristic white back of pectoral flipper	Scattergood (1949)
	Dorsal surface uniformly black, underparts mostly white. Broadness of flipper stripe is individually variable. Pale streaks about head and shoulders often occur	Omura and Sakiura (1956)
	Major area of body and trunk dark but flippers have white patch	Mitchell (1978)
North Atlantic	Colour above and on sides of lower jaw grey-black; below white. Dark colour of back descending obliquely behind the pectoral fins and occupying greater part of tail	True (1904)
	Considerable variation in form of flipper stripe, but always-present white borders may end distinctly or be preceded by grey shadows	Sergeant (1963)
Southern Hemisphere	No white flipper band	Williamson (1959)
	All minkes caught in 1967–1968 and 1968–1969 lacked white band on flipper	Ohsumi et al. (1970)
	Grey dorsally and white ventrally	Kato (1979a)
	Flipper band missing from most animals in Ross Sea, faint in remainder	Leatherwood et al. (1981)

stranded during the winter. However, Winn and Perkins (1976) report numerous winter sightings from the West Indies (18°08' N).

In spring and summer minke whales migrate northwards, some near shore, others apparently in pelagic waters. At least the coastwise portion passes Nova Scotia in May and reaches the northern Labrador coast by August (Katona *et al.*, 1977). Minkes continue to be common in the Gulf of St. Lawrence, along the Newfoundland coast, and off southwest Greenland throughout summer, although some whales move farther north to Davis Strait and Baffin Bay in late summer (Allen, 1916; Sergeant, 1963; Jonsgård, 1966; Kapel, 1980). Inshore records have been reported from eastern Canada (Sergeant, 1963; Sergeant *et al.*, 1970; Mitchell and Kozicki, 1975). There is a southward migration to temperate waters in winter (Moore and Palmer, 1955; Jonsgård, 1951; Fraser, 1953; Mitchell, 1975a) which, because of the paucity of nearshore records at this time, is thought to occur farther offshore.

In the northeast Atlantic, minkes migrate north into Norwegian and nearby Arctic waters in spring and are observed as far north as West Spitzbergen and the Barents and Kara seas in summer (Turner, 1893; Collett, 1912; Hentschel, 1937; Jonsgård, 1951, 1966; Fraser, 1979). Minkes have been occasional visitors to the waters off Sweden and Denmark (Collett, 1912) and are rarely seen in the Baltic Sea (Aguayo, 1978). The species is common in inshore northern and western coastal waters of British seas (47°–67° N) from July to October (Harmer, 1927; Fraser, 1934, 1946; Stephenson, 1951; Evans, 1980). As in the northwestern Atlantic, there appears to be a southward migration—at least some of it occurring near the European coast—to temperate waters in autumn (Jonsgård, 1966; Fraser, 1979). Although little is known of the species' southern limit in the eastern Atlantic, records have been reported for Holland (Weber, 1922; van Deinse, 1931; Oordt, 1926), France (Fischer, 1881; van Beneden, 1885; Legendre, 1943), Gibraltar (Collett, 1912), the Mediterranean and along the Italian coast (van Beneden, 1885; Carrucio, 1913; Monticelli, 1926; Cagnalaro *et al.*, 1984), and the Azores (Chaves, 1924). Nobre (1935) reported that the species was frequently sighted off Portugal in winter.

Data suggest that on both sides of the North Atlantic, sexual and age segregation occurs during summer, with males migrating farther north in open seas, females remaining in more southern and coastal areas, and immatures occurring slightly farther south (Jonsgård, 1951, 1966; Schwartz, 1962; Sergeant, 1963; Kapel, 1980).

The International Whaling Commission (IWC) recognises four stocks in the North Atlantic: Canadian east coast, west Greenland, central North Atlantic, and northeast Atlantic (IWC, 1980).

In the northeast Pacific, minke whales range from the Chukchi Sea and Pt Barrow, Alaska, south to Islas Revillagigedos, Baja California, Mexico

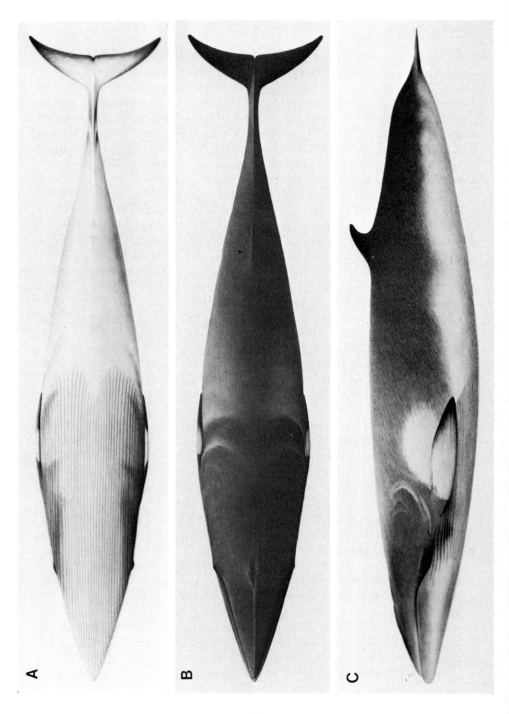

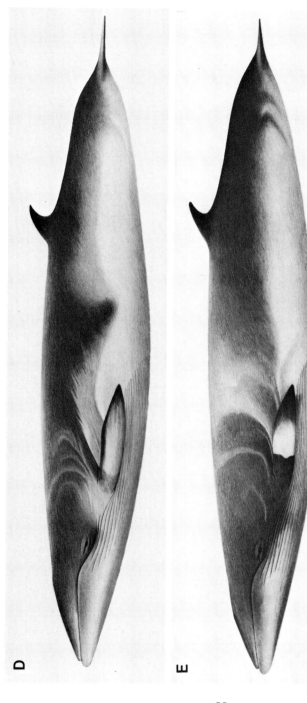

FIG. 1 General colour patterns of the minke whale. (A) Ventral view; (B) dorsal view; (C) lateral view, one Southern Hemisphere form; (D) lateral view, another Southern Hemisphere form; (E) lateral view, Northern Hemisphere form. (Drawings by L. Foster, courtesy of General Whale.)

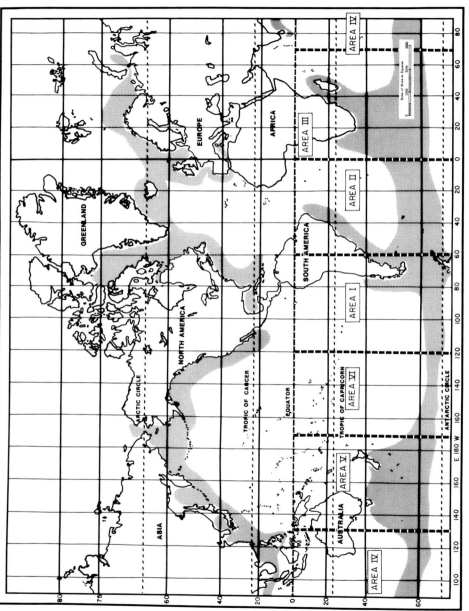

FIG. 2 General distribution of the minke whale. The shaded areas are supported by references in the text. Minke whales may occur in other areas and this map may not represent the total distribution of such a mobile marine organism.

(Scammon, 1869; Scattergood, 1949; Miller and Kellogg, 1955; Hall and Kelson, 1959; Rice, 1974; Everitt et al., 1980; Marquette et al., 1982; Leatherwood et al., 1982b; B. S. Stewart and S. Leatherwood, unpublished data). Minkes are occasionally sighted in the Sea of Cortez. The species is common in the Bering Sea, coastal Gulf of Alaska, Puget Sound, the San Juan Islands and the Strait of Juan de Fuca in spring and summer months (Scammon, 1874; Scheffer, 1973; Rice, 1974; Boran and Osborne, 1978; Braham et al., 1979; Everitt et al., 1980; B. S. Stewart and S. Leatherwood, unpublished data). Sighting records from northern Puget Sound and the San Juan Islands suggest that some whales may reside year-round in these areas (Everitt et al., 1980; Dorsey, 1983; Angell and Balcomb, 1982). The species is known to occur off central and northern California, but sightings are rare (Sullivan and Houck, 1979; Dohl et al., 1980). Although few records exist for southern California waters (Abbott, 1930; Fry, 1935; Norris and Prescott, 1961; Leatherwood et al., 1982b, B. S. Stewart, unpublished data), the species may be common year-round (Norris and Prescott, 1961; Dohl et al., 1980).

By individually identifying animals in the inland waters of Washington state, from dorsal fin shapes and colour patterns, Dorsey (1983) found that some minke whales had exclusive home ranges.

In the western Pacific this species ranges from the Okhotsk and Bering seas and at least as far north as Cape Serdtse-Kamen in the Chukchi Sea south to the Sea of Japan and the Yellow Sea. Records exist from tropical waters near the Philippines (Herre, 1925) and off Sri Lanka (Deraniyagala, 1948, 1963) but minke whales appear rarely to enter these waters. Minke whales are common near the Kurile Islands (45°–50° N) in summer (Matsuura, 1936) and are abundant in the Sea of Japan and the Yellow Sea in winter (Matsuura, 1936; Tago, 1937; Ohsumi, 1975; Gong, 1981). Whales in both the east and west Pacific appear to migrate north along the coasts in spring and summer and migrate south in autumn and winter, apparently farther offshore (Matsuura, 1936; Scattergood, 1949; Tomilin, 1957; Ohsumi, 1975; Rice, 1974; Mitchell, 1978). Near Japan, whales on their southward migration in autumn follow two courses, one into the Sea of Japan along the east coast of Honshu, and one down the Pacific coast of Honshu (Matsuura, 1936; Omura and Sakiura, 1956). During summer months, immature animals remain in more southern waters while predominantly mature animals travel to more northern feeding grounds; adult males and females may be geographically segregated at this time also (Omura and Sakiura, 1956). Inshore records along the Chukotka coast have been reported (Ivashin and Votrogov, 1981).

The IWC recognizes two stocks in the northwest Pacific: Okhotsk Sea/ West Pacific; and Sea of Japan, including the Sea of Japan, Yellow Sea, and East China Sea (IWC, 1980).

In the Southern Hemisphere, minkes are apparently circumpolar between the Antarctic continent and New Zealand, Argentina, central Chile, Brazil, Surinam, Angola, and Madagascar. The species is common off Antarctica in the austral summer (December–March), entering the pack ice and ice-free coastal waters (Lillie, 1915; Williamson, 1959; Doroshenko, 1979; Lockyer, 1981). Geographic abundance is not uniform during a particular season and such differences may not be inconsistent among seasons (Arsenev, 1960; Butterworth and Best, 1982). Minke whales apparently occur rarely in New Zealand and Tasmanian waters (Stead, 1910; Oliver, 1922; Pearson, 1936; Davies and Guiler, 1958; Mörzer-Bruyns, 1974). They are common off Brazil (July–November) and are present but uncommon off western South Africa and Angola. Minkes may be present in moderate numbers year-round off Durban, southeast Africa (Burmeister, 1867; Lahille, 1908; Marelli, 1918; Peringuey, 1921; Slipjer et al., 1964; Williamson, 1975; Best, 1967, 1974, 1977, 1982). Abundance in this area peaks from June to September when the population consists of adult animals, primarily males (Best, 1982). Off southeast Africa, minke whales have been observed as far north as 21° S, off both coasts of Madagascar, and eastwards towards Mauritius (Gambell et al., 1975). Minkes have also been reported north of the equator in the Gulf of Aden (Yukhov, 1969), and there are specimen records from Sri Lanka (6°–10° N) (Deraniyagala, 1948, 1963). Aguayo (1974) notes that although a few records exist for Chilean waters, minkes are rare visitors there.

Sightings and tag return data suggest that minkes migrate from temperate waters south to the Antarctic continent in spring and summer and migrate north towards South America, South Africa, and New Zealand in autumn and winter (Williamson, 1975; Lockyer, 1981; Best, 1982). A marlin spear found imbedded in the tissue of a minke whale caught in Antarctic waters also suggests movement between warm and cold waters (Ohsumi, 1973). Adult males appear to migrate farther south than females; immatures apparently do not travel as far south as either adult males or adult females (Ohsumi et al., 1970). On the Antarctic whaling grounds sexual segregation apparently occurs, at least temporally, with mature females arriving later in the season than mature males (Ohsumi et al., 1970; Ohsumi and Masaki, 1975; Masaki, 1979).

In the Antarctic, Arsenev (1960) recognized three population groups based on catch densities: Atlantic stock (50° W–40° E), Indian stock, (50°–140° E), and a third possible stock (160°–170° E). Doroshenko (1979) provisionally defined seven populations based on winter habitat (off Brazil, Africa, New Zealand, in the Pacific Ocean, Chile and Peru, in the Indian Ocean, and off Madagascar) but noted that morphological data support recognition of only four distinguishable populations in the southern hemisphere (Brazilian, Indian, New Zealand, Chilean–Peruvian). Winter distributions may have greater significance for management purposes. The IWC, however, currently manages the southern hemisphere minke whale stocks on the basis of the six arbitrarily assigned baleen whale areas (IWC, 1978; Fig. 2): area I, 60°–120° W; area II, 0°–60° W; area III, 0°–70° E; area IV, 70°–130° E; area V, 130° E–170° W; area VI, 170°–120° W.

Abundance and Life History

Whaling and history of exploitation

North Pacific. Historically minke whales were taken in very small numbers by natives of the Pacific Northwest of North America (Scammon, 1874; Collins, 1892; Waterman, 1920). Two minkes were reportedly taken off British Columbia in 1923 by a commercial shore-based whale fishery (Kellogg, 1931). American Eskimos at Savoonga on St. Lawrence Island still occasionally take minkes for subsistence purposes, although whaling efforts there are primarily directed at bowheads, *Balaena mysticetus*. Minkes were exploited in the coastal waters of Japan beginning several centuries ago (Omura and Sakiura, 1956; Matsuura, 1936). The Norwegian method of whaling using small catcher boats, introduced to Japan in about 1890, was used for minkes though they were not the primary species pursued (Ohsumi, 1975). As stocks of bowhead; right, *Eubalaena glacialis* and *E. australis;* blue, *Balaenoptera musculus;* fin, *B. physalus;* and sei, *B. borealis*, whales were depleted, the minke whale attracted the attention of pelagic whaling fleets. Modern catcher boats were first used in the 1920s and subsequently the coastal Japanese minke whale fishery expanded (Matsuura, 1936).

Russian pelagic whaling fleets began taking minkes in 1933 off the east coast of Kamchatka, in the Bering Sea, and in the Arctic Ocean (Matsuura, 1936). The Japanese pelagic vessels began exploiting minkes of the Okhotsk–West Pacific stock in 1930 and enjoyed a gradual increase in the annual catch until the beginning of the 1950s; the annual catch has been stable at about 400 individuals since (Ohsumi, 1981b).

The Republic of Korea has used small shore-based catcher boats to harvest animals reported as minke whales year-round in the waters off Korea since the late nineteenth century (Gong, 1981). Beginning in 1978, when Korea joined the IWC, that whaling season was limited to 6 months, from March to September (Brownell, 1981). The catch in Korean waters increased gradually from 170 in 1962 to 396 in 1969; 715 were taken in 1970; between 1971 and 1980 the catch has fluctuated between 500 and a maximum of 1033 (Brownell, 1981) (Table 4).

TABLE 4 Minke whales taken by year for each area[a]

Year	N. Atlantic	Japan	Korea	Antarctic[b]	Brazil	S. Africa	Canada	United States (Alaska)	S. Atlantic
1954	1527	365					32		
1955	1797	427			2		13		
1956	1425	532		41			57		
1957	1677	423		46			37		
1958	2243	512		493			42		
1959	1839	280		102	2		18		
1960	2087	253		203			11		
1961	3245	332		162			22		
1962	3286	238	170	2			18		
1963	3233	220	291	21	2		37		
1964	2732	301	384	101	44		54		
1965	2467	334	247	6	68		41		
1966	2172	365	301	14	352		28		
1967	2196	285	335	605	488		41		

and Norway, the IWC added the following ban in 1982: "The killing for commercial purposes of minke whales using the cold grenade harpoon shall be forbidden from the beginning of the 1982/83 pelagic and the 1983 coastal seasons" (IWC, 1982, p. 29). Japan and Russia announced in response that they could not comply with the ban on the cold harpoon, although Japan said that it would continue limited testing of an explosive-tipped harpoon.

Abundance

Based on analysis of catch per unit effort (CPUE) trends, and historical catches of minke whales, Ohsumi (1981b) estimated the 1981 stock of the Okhotsk Sea–West Pacific to number between 17 000 and 28 000. He also argues from the same data that the stock should be reclassified from a sustained management stock (SMS) to an initial management stock (IMS) and that the annual catch quota should be increased from 421 to more than 648. Wada (1976) estimated the entire North Pacific stock at 9000 individuals. However, because there are no reliable separate estimates available for the central and eastern North Pacific stocks, these have been classified by the IWC as IMSs with zero catch quotas.

The northeast Atlantic stock has been estimated, based on mark–recapture data, at 113 000, with 95% confidence limits of 70 000 and 186 000, and the exploitable stock has been estimated at 56 000 (Christensen and Rørvik, 1981). Estimates for other North Atlantic stocks are not available.

Laws (1977a) estimated both the initial and 1977 Antarctic stock sizes at 200 000 with a biomass in 1977 of 1 400 000 metric tons. Ohsumi (1979b) estimated the original Antarctic population at 52 100, and the 1978 population at 416 700 and concluded that minkes have become "overpopulated" in this area. Estimates of populations in various portions of the Antarctic are presented in Table 5. Some earlier estimates are summarized in Gaskin (1982, Table 8.2). Ohsumi (1979b, p. 420) recommended that the catch quota of Antarctic minkes be increased "to ensure recovery of the populations of blue, fin, and humpback whales."

Food habits

Omura and Sakiura (1956) reported that minke whales in the North Pacific feed on euphausiids, copepods, and sand lance and those in Okhotsk Sea take krill and, to a far lesser extent, fishes. We have observed them feeding on herring in northern Bristol Bay in summer (S. Leatherwood and B. S. Stewart, unpublished data).

In the North Atlantic minkes are known to feed on sand lance, sand eel, euphausiids, copepods, salmon, capelin, mackerel, cod, coal fish, whiting,

TABLE 5 Estimated minke whale populations[a]

Region	Area	Estimated abundance	Estimation method	Source
North Pacific	Okhotsk Sea–West Pacific stock	17 000–28 000	Catch per unit effort	Ohsumi (1982)
North Atlantic	Svalbard–Norway–British Isles stock	50 592 with 95% confidence interval ranging from 26 895 to 100 172	Mark recapture data	Christensen and Rørvik (1979)
Southern Hemisphere	Area I (south of 65° S)	35 000	Sighting and marking data	Miyashita (1982b)
	Area I (south of 65° S)	28 000	Sighting data	IWC (1982)
	Area III	55 000		Best (1974)
	Area III	62 500		Best (1977)
	Area III	>127 000	Tag–recapture data	Horwood (1981)
	Area III	96 000	Sighting data	Horwood (1981)
	Area III	71 644	Sighting data	IWC (1982)
	Area III	200 555	Tag–recapture data	Miyashita (1982a)
	Area III	257 790	Sighting data	IWC (1982)
	Area IV	23 000	Sighting data	Ohsumi et al. (1970), Ohsumi (1980b)
	Area IV	91 000 ± 38 000	Tag–recapture data	Best and Butterworth (1980)

Area IV	106 000 ± 44 000	Tag–recapture data	Best and Butterworth (1980)
Area IV	78 000	Sighting data	Best and Butterworth (1980)
Area IV	147 000	Sighting data	Chapman (1980)
Area IV	126 945	Tag–recapture data	Miyashita (1982a)
Area IV	57 951	Sighting data	IWC (1982)
Area IV	136 417	Marking data	IWC (1982)
Area IV	116 000 ± 36 000	Marking data	Brown and Best (1982)
Area IV	95 000 ± 30 000	Marking data	Brown and Best (1982)
Area V	79 847	Sighting data	IWC (1982)
Area V	94 109	Marking data	IWC (1982)
Area VI	159 000–185 000		Ohsumi (1981a)
Area VI	54 142		IWC (1982)
All areas	>70 000		Ohsumi et al (1970)
All areas	416 700		Ohsumi (1979b)
All areas	200 000		Laws (1977b)

[a] Estimates are of exploitable populations, total populations or recruited populations. Readers are referred to original sources for derivation and justification of these estimates.

sprat, wolffish, dogfish, pollack, haddock, and herring (Eschricht, 1849; Henking, 1901; Collett, 1912; Hentschel, 1937; Ruud, 1937; Jonsgård, 1951; Sergeant, 1963; Christensen, 1971, 1974; Mitchell, 1975a,b; Jonsgård, 1982; Larsen and Kapel, 1981). Krill are important food items off west Greenland (Larsen and Kapel, 1981), while capelin and cod are the dominant prey species in eastern Newfoundland waters (Sergeant, 1963).

Minke whales apparently do not feed while in Brazilian waters (Paiva and Grangeiro, 1965, 1970; Williamson, 1975). In Antarctic waters they feed predominantly on krill (*Euphausia superba*, *E. spinifera*, *E. crystallorphias*), although they also consume various species of myctophid fishes (Nemoto, 1959; Ohsumi *et al.*, 1970; Laws, 1977b; Ohsumi, 1979c, Best, 1982).

Growth and reproduction

Northern Hemisphere. In the southwest Sea of Japan breeding occurs from December through March. Calves are born approximately 10 months after conception and are 2.4–2.7 m long at birth (Matsuura, 1936). Omura and Sakiura (1956) and Mitchell (1975a,b) report that breeding occurs year-round in the eastern North Pacific but there are peaks in January and June; calving peaks in December and June. Females attain sexual maturity at 7.3 m and males at 6.7 to 7.0 m (see Table 6). Age at sexual maturity is unknown (Mitchell, 1975a,b).

In the North Atlantic, mating occurs from October to March and gestation is approximately 10 months. Mature females may give birth every year (Jönsgard, 1951; Stephenson, 1951; Sergeant, 1963; Christensen, 1975). Calving occurs from November to March and calves are 2.4–2.8 m long at birth, lactation lasts 4–5 months (Jonsgård, 1951; Sergeant, 1963; Mitchell, 1975b). Christensen (1975) reported pregnancy rates of 0.90 and 0.97 for the west Greenland and Barents Sea populations, respectively. Mitchell and Kozicki (1975) reported a pregnancy rate of 0.86 for the Newfoundland population.

Ohsumi *et al.* (1970) and Ohsumi and Masaki (1975) reported success in using growth layers in ear plug cores to determine ages of Antarctic minke whales. Sergeant (1963) and Mitchell and Kozicki (1975) reported moderate success when they applied this method to minkes in the northwest Atlantic where only a small percentage of plugs were readable. The use of ear plugs to age minkes from Icelandic waters was found unsatisfactory for a large part of the population (Sigurjonsson, 1980).

Age at sexual maturity has been estimated at 7.1 years in females and 6 years in males, based on growth layers in the tympanic bulla (see Table 6). The mean length at sexual maturity is estimated at 7.15 m for females and

6.75 m for males, while mean length at physical maturity is approximately 8.5 m for females and 7.9 m for males (Christensen, 1980, 1981).

Southern Hemisphere. In the Southern Hemisphere, mating occurs from June through December with a peak in August and September (Best, 1982). Gestation is about 10 months and peak numbers of births occur during late May and early June in warm waters north of the Antarctic convergence; there is evidence for a minimum 14-month calving cycle (Jönsgard, 1962; Williamson, 1975; Ivashin, 1976; Ivashin and Mikhalev, 1978; IWC, 1979; Masaki, 1979; Best, 1982). There is usually one calf. Of 10 675 pregnant females examined by Kato (1982), 60 (0.56%) carried twins and 3 (0.03%) carried triplets. Newborns are approximately 2.8 m; length at weaning is 5.7 m and length at 1 year is approximately 7–8 m (Williamson, 1975; Ivashin and Mikhalev, 1978; IWC, 1979; Lockyer, 1979; Masaki, 1979). Lactation lasts from 3 to 6 months (Williamson, 1975; IWC, 1979; Masaki, 1979; Best, 1982). Length and age at sexual maturity for females are approximately 7.9 m and 6–8 years, respectively (Ohsumi *et al.*, 1970; Ohsumi and Masaki, 1975; Masaki, 1979; Lockyer, 1981). Males are sexually mature when 5–8 years old and approximately 7.3 m long. Although pregnancy rates of 0.80–0.90 and 0.95–0.96 have been reported for the higher latitudes of Antarctica (Ohsumi *et al.*, 1970; Ohsumi and Masaki, 1975; Ohsumi, 1979a; Masaki, 1979), this figure may be affected by reproductive segregation of pregnant and nonpregnant females (Best, 1982). Best (1982) estimated the pregnancy rate of the (seasonal) population off southeast Africa at 0.78.

From 1974 through 1978, 333 minke whales were marked in the northeast Atlantic. Twenty-four marks were recovered from 1975 to 1980 (Christensen and Rørvik, 1978, 1980, 1981). A total of 2243 was marked in the Southern Hemisphere between 1971 and 1981 by scientists on international research cruises, primarily in areas III and IV. Twenty-five marks were recovered through 1981 (Brown, 1979; Best and Butterworth, 1980; Brown and Best, 1981; Horwood, 1981; Butterworth and Best, 1982; Brown and Wada, 1982).

Russian scientists marked an additional 253 minkes from 1957 through 1980 in Antarctic waters; three marks were recovered through 1980 (Brown and Best, 1981; Brown and Wada, 1982).

The small Discovery .410 tag has been the primary mark used for minke whales. The Russian program has used primarily a larger tag similar in size to the Discovery 12-bore calibre tag. The 12-bore calibre tag has proved unsuitable because of its potential for serious injury to minkes (Best, 1982). To date, population estimates of North Atlantic minke whales, based on mark–recaptures, have been presented only for the northeast Atlantic population (Christensen and Rørvik, 1981).

TABLE 6 Reproductive and growth parameters for minke whales

Area	Age at sexual maturity (years)		Age at physical maturity (years)		Length at sexual maturity (m)		Length at physical maturity (m)		Maximum length (m)	Maximum age (years)	Source
	Males	Females	Males	Females	Males	Females	Males	Females			
North Pacific	2	2			6.7–7.0	7.3					Omura and Sakiura (1956)
					6.7–7	7.3					Mitchell (1975a)
	7–8	7–8			6.7–7.0	7.3	8.2	8.2			Mitchell (1978)
					7	7.9			8.2		Everitt *et al.* (1980)
North Atlantic					6.7–7.0	7.3					Sergeant (1963)
					7.0	7.3					Mitchell (1975a)
			7.3 ear plug growth layers								Mitchell and Kozicki (1975)

									Reference
6				6.75	7.2				Christensen (1980)
7 7.1 bulla growth layers								≥33	Christensen (1980)
3.7 bulla growth layers									Larsen and Kapel (1981)
Southern Hemisphere									
7–8	18–22	18–22			8.3	8.8	9.2–9.8	<50	Ohsumi et al. (1970)
	18–20	20–22			8.5	9.0			Ohsumi and Masaki (1975)
			7.6	7.6					IWC (1978)
			7.1	7.9					Masaki (1979)
			7.7	7.9					da Rocha (1980)
6				8.0					Lockyer (1981)
7–9	18	18	7.9	8.1	8.6	8.9			Best (1982)
5–8									

Internal Anatomical Characteristics

Skull

Eschricht (1849), van Bambeke (1868), Carte and Macalister (1868), and van Beneden and Gervais (1880) have given detailed descriptions of skulls of European specimens. True (1904) presents skull measurements of 10 specimens from France, Great Britain, and the western Atlantic. Turner (1893) published measurements of five skulls preserved in the Museum of Edinburgh. The skull of the minke whale is typical of balaenopterids. The projection of the maxillary and mandibles are, however, less marked than in other balaenopterids. The maxillary is very narrow at the distal end and wide at the proximal end, resulting in a large, wide nasal cavity (Fig. 4). Few differences have been observed between skulls of minkes from the North Atlantic and those of the North Pacific (True, 1904; Cowan, 1939; Omura, 1957). Several differences, however, have been reported between the skulls of Northern and Southern hemisphere minke whales (Omura, 1975). The breadth of the Antarctic minke whale skull is narrower than that of Northern Hemisphere minkes. Antarctic minkes have larger rostrums than do North Pacific minkes; several differences also exist in rostrum breadth at parts along its length (Omura, 1975).

Organ systems

A limited body of literature exists on the internal anatomy and organ systems of the minke whale from various parts of its range (see Table 7).

Behaviour

Minke whales occur singly or in groups of two to three, although herds have been observed in high latitudes of the Northern and Southern hemispheres (Fig. 5) (Taylor, 1957; Christensen, 1975; Mitchell, 1978; Everitt *et al.*, 1980). Minkes in some areas approach boats (Kasuya and Ichihara, 1965; Beamish and Mitchell, 1973; Mitchell, 1975a; Horwood, 1981; Ivashin and Votrogov, 1982, B. S. Stewart, personal observation; see Fig. 6). When it

---

FIG. 4 (A) Lateral view of a skull and skeleton; (B) lateral view, (C) dorsal view, and (D) ventral view of a skull of a minke whale from the western North Atlantic. (From True, 1904, courtesy of J. G. Mead, Smithsonian Institution.)

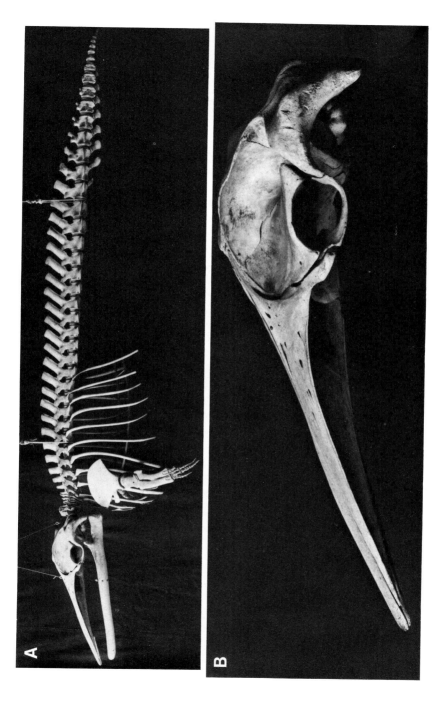

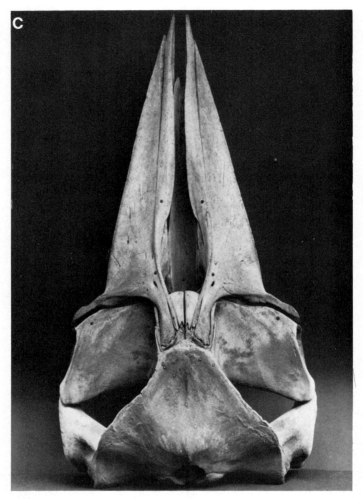

FIG. 4C

exists, this "seeking behaviour" (Mitchell, 1978) confounds efforts to estimate population size from sighting data by producing inflated estimates of observed density which are reflected in extrapolated population estimates (Leatherwood et al., 1982a). The tendency of minkes to segregate into age- and sex-dominated groups also makes it difficult to obtain the random samples required for accurate estimates of population size (Williamson, 1975). In some areas movements of minkes appear unaffected by boat presence (Hor-

Sound production

Sounds of minke whales have been recorded off Newfoundland (Beamish and Mitchell, 1973; Winn and Perkins, 1976), in the Ross Sea off Antarctica (Schevill and Watkins, 1972; Leatherwood et al., 1981), and in the St. Lawrence River (P. L. Edds, personal communication). Sounds recorded off Newfoundland were low frequency grunts (80–140 Hz, 165–320 msec duration), thumps (100–200 Hz, 50–70 msec duration), ratchets (850 Hz, 1–6 msec duration) and pinglike sounds and clicks at various frequencies (3.3–3.9 kHz, 5.5–7.2 kHz, 10.2–12 kHz) extending to over 20 kHz (Winn and Perkins, 1976). Pings and clicks were primarily 0.5–5 msec long with the exception of 3.3–3.8 kHz pings that were 16–20 msec duration (Winn and Perkins, 1976). Thompson et al. (1979) believed that potential signature information may occur within the "thump trains," since frequency composition and repetition rates appeared to be individually variable.

Beamish and Mitchell (1973) reported high frequency signals from a single minke whale off Newfoundland that emitted 200 clicks in 50 distinct series. The repetition rate was constant with a mean of 6.75 clicks/sec, with the principal energy between 4 and 7.5 kHz.

Schevill and Watkins (1972) recorded intense sounds in the Ross Sea of 65 dB re 1 dyne/cm^2 that swept down in frequency from 130–115 to 60 Hz, and lasted from 200–300 msec. All sounds were similar in duration, frequency sweep, and intensity (Schevill and Watkins, 1972). Leatherwood et al. (1981) recorded six types of underwater sounds in the Ross Sea believed to be those of minke whales, including a whistle series, a clanging bell series, a click, a screech, a low frequency grunt, and a frequency-modulated sweep.

Over 400 calls were recorded in the presence of minke whales in the St. Lawrence River (P. L. Edds, personal communication). Although fin whales were often present in the study area, careful analysis revealed differences in the majority of calls produced (Edds, 1980). Nearly 80% of minke whale vocalizations were frequency downsweeps with initial frequencies between 80 and 200 Hz. The frequency change during the downsweeps was between 10 and 70 Hz. Average duration was 0.4 sec (range 0.1–0.6 sec). The calls were of relatively low intensity (30–60 dB re 1 μbar). Sound production was infrequent and few of the sounds produced were in the higher frequency ranges reported previously for this species. Click series were not recorded (P. L. Edds, personal communication).

Parasites

Ohsumi et al. (1970) identified whale lice, *Cyamus balaenopterae*, from Antarctic minkes and noted that they were found only in the ventral grooves.

Best (1982) reported that of 160 minke whales examined off Durban, South Africa, 32 (20%) carried whale lice, mostly at the posterior ends of the ventral grooves or around the umbilicus. Larsen and Kapel (1981) reported on nematode parasites in the stomachs of three whales and liver parasites from one whale. Other internal parasites identified from the minke whale include three trematode species (*Fasciola skrajabini, Lecithodesmus goliath, Ogmogaster plicatus*), three nematode species (*Anisakis simplex, Crassicauda crassicauda, Porrocaecum decipiens*), three species of acanthocephalans (*Bolbosoma balaena, B. brevicolle, B. nipponicum*), and at least one species of cestode (*Tetrobothrius affinus*) (Baylis, 1932; Delyamure, 1955; Best, 1982). The diatoms *Cocconeis ceticola* and *Navicula* sp. were found on minkes off South Africa; *Cocconeis* was found on only one whale (Nemoto *et al.*, 1980). Ohsumi *et al.* (1970) observed *C. ceticola* on minkes from the Antarctic. Best (1982) collected a barnacle (*Penella* sp.) from the skin of a female minke whale killed off Durban, South Africa.

Captivity

Minke whales have been kept in captivity in Japan on three occasions. One whale was kept alive for nearly 3 months, another for 1 month and a third for 37 days before it escaped through holding nets (Kimura and Nemoto, 1956). The one attempt in North American to monitor a live-caught minke whale in captivity met with no success (Norris and Prescott, 1961).

Predation

Killer whales, *Orcinus orca*, have been reported to attack and kill minke whales in the Antarctic (Best, 1982) and in waters off British Columbia, Canada (Hancock, 1965). In Antarctic waters minke whale parts were identified in 84% of killer whale stomachs examined by Shevchenko (1975) and in 70 to 85% of stomachs examined by Doroshenko (1978). Budylenko (1981) estimated that minke whales constitute 85% of killer whales' diet in the Antarctic.

Acknowledgments

We thank S. Ohsumi, J. Horwood, P. Edds, and C. Lockyer for reviewing the manuscript. We also thank L. Foster for providing us with pencil drawings of the minke whale. This chapter was submitted to the volume editors 1 April 1983.

References

Abbott, C. G. (1930). California record of a sharp-headed finner whale. *J. Mammal.* **11,** 240–241.

Aguayo, A. (1974). Baleen whales off continental Chile. In "The Whale problem: A Status Report" (Ed. W. E. Schevill), pp. 210–217. Harvard Univ. Press, Cambridge, Massachusetts.

Aguayo, A. (1978). Smaller cetaceans in the Baltic Sea. *Rep. int. Whal. Commn* **28,** 131–146.

Allen, G. M. (1916). The whalebone whales of New England. *Mem. Boston Soc. Nat. Hist.* **8,** 107–322.

Angell, T., and Balcomb, K. C. (1982). "Marine Birds and Mammals of Puget Sound." Univ. of Washington Press, Seattle.

Arnason, U. (1972). The role of chromosomal rearrangement in mammalian speciation with special references to Cetacea and Pinnipedia. *Hereditas* **70,** 113–118.

Arnason, U., Benirschke, K., and Mead, J. G. (1977). Banded karyotypes of three whales: *Mesoplodon europaeus, M. carlhubbsi,* and *Balaenoptera acutorostrata. Hereditas* **87,** 189–200.

Arsenev, R. K. (1960). Distribution of *Balaenoptera acutorostrata* in the Antarctic. *Nor. Hvalfangst-Tid.* **49,** 380–382.

van Bambeke, C. (1868). Quelques remarques sur les squelettes de cétacés, conservés à la collection d'anatomie comparée de l'université de Gand. *Bull. Acad. R. Belg.* **26,** 20–61.

Bayliss, H. A. (1932). A list of worms parasitic in Cetacea. *Disc. Rep.* **6,** 393–418.

Beamish, P., and Mitchell, E. D. (1973). Short pulse length audio frequency sounds recorded in the presence of a minke whale (*Balaenoptera acutorostrata*). *Deep-Sea Res.* **20,** 375–386.

van Beek, J. G., and de la Mare, W. K. (1981). A statistical review of the biochemical evidence for minke whale stock identity in the Antarctic. *Rep. int. Whal. Commn* **31,** 114.

van Beek, J. G., and van Biezen, J. B. (1982). A review of morphological and biochemical research on population identification of southern minke whales. *Rep. int. Whal. Commn* **32,** 753–758.

van Beneden, P. J. (1885). Les cétacés des mers d'Europe. *Bull. Acad. R. Sci. Belg.* **54,** 707–732.

van Beneden, P. J., and Gervais, P. (1880). "Osterographie des cétacés vivants et fossiles." Paris.

Benham, W. B. (1901). An account of the external anatomy of a baby rorqual (*Balaenoptera rostrata*). *Trans. Proc. N.Z. Inst.* **34,** 151–155.

Best, P. B. (1967). Distribution and feeding habits of baleen whales off the Cape Province. *Invest. Rep. Div. Sea Fish. S. Afr.* **57,** 1–44.

Best, P. B. (1974). Status of the whale populations off the west coast of South Africa, and current research. In "The Whale Problem: A Status Report" (Ed. W. E. Schevill), pp. 53–81. Harvard Univ. Press, Cambridge, Massachusetts.

Best, P. B. (1977). Status of whale stocks off South Africa, 1975. *Rep. int. Whal. Commn* **27,** 116–121.

Best, P. B. (1979). External characters of minke whales landed at Durban. *Rep. int. Whal. Commn* **29**, 356.

Best, P. B. (1982). Seasonal abundance, feeding, reproduction, age, and growth in minke whales off Durban. *Rep. int. Whal. Commn* **32**, 759–786.

Best, P. B., and Butterworth, D. S. (1980). Report of the Southern Hemisphere minke whale assessment cruise 1978/79. *Rep. int. Whal. Commn* **30**, 257–283.

Boran, R. J., and Osborne, R. W. (1978). Orca survey: whale hotline cetacean sightings report. Unpublished report to Marine Mammal Div., Natl. Mar. Fish. Serv., Seattle, Washington.

Bourne, A. G. (1966). A study of the minke whale. *Oryx.* **8**, 229–232.

Braham H., Krogman, B., Marquette, W., Rugh, D., Johnson, J., Nerini, M., Leatherwood, S., Dahlheim, M., Sonntag, R., Carrol, G., Bray, T., Savage, S., and Cubbage, J. (1979). Bowhead whale preliminary research results, June–December 1978. Paper SC/31/Doc 15 presented to the Int. Whal. Commn Scientific Committee, June 1979 (unpublished).

Brown, R. (1868). Notes on the history and geographical relations of the Cetacea frequenting Davis Strait and Baffin's Bay. *Proc. Zool. Soc. London* **35**, 433–556.

Brown, S. G. (1975). Marking of small cetaceans using "Discovery" type whale marks. *J. Fish. Res. Board Can.* **32**, 1237–40.

Brown, S. G. (1979). The marking of minke whales in the Southern Hemisphere. *Rep. int. Whal. Commn* **29**, 359.

Brown, S. G., and Best, P. B. (1981). Recoveries of marks from minke whales in the Southern Hemisphere. *Rep. int. Whal. Commn* **31**, 357–359.

Brown, S. G., and Wada, S. (1982). Southern Hemisphere marking of minke whales. *Rep. int. Whal. Commn* **32**, 803–808.

Brownell, R. L., Jr. (1981). Review of coastal whaling by the Republic of Korea. *Rep. int. Whal. Commn* **31**, 395–402.

Budylenko, G. A. (1981). Distribution and some aspects of the biology of killer whales in the South Atlantic. *Rep. int. Whal. Commn* **31**, 523–526.

Burmeister, H. (1867). Preliminary description of a new finner whale (*Balaenoptera bonaerensis*).*Proc. Zool. Soc. London* **46**, 707–713.

Butterworth, D. S., and Best, P. B. (1982). Report of the Southern Hemisphere minke whale assessment cruise 1980/81. *Rep. int. Whal. Commn* **32**, 835–875.

Cagnalaro, L., Di Natale, A., and Notarbartolo di Sciara, G. (1984). "Cetacei. Guide per il Riconoseimento delle Specie Animali delle Acque Lagunari e Costiere Italiane". AQ/1/224/‡9, Consiglio Nazionale delle Ricerche, Rome.

Capellini, G. (1877). Sulla balenottera di Mondini, rorqual de la mer Adriatique di G. Cuvier. *Mem. R. Acad. Sci. Bologna*, pp. 413–448.

Carl, G. C. (1946). Sharp-headed finner whale stranded at Sidney, Vancouver Island, British Columbia. *Murrelet* **27**, 47–49.

Carrucio, A. (1913). Sulla *Balaenoptera acutorostrata* cattarata per la prima volfa nel mare laziale. *Bol. Soc. Zool. Ital. (Roma)* **2**, 157–173.

Carte, A., and Macalister, A. (1868). On the anatomy of *Balaenoptera rostrata*. *Philos. Trans. R. Soc. London*, pp. 201–261.

Chapman, D. G. (1980). Stratified estimate of area IV series B minke whale population from sightings data of assessment cruise 1978/79 and estimate of the variance. *Rep. int. Whal. Commn* **30**, 100–101.

Chaves, F. A. (1924). Cetaceos que aparacen nos mares dos açores. *Pesca Mar. (Lisboa)* **15**, 41–44.

Christensen, I. (1971). Minke whale investigations in the area Spitsbergen–Barents Sea in May–June 1972. *Fisk. Gang* **48**, 961–965.

Christensen, I. (1974). Minke whale investigations in the Barents Sea and off east and west Greenland 1973. *Fisk. Gang* **60**, 278–286.

Christensen, I. (1975). Preliminary report on the Norwegian fishery for small whales: Expansion of Norwegian whaling to Arctic and northwest Atlantic waters, and Norwegian investigations of the biology of small whales. *J. Fish. Res. Board Can.* **32**, 1083–1094.

Christensen, I. (1979). Norwegian minke whale fishery in 1976 and 1977. *Rep. int. Whal. Commn* **29**, 467–472.

Christensen, I. (1980). Catch and effort in the Norwegian minke whale fishery in the 1978 whaling season. *Rep. int. Whal. Commn* **30**, 209–212.

Christensen, I. (1981). Age determination of minke whales, *Balaenoptera acutorostrata*, from laminated structures in tympanic bullae. *Rep. int. Whal. Commn* **31**, 245–253.

Christensen, I., and Rørvik, C. J. (1978). Preliminary results from recent markings of minke whales in the northeast Atlantic. *Rep. int. Whal. Commn* **28**, 183–184.

Christensen, I., and Rørvik, C. J. (1979). Stock estimate of minke whales in the Svalbard–Norway–British Isles area from markings and recoveries 1974–1977. *Rep. int. Whal. Commn* **29**, 461–462.

Christensen, I., and Rørvik, C. J. (1980). Results from markings of minke whales in the northeast Atlantic. *Rep. int. Whal. Commn* **30**, 201–203.

Christensen, I., and Rørvik, C. J. (1981). Analysis of markings and recaptures of minke whales in the Barents Sea 1974–1979. *Rep. int. Whal. Commn* **31**, 255–258.

Collett, R. (1912). Norges Pattedyr (Norges Hvirveldyr I). H. Aschehoug, W. Nygaard, Kristiania (Oslo).

Collins, J. W. (1892). Report on the fisheries of the Pacific coast of the United States. *Rep. U.S. Fish. Commn 1888* **16**, 3–269

Cowan, I. M. (1939). The sharp-headed finner whale of the eastern Pacific. *J. Mammal.* **20**, 215–225.

Davies, J. L., and Guiler, E. R. (1958). A newborn piked whale in Tasmania. *J. Mammal.* **39**, 593–594.

van Deinse, A. B. (1931) "De Fossiele en Recente Cetacea van Nederland." H. J. Paris, Amsterdam.

Delyamure, S. L. (1955). "Helminthofauna of Marine Mammals (Ecology and Phylogeny)." Academy of Science U.S.S.R., Moscow. (Israel Program for Scientific Translations, Jerusalem, 1968.)

De Smet, W. M. A. (1979). The position of the testes in cetaceans. *In* "Functional Anatomy of Marine Mammals" (Ed. R. J. Harrison), Vol. 3, pp. 361–386. Academic Press, London.

Deraniyagala, P. E. P. (1948). Some mysticetid whales from Ceylon. *Spolia Zeylan.* **25**, 61–63.

Deraniyagala, P. E. P. (1963). Mass mortality of the new subspecies of little piked whale *Balaenoptera acutorostrata thalmaha* and a new beaked whale *Mesoplodon hotaula* from Ceylon. *Spolia Zeylan.* **30**, 80–84.

Dohl, T. P., Norris, K. S., Guess, R. C., Bryant, J. D., and Honig, M. W. (1980). "Summary of Marine Mammal and Seabird Surveys of the southern California Bight Area 1975–1978," Vol. II. Investigators Reports, Part II. Cetacea of the southern California Bight. National Technical Information Service (NTIS), Springfield, Virginia.

Doroshenko, N. V. (1978). On inter-relationship between killer whales (predator–prey) in the Antarctic. *Mar. Mamm., Abstr. Rep. 7th All-Union Meet. Moscow*, pp. 107–109.

Doroshenko, N. V. (1979). Populations of minke whales in the Southern Hemisphere. *Rep. int. Whal. Commn* **29,** 361–5.

Dorsey, E. (1983). Exclusive adjoining ranges in individually identified minke whales (*Balaenoptera acutorostrata*) in Washington state. *Can. J. Zool.* **61,** 174–181.

Edds, P. L. (1980). Variations in vocalizations of fin whales, *Balaenoptera physalus*, in the St. Lawrence River. M.S. thesis, Univ. of Maryland, College Park.

Erdman, D. S., Harms, J., and Flores, M. M. (1973). Cetacean records from the northeastern Caribbean regions. *Cetology* **17,** 1–13.

Eschricht, D. F. (1849). "Zoologisch–Anatomisch–Physiologische Untersuchungen über die nordischen Wallthiere". Voss, Leipzig.

Evans, P. G. H. (1980). Cetaceans in British waters. *Mammal. Rev.* **10,** 1–46.

Evans, P. G. H. (1982). Associations between seabirds and cetaceans: A review. *Mammal. Rev.* **12,** 187–206.

Everitt, R. D., Fiscus, C. H., and DeLong, R. L. (1980). "Northern Puget Sound Marine Mammals," U.S. Govt. Printing Office, Washington, D. C.

Fischer, P. (1881). Cétacés du sud-ouest de la France, *Actes Soc. Linn. Bordeaux* **35,** 5–220.

Flower, W. H. (1885). "List of the Specimens of Cetacea in the Zoological Department of the British Museum". British Museum, London.

Foote, D. C. (1975). Investigation of small whale hunting in northern Norway, 1964. *J. Fish Res. Board Can.* **32,** 1163–1189.

Fraser, F. C. (1934). Report on Cetacea stranded on the British coasts from 1927–1932. *Br. Mus. (Nat. Hist.)* **11,** 1–41.

Fraser, F. C. (1946). Report on Cetacea stranded on the British coasts from 1933 to 1937. *Br. Mus. (Nat. Hist.)* **12,** 1–56.

Fraser, F. C. (1953). Report on Cetacea stranded on the British coasts from 1938 to 1947. *Br. Mus. (Nat. Hist.)* **13,** 1–48.

Fraser, F. C. (1979). Royal fishes: the importance of the dolphin. In "Functional Anatomy of Marine Mammals" (Ed. R. J. Harrison), Vol. III, pp. 1–44. Academic Press, London.

Fruend, L. (1912). Walstudien. *Sb. Akad. Wiss. Wien* **121,** 1103–1182.

Fry, D. H. (1935). Sharp-headed finner whale taken at Los Angeles harbor. *J. Mammal.* **16,** 205–207.

Gambell, R., Best, P. B., and Rice, D. W. (1975). Report on the International Indian Ocean whale marking cruise. *Rep. int. Whal. Commn* **26,** 240–252.

Gaskin, D. E. (1972). "Whales, Dolphins, and Seals with Special Reference to the New Zealand Region". Heinemann, Auckland.

Gaskin, D. E. (1976). The evolution, zoogeography and ecology of Cetacea. *Oceanogr. Mar. Biol.* **14,** 247–346.

Gaskin, D. E. (1982). "The Ecology of Whales and Dolphins." Heinemann, London.

Gong, Y. (1981). Minke whales in the waters off Korea. *Rep. int. Whal. Commn* **31**, 241–244.

Gong, Y. (1982). A note on the distribution of minke whales in Korean waters. *Rep. int. Whal. Commn* **32**, 279–282.

Gray, J. E. (1846). On the cetaceous animals. *In* "The Zoology of the Voyage of H. M. S. *Erebus* and *Terror*, under the Command of Captain Sir James Clark Ross, during the Years 1839–1843. (Eds. J. Richardson and J. E. Gray), pp. 15–53. E. W. Janson, London.

Gray, J. E. (1874). On the skeleton of the New Zealand pike whale, *Balaenoptera huttoni* (*Physalus antarcticus* Hutton). *Ann. Mag. Nat. Hist.* **13**, 448–452.

Haley, D. (Ed.) (1978). Marine mammals of eastern North Pacific and Arctic waters. Pacific Search Press, Seattle.

Hancock, D. (1965). Killer whales kill and eat a minke whale. *J. Mammal.* **46**, 341–342.

Harmer, S. F. (1927). Report on Cetacea stranded on the British coasts from 1913–1926. *Br. Mus. (Nat. Hist).* **10**, 1–91.

Harris, G. W. (1950). Hypothalamo–hypophysial connexions in the Cetacea. *J. Physiol.* **111**, 361–367.

Hall, E. R., and Kelson, K. R. (1959). "The Mammals of North America". Ronald Press, New York.

Henking, H. (1901). Norwegen's Walfang. *Abh. Dtsch Seefisch. Ver.* **6**, 119–171.

Herre, A. W. (1925). A Philippine rorqual. *Science* **61**, 541.

Hentschel, E. (1937). Naturgeschichte der nordatlantischen Wale und Robben. *In* "Handbuch der Seefischerei Nordeuropas". (Eds. H. Lubbert and E. Ehrenbaum), Vol. 3, pp. 1–54. Stuttgart.

Heyerdahl, T., Jr. (1973). Sexual dimorphism and age criteria in the pelvic bones of the minke whale, *Balaenoptera acutorostrata* Lacépède. *Norw. J. Zool.* **21**, 39–43.

Horwood, J. W. (1981). Results from the IWC/IDCR minke whale marking and sighting cruise 1979/80. *Rep. int. Whal. Commn* **31**, 287–315.

Howell, A. B. (1934). Observations on the white whale. *J. Mammal.* **16**, 155–156.

Ivashin, M. V. (1976). On reproduction of the minke whale in the Indian Ocean section of the Antarctic. *Rep. int. Whal. Commn* **27**, 329–332.

Ivashin, M. V., and Mikhalev, Yu. A. (1978). To the problem of the prenatal growth of minke whales (*Balaenoptera acutorostrata*) of the Southern Hemisphere and of the biology of their reproduction. *Rep. int. Whal. Commn* **28**, 201–205.

Ivashin, M. V., and Votrogov, L. M. (1981). Minke whales, *Balaenoptera acutorostrata davidsoni*, inhabiting inshore waters off the Chukotka coast. *Rep. int. Whal. Commn* **31**, 231.

Ivashin, M. V., and Votrogov, L. M. (1982). Occurrence of baleen and killer whales off Chukotka. *Rep. int. Whal. Commn* **32**, 499–501.

IWC. (1978). Report of the special meeting on Southern Hemisphere minke whales. SC/30/Rep 4.

IWC. (1979). Report of the special meeting on Southern Hemisphere minke whales, Seattle, May 1978. *Rep. int. Whal. Commn* **29**, 349–358.

IWC. (1980). Draft Report of the minke whale sub-committee. IWC/32/4/Annex E.

IWC. (1982). Report of the special meeting on Southern Hemisphere minke whales, Cambridge, 22–26 June 1981. *Rep. int. Whal. Commn* **32,** 697–705.

IWS. (1981a). Whaling in the Antarctica season 1979/80 and outside the Antarctic in 1980. *Int. Whal. Stat.* **87,** 43.

IWS. (1981b). Whaling results for the various countries in the Antarctic in the season 1980/81. *Int. Whal. Stat.* **88,** 61.

Jansen, J., and Jansen, J. K. S. (1969). The nervous system of Cetacea. *In* "The Biology of Marine Mammals" (Ed. H. T. Andersen), pp. 176–249. Academic Press, New York.

Jonsgård, A. (1951). Studies on the little piked whale or minke whale (*Balaenoptera acutorostrata* Lacépède). *Nor. Hvalfangst-Tid.* **40,** 209–232.

Jonsgård, A. (1962). Population studies on the minke whale *Balaenoptera acutorostrata* Lacépède. *In* "The Exploitation of Natural Animal Populations" (Eds E. D. le Cren and M. W. Holdgate), pp. 159–167. Blackwell, Oxford.

Jonsgård, A. (1966). The distribution of Balaenopteridae in the North Atlantic Ocean. *In* "Whales, Dolphins, and Porpoises" (Ed. K. S. Norris), pp. 114–124. Univ. of California Press, Berkeley.

Jonsgård, A. (1974). On whale exploitation in the eastern part of the North Atlantic Ocean. *In* "The Whale Problem: A Status Report" (Ed. W. E. Schevill), pp. 97–107. Harvard Univ. Press, Cambridge, Massachusetts.

Jonsgård, A. (1982). The food of minke whales (*Balaenoptera acutorostrata*) in northern North Atlantic waters. *Rep. int. Whal. Commn* **32,** 259–269.

Jurasz, C. M., and Jurasz, V. P. (1979). Feeding modes of the humpback whale, *Megaptera novaeangliae*, in southeast Alaska. *Sci. Rep. Whales Res. Inst.* **31,** 69–83.

Kapel, F. O. (1978). Catch of minke whales by fishing vessels in west Greenland. *Rep. int. Whal. Commn* **28,** 217–226.

Kapel, F. O. (1980). Sex ratio and seasonal distribution of catches of minke whales in west Greenland. *Rep. int. Whal. Commn* **30,** 195–200.

Kasuya, T., and Ichihara, T. (1965). Some information on minke whales from the Antarctic. *Sci. Rep. Whales Res. Inst.* **19,** 37–43.

Kato, H. (1982). Some biological parameters for the Antarctic minke whale. *Rep. int. Whal. Commn* **32,** 935–945.

Kato, H. (1979b). Unusual minke whale with deformed jaw. *Sci. Rep. Whales. Res. Inst.* **31,** 101–103.

Kato, H. (1982). Some biological parameters for the Antarctic minke whale *Rep. int. Whal. Commn* **32,** 935–945.

Katona, S., Richardson, D., and Hazard, R. (1977). "A Field Guide to the Whales and Seals of the Gulf of Maine," College of the Atlantic, Bar Harbor, Maine.

Kellogg, R. (1931). Whaling statistics for the Pacific Coast of North America. *J. Mammal.* **12,** 73–77.

Kimura, S., and Nemoto, T. (1956). A note on a minke whale kept alive in aquarium. *Sci. Rep. Whales Res. Inst.* **11,** 181–189.

Lahille, F. (1908). Notas sobre un ballenato de 2.10 metros de largo. *An. Mus. Nac. Buenos Aires.* **16,** 375–401.

Lambertsen, R. H. (1983). Internal mechanism of rorqual feeding. *J. Mammal.* **64,** 76–88.

Larsen, F., and Kapel, F. O. (1981). Collection of biological material of minke whales off west Greenland 1979. *Rep. int. Whal. Commn* **31,** 279–287.

Laws, R. M. (1977a). Seals and whales of the southern oceans. *Philos. Trans. R. Soc. London* **279,** 81–96.

Laws, R. M. 1977b). The significance of vertebrates in the Antarctic marine ecosystem. *In* "Adaptation within Antarctic Ecosystems" (Ed. G. A. Llano), pp. 411–438. Smithsonian Institution, Washington, D.C.

Leatherwood, S., Caldwell, D. K., and Winn, H. E. (1976). Whales, dolphins and porpoises of the western North Atlantic: A guide to their identification. *NOAA Tech. Rep. NMFS Circ.* **396.**

Leatherwood, S., Thomas, J. A., and Awbrey, F. T. (1981). Minke whales off northwestern Ross Island. *Antarct. J.* **16,** 154–156.

Leatherwood, S., Awbrey, F. T., and Thomas, J. A. (1982a). Minke whale responses to a transiting survey vessel. *Rep. int. Whal. Commn* **32,** 795–802.

Leatherwood, S., Reeves, R. R., Perrin, W. F., and Evans, W. E. (1982b). Whales, dolphins, and porpoises of the eastern North Pacific and adjacent Arctic waters: A guide to their identification. *NOAA Tech. Rep. NMFS Circ.* **444.**

Leatherwood, S., Todd, F. S., Thomas, J. A., and Awbrey, F. T. (1982c). Incidental records of cetaceans in southern seas, January and February 1981. *Rep. int. Whal. Commn* **32,** 515–520.

Legendre, R. (1943). Notes cetologiques. A propos du *Balaenoptera acutorostrata* Lacépède observée à Concarnean. *Bull. Inst. Oceanogr.* **856,** 106.

Lillie, D. G. (1915). Cetacea. *Br. Antarct. ("Terra Nova") Exped. Nat. Hist. Rep., Zool.* **1,** 85–124.

Lockyer, C. (1979). Review (in minke whales) of the weight/length relationship and the Antarctic catch biomass, and a discussion of the implications of imposing a body length limitation on the catch. *Rep. int. Whal. Commn* **29,** 369–374.

Lockyer, C. (1981). Estimation of the energy costs of growth, maintenance, and reproduction in the female minke whale (*Balaenoptera acutorostrata*) from the Southern Hemisphere. *Rep. int. Whal. Commn* **31,** 337–343.

Marelli, C. A. (1918). Un ballenato hallado en la costa del Rio de la Plata. *Phys. Rev. Soc. Argent. Cien. Nat. (Buenos Aires)* **4,** 326–328.

Marquette, W. M., Braham, H. W., Nerini, M. K., and Miller, R. V. (1982). Bowhead whale studies, Autumn 1980–Spring 1981: Harvest, biology, and distribution. *Rep. int. Whal. Commn* **32,** 357–370.

Masaki, Y. (1979). Yearly changes of biological parameters for the Antarctic minke whale. *Rep. int. Whal. Commn* **29,** 375–395.

Matsuura, Y. (1936). On the lesser rorqual found in the adjacent waters of Japan. *Bull. Jpn. Soc. Sci. Fish.* **4,** 325–330.

Miller, G. A., Jr., and Kellogg, R. (1955). List of North American Recent mammals. *U.S. Natl. Mus. Bull. pp.* 1–954.

Mitchell, E. D. (1975a). Porpoise, dolphin, and small whale fisheries of the world: status and problems. *IUCN Monogr.* **3,** 31–42.

Mitchell, E. D. (1975b). Review of biology and fisheries for smaller cetaceans. *J. Fish. Res. Board Can.* **32,** 891–895.

Mitchell, E. D. (1978). Finner whales. *In* "Marine Mammals of Eastern North Pacific and Arctic Waters" (Ed. D. Haley), pp. 37–53. Pacific Search Press, Seattle.

Mitchell, E. D., and Kozicki, V. M. (1975). Supplementary information on minke whale (*Balaenoptera acutorostrata*) from Newfoundland fishery. *J. Fish. Res. Board Can.* **32,** 985–994.

Miyashita, T. (1982a). Estimation of the population size of minke whales in areas III and IV in 1980/81 using a mark recapture method. *Rep. int. Whal. Commn* **32,** 897–898.

Miyashita, T. (1982b). Estimation of the population size of minke whales in area I using sightings and marking data. *Rep. int. Whal. Commn* **32,** 911–917.

Monticelli, F. S. (1926). Sulla *Balaenoptera acutorostrata* Lacépède (1804) presa a Lacco Ameno. *Bol. Soc. Nat. Napoli* **37,** 8–9.

Moore, J. C. (1953). Distribution of marine mammals to Florida waters. *Am. Midl. Nat.* **49,** 117–158.

Moore, J. C., and Palmer, R. S. (1955). More piked whales from southern North Atlantic. *J. Mammal.* **36,** 429–433.

Mörzer-Bruyns, W. F. J. (1974). "Field guide of Whales and Dolphins". Mees, Amsterdam.

Nemoto, T. (1959). Food of baleen whales with reference to whale movements. *Sci. Rep. Whales Res. Inst.* **14,** 149–290.

Nemoto, T., Best, P. B., and Ishimaru, K. (1980). Diatom films on whales in South African waters. *Sci. Rep. Whales Res. Inst.* **32,** 97–103.

Nishiwaki, M. (1972). General biology. *In* "Mammals of the Sea, Biology and Medicine" (Ed. S. H. Ridgway), pp. 3–204. Thomas, Springfield, Illinois.

Nobre, A. (1935). Descriçao dos mamiferos marinhos de Portugal. *Fauna Mar. Port.* **1,** 1–21.

Norris, K. S., and Prescott, J. H. (1961). Observations on Pacific cetaceans in Californian and Mexican waters. *Univ. Calif. Publ. Zool.* **63,** 361–370.

Ohsumi, S. (1973). Find of marlin spear from the Antarctic minke whales. *Sci. Rep. Whales Res. Inst.* **25,** 237–239.

Ohsumi, S. (1975). Review of Japanese small-type whaling. *J. Fish. Res. Board Can.* **32,** 1111–1121.

Ohsumi, S. (1977). Catch of minke whales in coastal waters of Japan. *Rep. int. Whal. Commn* **27,** 164–166.

Ohsumi, S. (1979a). Interspecies relationships among some biological parameters in cetaceans and estimation of natural mortality coefficient of Southern Hemisphere minke whale. *Rep. int. Whal. Commn* **29,** 379–406.

Ohsumi, S. (1979b). Population assessment of the Antarctic minke whale. *Rep. int. Whal. Commn* **29,** 407–420.

Ohsumi, S. (1979c). Feeding habits of the minke whale in the Antarctic. *Rep. int. Whal. Commn* **29,** 473–476.

Ohsumi, S. (1980a). Minke whales in the coastal waters of Japan, 1978. *Rep. int. Whal. Commn* **30,** 307–312.

Ohsumi, S. (1980b). Estimation of population size of the minke whale in the Antarctic area IV by means of whale sightings. *Rep. int. Whal. Commn* **30,** 323–327.

Ohsumi, S. (1981a). Estimation of the population size of the minke whale in the Antarctic area VI by means of whale sightings. *Rep. int. Whal. Commn* **31,** 323–326.

Ohsumi, S. (1981b). Minke whales in the coastal waters of Japan, 1979. *Rep. int. Whal. Commn* **31,** 333–337.

Ohsumi, S. (1982). Minke whales in the coastal waters of Japan, in 1980 and a population assessment of Okhotsk Sea–West Pacific Stock. *Rep. int. Whal. Commn* **32,** 283–286.

Ohsumi, S., and Masaki, Y. (1975). Biological parameters of the Antarctic minke whale at the virginal population level. *J. Fish. Res. Board Can.* **32,** 995–1004.

Ohsumi, S., Masaki, Y., and Kawamura, A. (1970). Stock of the Antarctica minke whale. *Sci. Rep. Whales Res. Inst.* **22,** 75–125.

O'Leary, B. L. (1984). Aboriginal whaling from the Aleutian Islands to Washington state. *In* "The Gray Whale"(Eds M. L. Jones, S. L. Swartz, and S. Leatherwood), pp. 79–102. Academic Press, Orlando.

Oliver, W. B. R. (1922). A review of the Cetacea of the New Zealand seas. *Proc. Zool. Soc. London,* pp. 557–585.

Omura, H. (1957). Osteological study of the little piked whale from the coast of Japan. *Sci. Rep. Whales Res. Inst.* **12,** 1–21.

Omura, H. (1975). Osteological study of the minke whale from the Antarctic. *Sci. Rep. Whales Res. Inst.* **27,** 1–36.

Omura, H. (1976). A skull of the minke whale dug out from Osaka. *Sci. Rep. Whales Res. Inst.* **28,** 69–72.

Omura, H. (1978). Preliminary report on morphological study of the pelvic bones of the minke whale from the Antarctic. *Sci. Rep. Whales Res. Inst.* **30,** 271–279.

Omura, H. (1980). Morphological study of pelvic bones of the minke whale from the Antarctic. *Sci. Rep. Whales Res. Inst.* **30,** 25–37.

Omura, H., and Kasuya, T. (1976). Additional information on skeleton of the minke whale from the Antarctic. *Sci. Rep. Whales Res. Inst.* **28,** 57–68.

Omura, H., and Sakiura, H. (1956). Studies on the little piked whales from the coast of Japan. *Sci. Rep. Whales Res. Inst.* **11,** 7–18.

van Oordt, E. D. (1926). Over eenige aan te kast uan Nederland waargenomen cetaceen sooten. *Zool. Meded. Leiden* **9,** 211–214.

Paiva, M. P., and Grangeiro, B. F. (1965). Biological investigations on the whaling season 1960–1963, off northeastern coast of Brazil. *Arq. Est. Biol. Mar. Univ. Ceará* **5,** 29–64.

Paiva, M. P., and Grangeiro, B. F. (1970). Investigations on the whaling seasons 1964–1967, off northeastern coast of Brazil. *Arq. Est. Biol. Mar. Univ. Ceará* **10,** 111–126.

Pearson, J. (1936). The whales and dolphins of Tasmania. Part I. External characters and habits. *Papers Proc. R. Soc. Tas., 1935* , pp. 163–192.

Peringuey, L. (1921). A note on the whales frequenting South American waters. *Trans. R. Soc. S. Afr.* **9,** 73–76.

Perrin, J. B. (1870). Notes on the anatomy of *Balaenoptera rostrata*. *Proc. Zool. Soc. London* pp. 805–817.

Rice, D. W. (1974). Whales and whale research in the eastern North Pacific. *In* "The Whale Problem: A Status Report" (Ed. W. E. Schevill), pp. 170–195. Harvard Univ. Press, Cambridge, Massachusetts.

Rice, D. W. (1977). A list of the marine mammals of the world. *NOAA Tech. Rep. NMFS SSRF* **711,** 1–75.

da Rocha, J. M. (1980). Progress report on Brazilian minke whaling. *Rep. int. Whal. Commn* **30,** 379–384.

Rowlatt, U. (1981). The cardiac ventricles of a baleen whale (*Balaenoptera acutorostrata*: minke whale) and a toothed whale (*Hyperoodon ampullatus:* bottlenose whale). *J. Morphol.* **168,** 85–96.

Ruud, J. T. (1937). Vaagehvalen, *Balaenoptera acutorostrata* Lacépède. *Nor. Hvalfangst-Tid.*, pp. 193–194.

Saemundsson, B. (1939). "The Zoology of Iceland". Vol. 4, Mammalia. Munsksgaard, Copenhagen.

Satake, Y., and Omura, H. (1974). A taxonomic study of the minke whale in the Antarctic by means of hyoid bone. *Sci. Rep. Whales Res. Inst.* **26,** 15–24.

Scammon, C. M. (1869). On the cetaceans of the western coast of North America. *Proc. Acad. Nat. Sci. (Philadelphia)* pp. 13–63.

Scammon, C. M. (1873). On a new species of *Balaenoptera*. *Proc. Calif. Acad. Sci.* **4,** 29–270.

Scammon, C. M. (1874). "The Marine Mammals of the Northwestern Coast of North America". Carmany and Co., San Francisco.

Scattergood, L. W. (1949). Notes on the little piked whale. *Murrelet* **30,** 3–16.

Scheffer, V. B. (1973). Marine mammals in the Gulf of Alaska. *In* "A Review of the Oceanography and Renewable Resources of the Northern Gulf of Alaska" (Ed. D. H. Rosenburg), pp. 175–207. Inst. Marine Science, Univ. of Alaska, Fairbanks.

Schevill, W. E., and Watkins, W. A. (1972). Intense low frequency sounds from an Antarctic minke whale, *Balaenoptera acutorostrata*. *Breviora* **388,** 1–8.

Schmidly, D. J. (1981). Marine mammals of the southeastern United States coast and the Gulf of Mexico. *U.S. Fish. Wildl. Serv., Biol. Serv. Program, Tech. Rep.* **FWS/OBS-80/41,** 1–163.

Schwartz, F. J. (1962). Summer occurrence of an immature little piked whale, *Balaenoptera acutorostrata*, in Chesapeake Bay, Maryland. *Chesapeake Sci.* **3,** 206–209.

Sergeant, D. E. (1963). Minke whales, *Balaenoptera acutorostrata* Lacépède of the western North Atlantic. *J. Fish Res. Board Can.* **20,** 1489–1504.

Sergeant, D. E., Mansfield, A. W., and Beck, B. (1970). Inshore records of Cetacea for eastern Canada, 1949–1968. *J. Fish. Res. Board Can.* **27,** 1903–1915.

Schevchenko, V. I. (1975). The nature of inter-relationships between killer whales and other cetaceans. *Mar. Mamm. Rep. 6th All-Union Meet.* **2,** 173–175.

Sigurjonsson, J. (1980). A preliminary note on ear plugs from Icelandic minke whales. *Rep. int. Whal. Commn* **30,** 193–194.

Sigurjonsson, J. (1982). Icelandic minke whaling 1914–1980. *Rep. int. Whal. Commn* **32,** 287–295.

Simonsen, V., Kapel, F., and Larsen, F. (1982). Electrophoretic variation in the minke whale, *Balaenoptera acutorostrata* Lacépède. *Rep. int. Whal. Commn* **32**, 275–278.

Singarajah, K. V. (1981). Observations on the occurrence and behavior of minke whales off the coast of Brazil. Unpublished paper SC/JN81/MIS1 presented to the Int. Whal. Commn Scientific Committee, June 1981.

Singarajah, K. V. (1983). Behavioral observations and estimates of relative abundance of minke whales based at Costinha Land Station, Brazil. *Proc. Symp. Mar. Mamm. Indian Ocean, Sri Lanka*, 1–12.

Slijper, E. J., van Utrecht, W. L., and Naaktgeboren, C. (1964). Remarks on the distribution and migration of whales based on observations from Netherlands ships. *Bijdrgen Dierkunde* **34**, 1–98.

Stead, D. G. (1910). "A Brief Review of the Fisheries of New South Wales". Sydney.

Stephenson, W. (1951). The lesser rorqual in British waters. *Dover Mar. Lab. Rep.* **3**, 7–48.

Struhsaker, P. (1967). An occurrence of the minke whale, *Balaenoptera acutorostrata*, near the northern Bahama Islands. *J. Mammal.* **48**, 483.

Sullivan, R. M., and Houck, W. J. (1979). Sightings and strandings of cetaceans from northern California. *J. Mammal.* **60**, 828–833.

Tago, K. (1937). Cetacea found in Japanese waters. *C. R. Int. Congr. Zool.* **12**, 2193–2228.

Taylor, R. J. F. (1957). An unusual record of three species of whale being restricted to pools in Antarctic sea ice. *Proc. Zool. Soc. London* **129**, 325–331.

Thompson, T. J., Winn, H. E., and Perkins, P. J. (1979). Mysticete sounds. *In* "Behavior of Marine Animals, Vol. 3: Cetaceans" (Eds H. E. Winn and B. L. Olla), pp. 403–431. Plenum, New York.

Tomilin, A. G. (1957). "Mammals of the USSR and Adjacent Countries. Vol. IX: Cetacea" (Ed. V. G. Heptner), Nauk SSSR, Moscow. English Translation, 1967, Israel Program for Scientific Translations, Jerusalem.

True, F. W. (1904). The whalebone whales of the western North Atlantic. *Smithson. Contrib. Knowl.* **33**, 192–210.

Turner, W. (1893). The lesser rorqual (*Balaenoptera rostrata*) in the Scottish seas with observations on its anatomy. *Proc. R. Soc. Edinburgh* **19**, 36–75.

Wada, S. (1976). Indices of abundance of large-sized whales in the 1974 whaling season. *Rep. int. Whal. Commn* **26**, 382–391.

Wada, S. (1982). Analysis of the biochemical data by G-statistics. Appendix 4. *Rep. int. Whal. Commn* **32**, 707.

Wada, S., and Numachi, K. (1974). External and biochemical characters as an approach to stock identification for the Antarctic minke whale. *Rep. int. Whal. Commn* **29**, 421–432.

Wang, P. (1978). Studies on the baleen whales in the Yellow Sea. *Acta Zool. Sin.* **24**, 269–277.

Waterman, T. T. (1920). The whaling equipment of the Makah Indians. *Univ. Wash. Publ. Pol. Sci.* **1**, 1–67.

Weber, M. (1922). "Flora en fauna der Zuiderzee". Dierkund. Veraeniging, The Helder, Netherlands.

Williamson, G. R. (1959). Three unusual rorqual whales from the Antarctic *Proc. Zool. Soc. London* **133,** 135–144.

Williamson, G. R. (1961). Two kinds of minke whales in the Antarctic. *Nor. Hvalfangst-Tid.* **50,** 133–141.

Williamson, G. R. (1975). Minke whales off Brazil. *Sci. Rep. Whales Res. Inst.* **27,** 37–59.

Winn, H. E., and Perkins, P. J. (1976). Distributions and sounds of the minke whale, with a review of mysticete sounds. *Cetology* **19,** 1–12.

Yukhov, V. L. (1969). Observations of cetaceans in the Gulf of Aden and the northwestern part of the Arabic Sea. *In* "Morskie Mlekopitayuschie", 3rd All Union Conf. Mar. Mammals. Fisheries Research Board of Canada Translation Service No. 1510.

Zemsky, V. A., and Tormosov, D. D. (1964). Small rorqual (*Balaenoptera acutorostrata*) from the Antarctic. *Nor. Hvalfangst-Tid.* **53,** 302–305.

5

Bryde's Whale

Balaenoptera edeni Anderson, 1878

William C. Cummings

Genus and Species

Anderson (1878) examined a balaenopterid whale that had been stranded in Thaybyoo Creek, near the Gulf of Martaban, Burma. After careful study of this specimen, he named it *Balaenoptera edeni*. The common name, Bryde's whale, was given in honor of a Norwegian consul, Johan Bryde, who built the first two whaling stations in South Africa. Whaling history and early findings on the natural history of Bryde's whale are obscured because whalers usually did not recognise *B. edeni* as a separate species. Bryde's whale is very similar to the sei whale, *B. borealis*, in colour, shape, and even behaviour, making the two species difficult to distinguish, especially in the field. Records on Bryde's whale were frequently included under those of the sei whale, a practice carried out to some extent into the 1970s.

Olsen (1913) recognised the two as different species, but he called Bryde's whale *Balaenoptera brydei*. Five years later, Andrews (1918) studied the skeletal structure and he also concluded that there were two different species. In fact, Lonnberg (1931) thought that the osteology of Bryde's whale was different from all other members of the genus. As further verification, Junge (1950) examined the skeleton of a rorqual washed ashore on Pulu Sugi, near Singapore, and after comparing his data with those already recorded from *B. brydei* Olsen and *B. edeni* Anderson, he concluded that the two species names were synonymous. Omura (1959) and Best (1960) made similar observations, agreed with Junge's position, and accepted *Balaenoptera edeni* as the species name, using "Bryde's whale" as the vernacular.

Recent data show considerable variation in the characteristics of Bryde's whale from one locality to another (Gaskin, 1972), and Best (1975) listed five different populations on the basis of morphological "forms." Omura (1977) reported that two forms may exist off Japan. Although these may be only regional differences, the taxonomic status eventually could be revised to reflect inshore versus offshore forms, perhaps even subspecies. Specimens considered to be intermediate between Bryde's and sei whales have been encountered (Mead, 1974), although in most regions the populations are considered to be distinct.

External Characteristics

Size

The second smallest of the balaenopterids, Bryde's whale reaches an average length of 13 m and a maximum of 15.5 m. Females are slightly larger than males of the same age (Olsen, 1913). Bryde's whale (Fig. 1) is typically elongated and somewhat less powerfully built than sei and minke whales. I have observed that the general body shape is more like that of the fin whale.

Colour

The colour of Bryde's whale is variable, but usually the dorsal surface is bluish black and the ventral area is white or yellowish. A dark bluish grey area near the throat extends laterally and posteriorly to the flippers. On many individuals of some populations a grey band occurs across the belly, just anterior to the navel. The demarcation between dorsal and ventral surfaces is not very distinct (Norman and Fraser, 1949). Also in some populations, whitish grey oblong spots, about 3 × 7 cm, occur over much of the body. Olsen (1913) believed such spots to be areas of attachment by the parasite *Penella*. On the other hand, Chittleborough (1959) rarely saw parasites on *B. edeni*. Whales captured off the coast of Japan bore scars thought to be caused by the healing of open wounds of unknown origin (Omura, 1962). Best (1974) described oval scars that possibly were due to the shark *Istitius braziliensis* and scratches on the tail region that could have been produced by

FIG. 1 Head of a Bryde's whale. (Courtesy of S. Sinclair.)

rubbing against a rocky bottom. In fact, these characters are used by Best, among others, in differentiating inshore from offshore forms.

Appendages, eyes, and grooves

The flippers of Bryde's whale are slender and somewhat pointed, measuring 8–10% of total body length. They are dark bluish grey on both the dorsal and ventral surfaces. The breadth of the tail flukes is about 20% of the total body length. The pointed, falcate dorsal fin (Fig. 2) is up to 46 cm long (Leatherwood et al., 1976) and it somewhat resembles that of the finback whale.

The eyes of Bryde's whale are relatively large compared with those of other balaenopterids. The ventral grooves, about 45 in number, terminate at or behind the umblicus, unlike the grooves of the sei whale, which stop at the region of the midbody (Omura, 1962). Best (1960) described a prominent single median groove which runs from the umbilicus to the genital aperture. The penis is located in a darkly colored, 1–1.5 m furrow, two-thirds of which extends anterior to the genital aperture (Olsen, 1913).

Baleen

The baleen of Bryde's whale is generally distinctive in size, shape, and texture. However, Best (1960), in comparing specimens from the South African area, described baleen that was exceptionally narrow and more like

FIG. 2 Dorsal fin and surrounding portion of a Bryde's whale as viewed underwater. (Photo by W. Cummings.)

that from the sei whale. Normally the plates are wide, about 19 cm, and approximately 50 cm long (excluding bristles). The inner margin is somewhat concave. The number of well-developed plates ranges from 250 to 280, but if rudimentary plates are included, the total may reach 350 (Olsen, 1913). Kawamura (1978) found that South Pacific Bryde's whales have a smaller total filtering area than North Pacific animals. Kawamura and Satake (1976) presented numerous excellent photos of the baleen (Fig. 3). The greyish or brownish bristles are thick, stiff, and uncurled, unlike the white, fine, and curled baleen of sei whales (Omura, 1977). Most of the baleen is dark grey, but anterior plates are partially white. On many individuals, this light area has grey stripes (Norman and Fraser, 1949). Variability in the colour of baleen was noted by Mead (1974).

Lateral ridges

A unique external feature of *B. edeni* is two lateral ridges that run from just behind the tip of the snout back as far as the blowholes. These lateral ridges (see Fig. 1) are situated between the median line and the outer margin, one on each side of the usual balaenopterid median ridge. Omura (1962) re-

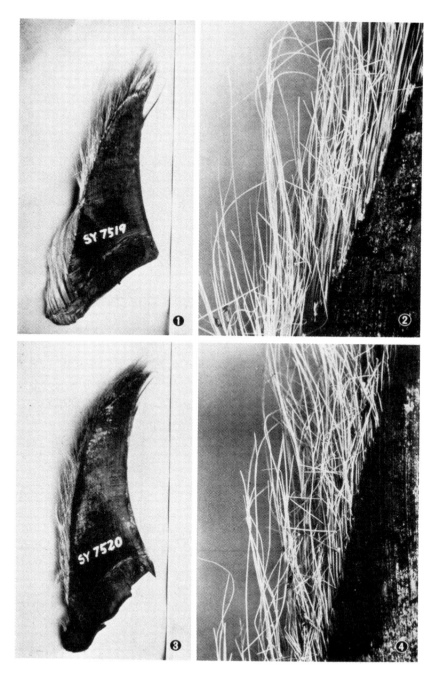

FIG. 3 Baleen plates and bristles of a female (1, 2) and a male (3, 4) Bryde's whale. (After Kawamura and Satake, 1976; courtesy of A. Kawamura.)

ported that these ridges are present on Bryde's whales off Brazil, but not elsewhere. Apparently this is incorrect, because Rice (1974) found them on a stranded animal in Florida, W. C. Cummings and P. O. Thompson (unpublished) found them on a living animal in the Gulf of California, and G. N. di Sciara (personal communication) noted them off Venezuela. The ridges are 1–2 cm high, but their posterior ends disappear from the surface and change into grooves of varying lengths. On some *B. edeni* these grooves may begin at about midlength of the snout, whereas on others they may be totally absent. Omura (1962) noted several hairs on each ridge. There are also two rows of hairs near the tip of the lower jaw, an average of 12 hairs in each row (Olsen, 1913). Except for a small number of Bryde's whales that may not have them, the presence of the head ridges is useful in identifying *B. edeni* at sea, but only under conditions of excellent visibility.

Internal Anatomical Characteristics

Little has been published on the internal anatomy of Bryde's whale, except for extensive studies conducted on its skeleton by Anderson (1879), Andrews (1918), Lonnberg (1931), Junge (1950), Omura (1959), Soot-Ryen (1961), Cagnolaro and Notarbartolo di Sciara (1979), and Omura *et al.* (1981).

The skull of Bryde's whale (Fig. 4) is very broad and short compared with that of *B. borealis*. The relatively short rostrum is pointed at the end and dorsally and ventrally flattened. Flatness of the skull is a conspicuous character in differentiating Bryde's and sei whales (Omura *et al.*, 1981). The sides are nearly straight posteriorly but slightly curved anteriorly. The nasals, which taper backwards, are separated by a triangular process of the frontals (Lonnberg, 1931) and they are concave or straight along the front margin (Omura *et al.*, 1981). The palatine bones of a fully grown Bryde's whale are about 55 cm long and the pterygoid, 18 cm. The palatines do not extend as far back as in the sei whale (Omura *et al.*, 1981). The curved and robust mandible of Bryde's whale (Fig. 5) is more pronounced than that of any other balaenopterid. A prominent, deep groove appears on the inner side of the mandible (Junge, 1950), but it is not as well developed as in the sei whale (Omura *et al.*, 1981).

The cervical vertebrae are free. The breadth of the atlas is about 17% of the skull's length. The thoracic vertebrae appear to have short spinous processes.

Each species of *Balaenoptera* appears to vary with regard to zygapophyses, a characteristic of some taxonomic value. Furthermore, Lonnberg (1931) thought that development and shape of the zygapophyses may affect bodily movements that presumably may be characteristic for each species. The total number of vertebrae may range from 49 to 55 in Bryde's whale, and there are as many as 17 chevron bones, the first being attached to the thirty-fourth vertebra (Omura, 1959; Cagnolaro and Notarbartolo di Sciara, 1979; Omura et al., 1981). While variability in vertebra number has been noted, Omura et al. (1981) stated that the number in Bryde's whale is 54–55, as compared with 56–57 in the sei whale. In addition to this characteristic feature, these authors reported that the most remarkable vertebral difference between the two species is in the strongly backward inclination of the spinous process in the Bryde's whale.

Of the 13 pairs of ribs, only the first is double headed. With the exception of this pair, the ribs are relatively thin and broad (Lonnberg, 1931; Omura et al., 1981). Humerus, radius, and ulna are typical of the Balaenopteridae (Junge, 1950). Nishiwaki (1972) reported that the flippers have four fingers of I:6, II:5, IV:5, and V:3. The sternum of *B. edeni* is cross shaped. The scapula is typically balaenopterid in shape. The well-developed acromion broadens towards the end, and the coracoid is relatively erect (Junge, 1950).

In summary, the skeleton of Bryde's whale appears to be variable and often not very useful in distinguishing this species from other balaenopterids (Mead, 1974). Thin ribs, small limbs, short head, and straight rostrum lend to an overall appearance of slenderness. The best single reference to the osteology of this species probably is Omura et al. (1981).

Distribution

Long migrations probably are not typical of Bryde's whales, although there is an indication of limited shifts towards the equator in winter and towards more temperate waters in summer (Best, 1975). Two forms are found off the west coast of South Africa (Best, 1974). One of these, located within 37 km of the coast, is considered to reside there throughout the year. This inshore population has baleen plates that are similar in shape to those of the sei whale. The second population, comprising 9000–12 000 whales (Best, 1975), generally occurs 93 km or more from the coastline, and appears in autumn and spring. The offshore group's baleen is generally broader than that of the inshore group, and the offshore whales are slightly larger in size. The offshore population possibly undertakes its north–south migration to follow shoals of fish on which it feeds throughout the year (Best, 1960). Two populations, each with recognisable morphological features, may also be present in other regions of the Southern Hemisphere, and there may be a third "form" in the eastern Indian Ocean (Best, 1975).

In the Western Pacific, *B. edeni* occurs from Japan to New Zealand, and in the Eastern Pacific, from Baja California (Mexico) to Chile (Anonymous,

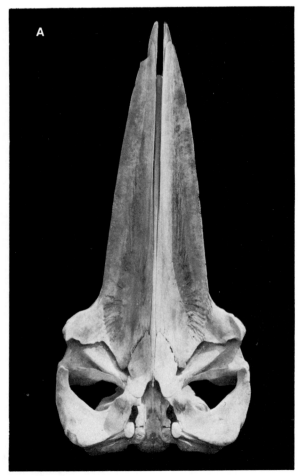

FIG. 4 Two views of skull of Bryde's whale (A) Ventral view; (B) dorsal view.

1973; Clarke and Aguayo, 1965; Clarke *et al.*, 1968; Aguayo, 1974). In the northwest Pacific Bryde's whales move from the Bonin Islands north to the coast of Japan, a seasonal migration of only about 540 km.

In the Atlantic this species is seen from Virginia, the Gulf of Mexico, and the Caribbean southward to Brazil, and from Morocco southward to the Cape of Good Hope. Bryde's whales were radio-tagged off Venezuela by Watkins *et al.*(1979) during their passage through the area. In the Indian

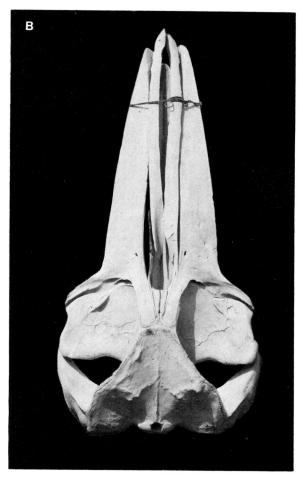

(After Omura et al., 1981; courtesy of H. Omura.)

Ocean their north–south range is from the Persian Gulf to the Cape of Good Hope, and from Burma to Western Australia. They generally remain in waters warmer than 15° or 20°C around the world, so are restricted to a belt from about 40° N to 40° S, except where there are warm water projections, such as the Kuroshio current (Omura and Nemoto, 1955). Bryde's whale is essentially a tropical and subtropical species, but it can be found in slightly cooler waters (Fig. 6).

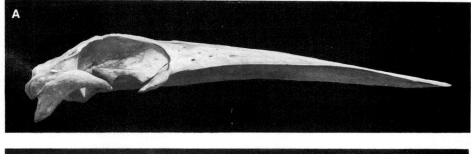
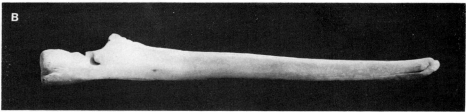
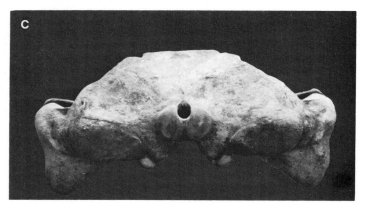

FIG. 5 Three views of skull of Bryde's whale. (A) Lateral view; (B) medial view of left mandible; (C) posterior view. (After Omura *et al.*, 1981; courtesy of H. Omura.)

Whaling and Abundance

Bryde's whale has not been of major importance to whalers. In the past, the species was taken by shore stations in Japan, South Africa, and a few other places, but the records may not be entirely accurate because of the confusion in separating Bryde's from sei whales. In 1975 the International Whaling Commission (IWC) separated the two species in setting catch limits (Anony-

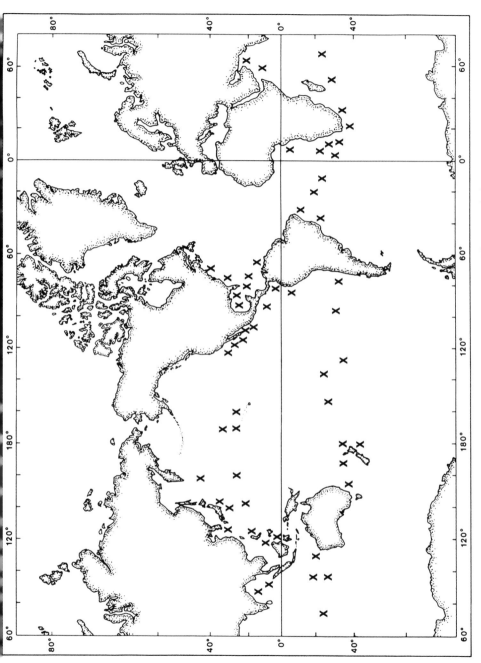

FIG. 6 Probable present general distribution of Bryde's whales.

mous, 1978). The IWC has permitted no quota of Bryde's whales in the southern oceans (since 1974), the North Atlantic, and the eastern North Pacific, pending a satisfactory estimate of population size. It established a catch limit of 454 for the western North Pacific, west of 160° W, for the 1979 season (Bollen, 1978). In 1975 IWC's Scientific Committee agreed that Bryde's whale may be designated as an Initial Management Stock with a permitted catch quota of 90% of the maximum sustainable yield. Since 1970, the harvest of Bryde's whales has increased somewhat, because of diminishing stocks of other catchable species and because Soviet and Japanese whaling in the western North Pacific has moved farther south into the warmer regions occupied by this species. Bryde's whales have been virtually unexploited in the eastern North Pacific (Rice, 1974). In addition to Bryde's whale, the IWC has allowed minke, finback, sei, and sperm whales to be taken in the North Pacific region.

A harvest of Bryde's whales was permitted by the IWC only in the western North Pacific, because there were no satisfactory estimates of population size in other areas. However, even for the western North Pacific, population estimates apparently have been based on very limited data, as indicated by the variation in estimated numbers for pelagic combined with coastal stocks for this area (about 10 000, Best, 1975; 18 000–49 000, Ohsumi, 1978; 15 000, Privalikhin and Berzin, 1978; and 17 840, Tillman, 1978). Best (1975) estimated a population of 20 000–30 000 Bryde's whales in the entire North Pacific. Nishiwaki (1972) expressed the belief that, since the migration of this whale is generally not extensive and the species is confined to small areas, the total world population may be relatively small. It probably has not changed much in recent times.

Behaviour

Breathing, diving, and schooling

The blows of *B. edeni* are not high. They are generally quite lean, and may give the impression of being bushy when wind is present. Generally four or five short breaths are taken before making prolonged dives. Olsen (1913) reported that Bryde's whales could be seen for a long time just under the surface before they came up to breathe, and that they remained submerged for unusually long periods during deeper dives. W. C. Cummings and P. O. Thompson (unpublished) have seen Bryde's whales coming to the surface as often as once every 1 min and diving as long as 20 min. Bryde's whales rarely if ever show their flukes when beginning a dive. The apparent attraction of an individual to a ship (returning 11 times in 91 min) was reminiscent of the behaviour of minke whales, *B. acutorostrata,* in the North Pacific (W. C. Cummings and P. O. Thompson, personal observation). Nishiwaki (1972) reported that Bryde's whales may form dense schools, with tens or sometimes over a hundred whales seen at one time. In the Gulf of California, I have observed only individuals or small groups.

Sound production

W. C. Cummings and P. O. Thompson (unpublished) recorded underwater sounds from *B. edeni* in the Gulf of California. This whale produces powerful low-frequency moans that average 0.4 sec in duration (ranging from 0.2 to 1.5 sec), with most of the energy concentrated at 124 Hz and very little above 250 Hz. These low-frequency sounds commonly exhibited a modulation in frequency of as much as 15 Hz. Shifts may be downwards, upwards, or downwards and then upwards (Fig. 7). The sounds from one whale were produced at random intervals lasting from 0.2 to 9 min. Source levels were about 156 dB, re 1 μPa at 1 m.

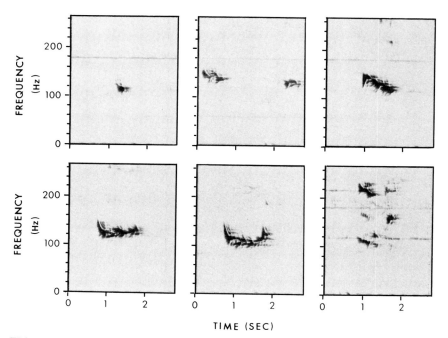

FIG. 7 Sonagrams of a variety of underwater moaning sounds recorded from a Bryde's whale. Analyzing filter bandwidth, 3 Hz. (After W. C. Cummings and P. O. Thompson, unpublished report.)

Bryde's whales produced these moaning sounds only while submerged, the sounds varying widely in duration and frequency structure. With one or two possible exceptions among other baleen species, the Bryde's whale is characteristic of the group in that it has not been noted to produce sounds that we normally associate with echoranging, such as trains of broadband impulses reaching into the kHz region of the spectrum. The only airborne sounds that we recorded from Bryde's whales were low-level exhalations that were nearly obscured by splashing noises as the rising whales broke the surface (Fig. 8).

Feeding

The coarseness of fringe on the whalebone plates (Fig. 3) of *B. edeni* apparently is associated with diet and feeding habits. This species frequently feeds on pelagic fishes such as pilchard, mackerel, and herring (Chittleborough, 1959; Best, 1960) and has been seen to eat the cephalopod *Lycoteuthis diadema*, (Best, 1974). Omura (1962) reported that the stomachs of many Bryde's whales taken off the coast of Japan contained large quantities of pelagic

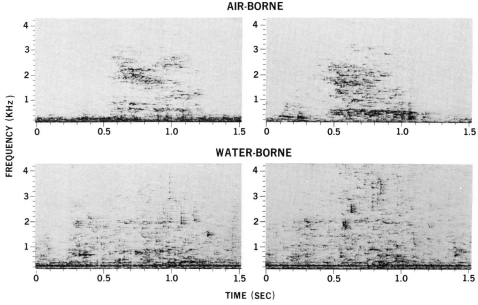

FIG. 8 Sonagrams of two blow sounds from a surfacing Bryde's whale as received in air and underwater, the latter almost masked by accompanying splash sounds. Analyzing filter bandwidth, 20 Hz). (After W. C. Cummings and P. O. Thompson, unpublished report.)

crustaceans. Olsen (1913) had also noted this, finding mainly euphausiids. In a sample of 459 Bryde's whales taken in the South Pacific and Indian oceans, Kawamura (1980) found that whales in these locations largely preyed upon euphausiids instead of fish.

Gaskin (1972), while marking whales in the Hauraki Gulf during September–October 1963 had the opportunity of seeing a small group of Bryde's whales feeding on surface fish. Two of the whales raced side by side among the fish at about 10 knots and then suddenly lunged vertically out of the water. Sliding back into the sea, they started moving rapidly forward again with snouts clear of the surface. Fish could be seen jumping clear of the whales' closing mouths. Olsen (1913) commented on the voracity of Bryde's whales that were seen hunting among sharks, some of which were found in the whales' stomachs. He also reported that penguins had been found in the stomach of a Bryde's whale, but that the whale was most likely feeding on fish and had caught the penguins inadvertently.

Reproduction

Data from 120 Bryde's whales taken off the coast of Japan indicates that females attain sexual maturity at a length of approximately 12 m (Omura, 1962). Assuming that one lamination of the ear plug occurs each year, females reach sexual maturation in 10 years and males become sexually mature between 9 and 13 years of age, at a length of 12 m. Omura's findings agree with those of Best (1960) who worked with Bryde's whales taken off South Africa. Chittleborough (1959) noted that sexual maturity of Bryde's whales off the west coast of Australia was attained at lengths slightly less than 12 m.

From records on the appearance of foetuses, Best (1960) reported that *B. edeni* bred throughout the year in South African waters. He later (1975) determined that the offshore population there bred only in the autumn. The gestation period is thought to be about 1 year, but females may not necessarily give birth every 2 years. Newborn calves are about 3.4 m in length. Ellis (1980) reviewed the biology and provided excellent colour plates of *B. edeni*.

Conclusion

Even though Bryde's whale has been pursued by the whale fisheries, especially in the northwest Pacific, it is one of the least known of the large whales. This species is distributed around the warm waters of the world, but its population size is virtually unknown. Population estimates in two regions disagree and imply a wide margin of error. The possible racial or "form" characteristics noted earlier are intriguing, and it is doubtful whether we have heard the last word on the species' taxonomic status. The basic reason for the lack of biological information on Bryde's whale is the difficulty in identifying the species from sea or aircraft, even under the best conditions and by the experienced eye.

Acknowledgement

I thank M. E. Dahlheim for considerable help with the earlier literature surveys and for typing previous versions; J. C. Cummings, T. Rydlinski, and C. L. Gray for their valuable assistance with the manuscript; H. Omura for recently published information, and Dr G. N. di Sciara, J. S. Leatherwood, W. E. Schevill, and Dr W. A. Watkins, for their helpful comments. Partial support was received from the Office of Naval Research (Contract N00014-78-C-0419), Dr R. C. Tipper, and Dr B. Zahuranec, Program Managers.

References

Aguayo, L. A. (1974). Baleen whales off continental Chile. *In* "The Whale Problem. A Status Report" (Ed. W. E. Schevill), pp. 209–217. Harvard Univ. Press, Cambridge, Massachusetts.

Anderson, J. (1878). "Anatomical and Zoological Researches. Comprising an Account of the Zoological Results of the Two Expeditions to Western Yunnan in 1868 and 1875", pp. 551–564. B. Quaritch, London.

Andrews, R. C. (1918). A note on the skeleton of *Balaenoptera edeni* in the Indian Museum, Calcutta. *Rec. Indian Mus.* **15,** 105–107.

Anonymous. (1973). Current status of stocks and life histories of species. Report of the Secretary of Commerce. Department of Commerce, National Oceanic and Atmospheric Administration, Marine Mammal Protection Act. *Fed. Regist.* **38**(147), 20580 (8-1-73).

Anonymous. (1978). The Soviet whaling industry. *Mar. Fish. Rev.* **40**(11), 33–36.

Best, P. B. (1960). Further information on Bryde's whale (*Balaenoptera edeni* Anderson) from Saldanha Bay, South Africa. *Nor. Hvalfangst-Tid.* **49,** 201–215.

Best, P. B. (1974). Status of the whale populations off the west coast of South Africa and current research. *In* "The Whale Problem. A Status Report," (Ed. W. E. Schevill), pp. 53–81. Harvard Univ. Press, Cambridge, Massachusetts.

Best, P. B. (1975). "Status of Bryde's whale (*Balaenoptera edeni* or *B. brydei*)". (FAD

Advisory Committee on Marine Resources Research, Marine Mammal Symposium).

Bollen, A. G. (1978). "IWC—Chairman's report of the thirtieth meeting, 26–30 June 1978", Int. Whal. Commn.

Cagnolaro, L., and Notarbartolo di Sciara, G. (1979). Su di uno scheletro di *Balaenoptera edeni* Anderson, 1978, spiaggiato sulle coste caraibiche del Venezuela (Cetacea Balaenopteridae). *Natura (Milan)* **70**(4), 265–274. (In Italian)

Chittleborough, R. G. (1959). *B. brydei* Olsen on the west coast of Australia. *Nor. Hvalfangst-Tid.* **48,** 62–66.

Clarke, R., and Aguayo, L. A. (1965). Bryde's whale in the south east Pacific. *Nor. Hvalfangst-Tid.* **54**(7), 141–148.

Clarke, R., Aguayo, L. A., and Paliza, G. (1968). Sperm whales of the southeast Pacific. Parts 1 and 2. *Hvalradets Skr.* **51,** 1–80.

Ellis, R. (1980). "The Book of Whales". Knopf, New York.

Gaskin, D. C. (Ed.) (1972). "Whales, Dolphins, and Seals, with Special Reference to the New Zealand Region". St. Martin's Press, New York.

Junge, G. C. A. (1950). On a specimen of the rare fin whale, *Balaenoptera edeni* Anderson, stranded on Pulu Sugi near Singapore. *Zool. Verhandelingen* **9,** 1–26.

Kawamura, A. (1978). On the baleen filter area in the South Pacific Bryde's whales. *Sci. Rep. Whales Res. Inst.* **30,** 291–300.

Kawamura, A. (1980). Food habits of the Bryde's whales taken in the South Pacific and Indian oceans. *Sci. Rep. Whales Res. Inst.* **32,** 1–23.

Kawamura, A., and Satake, Y. (1976). Preliminary report on the geographical distribution of the Bryde's whale in the North Pacific with special reference to the structure of filtering apparatus. *Sci. Rep. Whales Res. Inst.* **28,** 1–35.

Leatherwood, S., Caldwell, D. K., and Winn, H. E. (1976). Whales, dolphins and porpoises of the western North Atlantic. A guide to their identification. *NOAA Tech. Rep. NMFS Circ.* **396.**

Lonnberg, E. (1931). The skeleton of *Balaenoptera brydei* O. Olsen. *Ark. Zool.* **23,** 1–23.

Mead, J. G. (1974). Records of sei and Bryde's whales from the Atlantic coast of the United States, the Gulf of Mexico and the Caribbean. *Rep. int. Whal. Commn (Spec. Issue I)* **36,** 113–116. 1977.

Nishiwaki, M. (1972). General biology. *In* "Mammals of the Sea, Biology and Medicine" (Ed. S. H. Ridgway), pp. 3–204. Thomas, Springfield, Illinois.

Norman, J. R., and Fraser, F. C. (1949). "Field Book of Giant Fishes". Putnam, New York.

Ohsumi, S. (1978). Bryde's whales in the North Pacific in 1976. *Rep. int. Whal. Commn* **28,** 277–287.

Olsen, O. (1913). On the external characteristics and biology of Bryde's whale (*Balaenoptera brydei*) a new rorqual from the coast of South Africa. *Proc. Zool. Soc. London*, pp. 1073–1090.

Omura, H. (1959). Bryde's whale from the coast of Japan. *Sci. Rep. Whales Res. Inst.* **14,** 1–33.

Omura, H. (1962). Further information on Bryde's whales from the coast of Japan. *Sci. Rep. Whales Res. Inst.* **16,** 7–18.

Omura, H. (1977). Review of the occurrence of Bryde's whale in the northwest Pacific. *Rep. int. Whal. Commn. (Spec. Issue I)*, 88–91.

Omura, H., and Nemoto, T. (1955). Sei whales in the adjacent waters of Japan, relation between movement and water temperature of the sea. *Sci. Rep. Whales Res. Inst.* **10,** 79–87.

Omura, H., Kasuya, T., Kato, H., and Wada, S. (1981). Osteological study of the Bryde's whale from the central South Pacific and eastern Indian Ocean. *Sci. Rep. Whales Res. Inst.* **33,** 1–26.

Privalikhin, V. I., and Berzin, A. A. (1978). Abundance and distribution of Bryde's whale *Balaenoptera edeni* in the Pacific Ocean. *Rep. int. Whal. Commn* **28,** 301–302.

Rice, D. W. (1974). Whales and whale research in the eastern North Pacific. *In* "The Whale Problem, A Status Report" (Ed. W. E. Schevill), pp. 170–195. Harvard Univ. Press, Cambridge, Massachusetts.

Soot-Ryen, T. (1961). On a Bryde's whale stranded in Curaçao. *Nor. Hvalfangst-Tid.* **50**(6), 323–332.

Tillman, M. F. (1978). Modified DeLury Estimates of the North Pacific Bryde's whale stock. *Rep. int. Whal. Commn,* **28,** 315–317.

Watkins, W. A., Notarbartolo di Sciara, G., and Moore, K. E. (1979). Observations and radio tagging of *Balaenoptera edeni* near Puerto La Cruz, Venezuela (Ref. No. WHOI-79-78). Woods Hole Oceanographic Inst., Woods Hole, Massachusetts.

6

Sei Whale

Balaenoptera borealis Lesson, 1828

Ray Gambell

Genus and Species

Taxonomy

The sei whale *Balaenoptera borealis* is named from the Latin word *borealis* meaning northern. Lesson (1828) reviewed the various balaenopterid species described by Lacépède, and noted that the specimen from Holstein on the German coast described by Rudolphi in 1822 as *Balaena rostrata* was identified by Cuvier in 1823 as a rorqual (from the Norwegian *rorhval*, the whale with pleats, as distinct from the right whales, *Balaena* spp., which lack throat grooves). Cuvier called this whale *rorqual du nord*, and it is this name which was latinised by Lesson. Because of the size difference between sei whales inhabiting the Northern and Southern hemispheres, the slightly smaller northern form is distinguished as *B. b. borealis* from the southern *B. b. schlegelii*, after a specimen from Java described by Flower (1865).

TABLE 1 Some common names given to the sei whale[a]

English	Sei whale, coalfish whale, pollack whale, Rudolph's rorqual, sardine whale, Japan finner
Norwegian	Seihval
Japanese	Kaguo-kuzira, iwasikurira, iwashi kujira
Russian	Seival, saidianoi kit, ivasevhyi polosatik, ivasevyi kit
German	Seiwal
Danish	Sejhval
Icelandic	Sandereydur
French	Rorqual du Nord, rorqual de Rudolph, baleine noire
Dutch	Noordse vinvis
Czech	Plejtvak severny
Eskimo	Komvokhgak
Aleutian	Agalagitakg
Spanish	Ballena boba, ballena sei[b]
Swedish	Sejval

[a]From Klinowska (1980).
[b]A. Mignucci-Giannoni, Colorado State University (personal communication).

Common names

The common name sei comes from the Norwegian *seje*, the pollack or coalfish, which occurs at the same time as the whales appear off the coast of Finmark (although the whales do not eat the fish, both whales and fish consume the plankton which is available then). Other common names in English and other languages are given in Table 1.

External Characteristics and Morphology

Sei whales (Fig. 1) grow to a maximum length of 20 m in the Southern Hemisphere and 18.6 m in the North Pacific, but perhaps only 17.3 m in the North Atlantic. The females reach a slightly larger size than the males. According to the description by Ivashin *et al.* (1972) the body shape is rather more bulky than that of the fin whale, but still laterally compressed in the caudal region. The head is intermediate in shape between that of the blue and fin whales and occupies 20–25% of the body length, the percentage increasing with age. From the side, the head appears slightly arched because of a median crest rising towards the position of the blowholes, which are located at the highest point of the head. There are 318–340 baleen plates in each row in the North Atlantic and 296–402 in the Southern Hemisphere whales.

SEI WHALE

The ventral grooves vary in number: 38–56 in the North Atlantic, 32–60 in the North Pacific, and 40–62 in the Southern Hemisphere. They do not reach the level of the umbilicus. The dorsal fin is placed slightly less than two-thirds of the way along the back. It is relatively larger than that of the fin whale and different in shape. Its height is 3–4.5% of the body length (up to 60 cm) and more upright. The fin is slender and the leading edge forms an angle about 45° with the body. The point is directed well backwards so that the front edge is curved like a bow. The flippers are pointed and rather small, about 9% of the body length. The tail flukes are also relatively small.

In colour, the sei whale is dark grey with a bluish tinge which extends from the back down to the sides, although the colour may be somewhat lighter ventrally. There are often small lighter coloured scars on the body giving it a galvanised metallic look. There is usually a white patch of greater or lesser extent in the area of the ventral grooves. The undersides of the flippers and flukes are the same colour or only slightly paler than the rest of the body. The baleen plates are dark grey with fine paler fringes.

Weight

The weights of 16 sei whales obtained during commercial whaling operations were used by Lockyer (1976) to calculate the weight:length relationship. A factor of 6% for blood and other fluid losses was added during the piecemeal weighing process, giving the calculated curve of body weight against length shown in Fig. 2.

Distribution

Sei whales are widely distributed in all oceans, although they do not go so far towards the polar waters as do the other rorquals. They move from the temperate zones, which they occupy in the respective winter months of the

FIG. 1   External form of sei whale. (From Nishiwaki, 1972.)

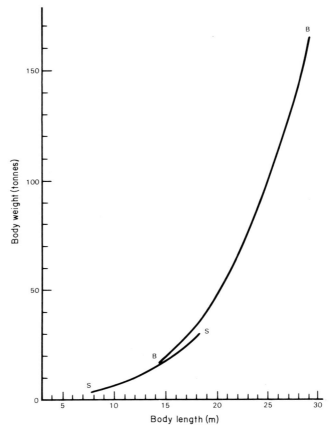

FIG. 2 Calculated total body weight curve for sei whale compared with that of blue whale B, Blue whale; S, sei whale. (Redrawn from Lockyer, 1976.)

Northern and Southern hemispheres, to the higher latitudinal waters of each hemisphere, where most feeding takes place, for the summer months. These movements are apparent from the seasonal pattern of whaling operations, sightings, and recoveries of marked whales, reviewed by Gambell (1976). However, much less is known of these migrations for sei whales than of those of the larger baleen whales. Also, the movements of sei whales in some parts of their range are somewhat more irregular than the movements of blue and fin whales, so that the sei whales' appearance in large numbers on a particular whaling ground, such as South Georgia or Iceland, was described as a "sei whale year."

In the North Pacific in summer, sei whales can be found from California to the Gulf of Alaska on the east, across the Bering Sea and down to the coasts of Japan and Korea in the west. In winter, the centers of abundance move southwards to around 20° N. The relationships of the sei whales in the North Atlantic are not so well known; they occur off Nova Scotia, Labrador, and move up to west Greenland on the west. On the east, the winter stocks off Spain, Portugal, and northwest Africa probably move towards Norway and more northern waters in summer, but the wintering grounds of the sei whales found in the Denmark Strait in summer are not known.

In the Southern Hemisphere the movements of the sei whales are broadly similar to those of the blue and fin whales, except the southward movement to the Antarctic feeding grounds does not penetrate to such high latitudes. The main summer concentrations seem to be between 40 and 50° S. In winter, sei whales are present off Brazil and the west and east coasts of South Africa and Australia, but the wintering grounds in the open oceans are unknown.

Sei whales migrate into and out of the Antarctic rather later than the blue and fin whales, but, like them, the pregnant females are amongst the first in the succession. There is also evidence of segregation by age, so that in both the Antarctic and the North Pacific there is a greater proportion of the older and bigger sei whales in the higher latitudes and a greater number of the smaller and younger whales in the lower latitudes.

Stocks

The evidence for stock separation of sei whales in both the Northern and Southern hemispheres is sparse. Because the seasons are opposite in the two hemispheres, with the whales migrating in the same direction at the same time of year, they do not converge towards the equator simultaneously. It is believed that they remain essentially independent, although with the possibility of some small interchange through whales out of step with the main movements. The following summary of the stock units recognised is derived from Gambell (1976).

Based primarily on the fact that catches have been made from land stations in Canada, Iceland, and Europe, as well as pelagically, the North Atlantic stocks are broadly divided into eastern, central, and western components for management (Fig. 3). The North Pacific sei whales have, by analogy with the fin whales, been divided arbitrarily into eastern and western stocks at 180°; although there have been suggestions of three stocks, the evidence is not conclusive (Masaki, 1977).

The six Antarctic management areas, derived from the humpback, blue,

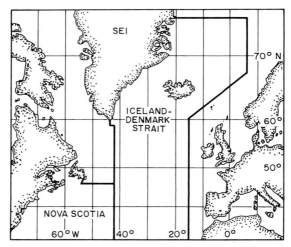

FIG. 3   Stock divisions adopted for management of sei whales in the North Atlantic.

and fin whale distributions, have been used for the southern sei whales (see Chapter 7, Fig. 5a). There is some evidence from the pattern of catches that sei whales in the Atlantic sector can be divided into eastern and western components.

Abundance and Life History

Whaling

Although Japanese whalers caught sei whales in their net fisheries from about the seventeenth century the main exploitation of this species required the development of modern whaling based on the combination of the explosive grenade harpoon fired from a fast steam-powered catching vessel and pioneered by Svend Foyn in 1864 (Tonnessen and Johnson, 1982). Once the capacity to pursue and hold on to the fast swimming rorquals was achieved, the larger blue and fin whales were the preferred species, but the later arriving sei whales were taken at the end of the season in the North Atlantic fisheries from the 1880s onwards and the land station operations in the Antarctic from the early 1900s. As the stocks of the larger whales were reduced in both areas, the sei whales assumed a greater importance. There were significant catches first off Norway, Iceland, and finally Nova Scotia around 1970 in the North Atlantic. All sei whales are now protected from commercial catching in the North Atlantic except the Iceland–Denmark Strait stock. In the Southern Hemisphere the peak of catching occurred in

TABLE 2 Organ weights (tonnes) of the largest and smallest male and female sei whales weighed from the coast of Japan[a]

Organ	Female (14.5 m)	Male (12.4 m)	Female (11.5 m)	Male (11.5 m)
Blubber	3.148	2.421	1.704	1.273
Meat	8.958	6.058	4.724	5.131
Internal organs	1.387	1.068	0.840	0.704
Heart	0.060	0.037	0.041	0.029
Lungs	0.156	0.089	0.071	0.071
Stomach	0.134	0.097	0.099	0.068
Intestine	0.260	0.209	0.268	0.223
Kidney	0.060	0.060	0.045	0.034
Liver	0.185	0.130	0.130	0.104
Bone	1.741	1.568	1.205	1.269
Skull	0.480	0.516	0.383	0.352
Vertebrae	0.760	0.618	0.532	0.600
Ribs	0.197	0.193	0.141	0.171
Jaw	0.178	0.148	0.089	0.097
Total weight	15.557	11.284	8.583	8.527

[a] From Omura (1950).

absent from the flipper skeleton, although embryos may show rudiments. The carpal formula is variable: I—3/4, II—5/7, IV—4/6, V—2/3.

Organ weights

The weights of the chief internal organs in the largest and smallest sei whales of each sex which have been recorded in the literature are shown in Table 2.

Kasuya (1966) found the sei whale to have 44 chromosomes. Reports of sei whale osteology are found in Andrews (1916), Kellogg (1928), and Tomilin (1957). Hosokawa (1950) studied the larynx and laryngeal sacs while Nakai and Shida (1948) discussed the "sinus-hairs" present in the skin of the rostrum. Shulte (1916) described extensive dissections of a sei whale foetus.

A good overall review of the cetacean central nervous system is found in Breathnach (1960). Specific works on the sei whale brain are those of Pilleri (1965) and Kraus and Pilleri (1969). R. H. Burne (Fraser, 1952) made anatomical comparisons including the sei whale and published a cast of the nostrils.

Pivorunas (1976, 1977) has examined the anatomy of the baleen plates and the ventral pouch. Baleen rows of the sei whale are shown in comparison with one from a fin whale in Fig. 6.

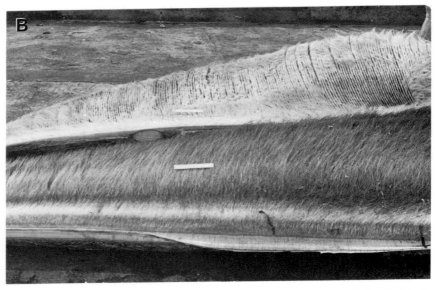

FIG. 6 (A) Baleen rows from one fin whale (in front) and three sei whales (behind); (B) the streaking seen in the coarsely fringed fin whale baleen (nearest baleen) was normally found in all fin whales examined by Pivorunas (1976) and contrasts with the light fringe of the sei whale (behind). White rule is 15.2 cm (6 inches). This material was from whales taken commercially at Blandford, Nova Scotia, in 1971. (Photography courtesy of A. Pivorunas.)

Behaviour

Sei whales usually are found in small groups of up to half a dozen individuals, although on the feeding grounds there may be larger numbers associated together.

The blow is similar to that of the blue and fin whales, but not so tall, reaching 3 m. Sei whales normally do not dive deeply, and come to the surface at a shallow angle, so that the head and back break the surface together. The whale seems to slip below the surface on the next dive, rather than arching over. The breathing pattern is rather more regular in the sei whale than some other rorquals, with blows at 20- to 30-sec intervals followed by submersion for up to 15 min or longer. It is often possible to follow the slicks of the sei whale made by the beat of the tail on its shallow dives. Sei whales were thought by the whalers to be the fastest of the rorquals, at least over short distances.

The sei whale feeds primarily by skimming plankton out of the water as it swims forward with its mouth open. The food organisms are caught on the particularly fine fringe fibres of the baleen plates (Fig. 6). However, they can also gulp single mouthfuls of water containing food, straining the water through the baleen plates and sieving off the food by reducing the volume of the mouth.

Reproduction

In the Southern Hemisphere 95% of sei whales are conceived in the 4 months between April and August (Gambell, 1968). The single calf is born 12 months later, although the gestation period in the North Pacific is reported to be $10\frac{1}{2}$ months (Masaki, 1976). At birth the northern sei whale is 4.4 m long and weighs 0.65 tonnes (Laws, 1959); southern sei whales are 4.5 m at birth (Matthews, 1938). The lactation period extends over 6 months and the calves are weaned on the higher latitudinal feeding grounds when they are about 9 m in length. The recent mothers then pass through a resting phase before coming into oestrus at the next winter breeding season about 6 months later. This gives a possible 2-year breeding cycle, although some females may not become pregnant again until the third year.

Sei whales reach sexual maturity at a length of 13.5 m in males and 13.9 m in southern females; the corresponding lengths are 12.8 and 13.3 m, respectively, in the North Pacific. It appears that sei whales in the Southern Hemisphere have responded to the decline in the abundance of the larger blue and fin whales by growing faster, even in the years before they were directly exploited themselves in significant numbers. Evidence from study of the transition layer in the earplug (Lockyer, 1974) indicates that the mean age at sexual maturity for both males and females in the year classes up to 1935 was a little over 11 years; by 1945 this had fallen to just under 10 years. Direct analysis of the ovaries and ages of female sei whales sampled in the early 1970s showed that the average age at sexual maturity had decreased by then to 7 years in some areas (Masaki, 1978). A similar decline has been reported in the North Pacific, from 10 years prior to 1930 to 6–7.5 years in 1960 (IWC, 1977).

The pregnancy rates of southern sei whales have also changed apparently in response to the reduced abundance of the larger rorquals. Since 1946 the proportion of nonlactating mature females pregnant in the catches has more than doubled, from about 25 to nearly 60% in more recent years (Gambell, 1973).

Parasites

Matthews (1938) noted that large ectoparasites are uncommon on sei whales. *Penella* and *Xenobalanus globicipitis* have been recorded in the Southern Hemisphere and North Pacific. *Balaenophilus*, on the other hand, commonly infests the baleen, as does *Haematophagus*. Diatoms occur on the skin in the Antarctic, commonly *Cocconesis ceticola*, together with *Lycmophora lynbyei*, *Fragilara antarctica*, and *Navicula* sp..

Internal parasites are very prevalent. The acanthocephalan *Bolbosoma turbinella* is common in the intestine with the cestodes *Tetrabothrius affinis* and *Balaenoptera turbinella*. Nematode worms occur frequently in the kidney and in the erectile tissue of the penis and the urethra of males.

In the North Pacific, sei whales are often heavily infected with the stomach worm *Anisaleis simplex* and the liver fluke *Lecithodesmus spinosus*, and 7% are infected with a disease which causes loss of the baleen plates (Rice, 1977).

According to Nishiwaki (1972) whalers often reported killer whales attacking sei whales.

References

Allen, K. R. (1980). "Conservation and Management of Whales". Univ. of Washington Press, Seattle.
Andrews, R. C. (1916). The sei whale (*Balaenoptera borealis* Lesson), history, habits, external anatomy, osteology and relationship. *Mem. Am. Mus. Nat. Hist. New Ser.* **1,** 291–388.

Breathnach, A. S. (1960). The cetacean central nervous system. *Biol. Rev.* **35,** 187–230.

Cuvier, G. (1823). "Recherches sur les Ossements Fossiles".

Flower, W. H. (1864). Notes on the skeletons of whales in the principal museum of Holland and Belgium. *Proc. Zool. Soc. London,* p. 408

Fraser, F. C. (1952). "Handbook of R. H Burne's Cetacean Dissections". Brit. Mus. (Nat. Hist.), London.

Gambell, R. (1968). Seasonal cycles and reproduction in sei whales of the Southern Hemisphere. *Disc. Rep.* **35,** 31–134.

Gambell, R. (1973). Some effects of exploitation on reproduction in whales. *J. Reprod. Fertil. Suppl.* **19,** 531–551.

Gambell, R. (1976). Population biology and the management of whales. *Appl. Biol.* **1,** 247–343.

Horwood, J. W. (1980). Population biology and stock assessment of Southern Hemisphere sei whales. *Rep. int. Whal. Commn* **30,** 519–530.

Hosokawa, H. (1950). On the cetacean larynx, with special remarks on the laryngeal sack of the sei whale and the arytenoepiglottideal tube of the sperm whale. *Sci. Rep. Whales Res. Inst.* **3,** 23–62.

Ivashin, M. V., Popov, L. A., and Tsapko, A. S. (1972). "Morskie Mlekopttayuschie". Pischevaya Promychlennosk, Moscow.

IWC. (1977). Report of the special meeting of the scientific committee on sei and Bryde's whales. *Rep. int. Whal. Commn (Spec. Issue I),* 1–9.

Kasuya, T. (1966). Karyotypes of a sei whale. *Sci. Rep. Whales Res. Inst.* **20,** 83–88.

Kawamura, A. (1973). Food and feeding of sei whales caught in the waters south of 40°N in the North Pacific. *Sci. Rep. Whales Res. Inst.* **25,** 219–236.

Kawamura, A. (1974). Food and feeding ecology in the southern sei whale. *Sci. Rep. Whales Rep. Inst.* **26,** 25–144.

Kellogg, R. (1928). The history of whales—their adaptation to life in the water. *Q. Rev. Biol.* **3,** 29–76, 174–208.

Klinowska, M. (1980). "A World Review of the Cetacea". Nature Conservancy Council, London.

Kraus, C., and Pilleri, G. (1969). Zur Histologie der Grosshirnrinde von *Balaenoptera borealis. Investigations on Cetacea* **1,** 151–170.

Laws, R. M. (1959). The foetal growth rates of whales with special reference to the fin whale. *Balaenoptera physalus* Linn. *Disc. Rep.* **29,** 281–308.

Lesson, R. P. (1828). "Compléments des oeuvres de Buffon ou histoire naturelle des animaux rares". Paris.

Lockyer, C. (1974). Investigation of the ear plug of the southern sei whale *Balaenoptera physalus* as a valid means of determining age. *J. Cons. Cons. Int. Explor. Mer* **36**(1), 71–81.

Lockyer, C. (1976). Body weights of some species of large whales. *J. Cons. Cons. Int. Explor. Mer.* **34**(2), 276–294.

Masaki, Y. (1976). Biological studies on the North Pacific sei whale. *Bull. Far Seas Fish. Res. Lab.* **14,** 1–104.

Masaki, Y. (1977). The separation of the stock units of sei whales in the North Pacific. *Rep. int. Whal. Commn (Spec. Issue 1)*, 71–79.

Masaki, Y. (1978). Yearly changes in the biological parameters of the Antarctic sei whale. *Rep. int. Whal. Commn* **28**, 421–429.

Matthews, L. H. (1938). The sei whale, *Balaenoptera borealis*. *Disc. Rep.* **17**, 183–290.

Nakai, J., and Shida, T. (1948). Sinus-hairs of the sei-whale (*Balaenoptera borealis*). *Sci. Rep. Whales Res. Inst.* **1**, 41–47.

Nemoto, T. (1959). Food of baleen whales with reference to whale movements. *Sci. Rep. Whales Res. Inst.* **14**, 149–290.

Nishiwaki, M. (1972). General biology. *In* "Mammals of the Sea: Biology and Medicine" (Ed. S. H. Ridgway), pp. 3–204. Thomas, Springfield, Illinois.

Omura, H. (1950). On the body weight of sperm and sei whales located in the adjacent waters of Japan. *Sci. Rep. Whales Res. Inst.* **4**, 1–13.

Pilleri, G. (1965). The brain of the southern sei whale (*Balaenoptera borealis* Lesson). *Experientia* **21**, 703–708.

Pivorunas, A. (1976). A mathematical consideration on the function of baleen plates and their fringes. *Sci. Rep. Whales Res. Inst.* **28**, 37–55.

Pivorunas, A. (1977). The fibrocartilage skeleton and related structures of the ventral pouch of balaenopterid whales. *J. Morphol.* **151**, 299–314.

Rice, D. W. (1977). Synopsis of biological data on the sei whale and Bryde's whale in the eastern North Pacific. *Rep. int. Whal. Commn (Spec. Issue 1)*, 92–97.

Rudolphi, D. K. A. (1822). *Abh. Akad. Wiss. Berlin*, 1820–1821, pp. 27–40.

Schulte, H. v. W. (1916). Anatomy of a fetus of *Balaenoptera borealis* Lesson. *Mem. Am. Mus. Nat. Hist.* **1**, 389–502.

Tomilin, A. G. (1957). "Mammals of the U.S.S.R. and Adjacent Countries. Vol. IX: Cetacea" (Ed. V. G. Heptner), Nauk SSSR, Moscow. English Translation, 1967, Israel Program for Scientific Translations, Jerusalem.

Tonnessen, J. N., and Johnson, A. D. (1982). "The History of Modern Whaling". Hurst, London.

7

Fin Whale

Balaenoptera physalus(Linnaeus, 1758)

Ray Gambell

Genus and Species

Taxonomy

The fin whale *Balaenoptera physalus* was named by Linnaeus from the Greek *physalis*, a wind instrument or a kind of toad which puffs itself up (Fig. 1). Lacépède in 1804 established a separate genus *Balaenoptera* to distinguish the rorquals (after the Norwegian *rorhval*, the whale with pleats) from the right whales, which do not have throat grooves, and called the whale *B. rorqual*. Racovitza (1903) synonymised *B. rorqual*, *B. australis* (Gray, 1846), a southern whale described on the voyages of HMS *Erebus* and *Terror*, and *B. patachonicus* (Burmeister, 1865), an Argentinian whale, under the single species name *B. physalus*. However, because the form found in the Southern Hemisphere grows slightly larger than that found in the Northern Hemisphere, some authorities recognise a northern subspecies *B. p. physalus* distinct from the southern *B. p. quoyi* (Fischer, 1829). The latter name was given to an animal captured in the Falkland Islands.

FIG. 1 Feeding fin whale with ventral forebody distended as viewed from the air. (Photo by Gary Carter, courtesy of Stephen Leatherwood.)

Common names

The fin whale is widely distributed and has been known and hunted over most of its range, so that there are many local names both in English and other languages. Some of these are listed in Table 1. Although all rorquals have dorsal fins, the common names based on the rather prominent back fin are widespread as are those reflecting the throat grooves, while the ridged tail-stock has given rise to the name razorback.

External Characteristics and Morphology

The fin whale is the second largest whale in size, exceeded only by the blue whale. It can grow up to 25 and 27 m for males and females, respectively, in the Southern Hemisphere, and 22 and 24 m, respectively, in the Northern Hemisphere. Like all the rorquals, it is very streamlined in appearance, and the following features are described following Ivashin *et al.* (1972). There is a series of 50 to 100 folds or grooves which extend from the chin backwards

TABLE 1 Some common names given to the fin whale[a]

English	Fin whale, finback, common rorqual, herring whale, true fin whale, finfish, gibbar, finner, razor back, common fin whale, common fin back
Norwegian	Finefisk, finhval, loddehval, rorhval, sildehval, sildror, tuequal, finnhval, storhval (used for other large whales also)
Japanese	Nagasu-kujira
Russian	Finval, seldianoi polostaik, seldianoi kit, obyknovennyi polosatik, nastoiashchii polosatik, kiit
German	Schnabelwal, Finnwal, Finnfisch
Swedish	Finnfisk, sillval, sillhval, rorval
Icelandic	Sildrek, seldreki, langreydur, hunfubaks, furehvaler, finnhvaler
French	Baleine americaine, vraie baleine, rorqual commun
Dutch	Vinvis, gewone vinvis
Eskimo	Tykyshkok, keporkarnak, vapaklichan, niltkokkein uiiut, nilchoken biuu
Aleutian	Mangidakh, mangidadakh (juveniles)
Czech	Plejtvak mysok
Spanish	Ballena de Aleta,[b] ballena boba (also *Balaenoptera borealis*)
Italian	Capidolio (*NB*, Capidoglio or Capodoglia = *Physeter catodon*)
Lapp	Reider
Danish	Rorhval

[a]After Klinowska (1980).
[b]A. Mignucci-Giannoni, Colorado State University (personal communication).

under the verntral region to the umbilicus, or slightly beyond in the midline. The head occupies 20–25% of the body length, becoming larger in the adults. It is markedly triangular in dorsal view, flat with a median crest. The eyes are small and lie above the corner of the mouth. The lower jaw is large and strongly convex laterally. When the mouth is closed the jaw protrudes 10–20 cm beyond the tip of the snout.

The body is somewhat fuller in shape than that of the blue whale. It is rounded in front but greatly compressed laterally in the caudal part, with a distinct ridge along the back behind the dorsal fin. The tail flukes are broad with a conspicuous indentation in the centre. The dorsal fin is set about two-thirds of the way along the back (Fig. 2). It is large, 2.1–2.5% of the body length or up to 60 cm tall and forms an angle of < 40° with the back. The flippers are lancetlike and relatively small, their length from tip to axilla generally 7.5–9.9% of the body length.

The colour of the fin whale is dark grey above, grading into white ventrally. The undersides of the flippers and flukes are also white. The dark

FIG. 2 Drawings of fin whale. (A) Left side; (B) right side. Note the asymmetrical colour pattern. (Drawings by Peter Folkens, The Oceanic Society.)

colour on the head is strikingly asymmetrical (Fig. 2). While the left side is pigmented evenly, the front of the lower jaw on the right is white. The 260–480 baleen plates on each side share this uneven pigmentation, so that those on the front 20–30% of the right side are white or yellow and the remainder and those on the left side are dark blue–grey, with some streaks of white or yellow. The fringe fibres on the baleen plates are uniformly yellowish white. Some animals show a pale grey chevron mark at the back of the head, the lateral arms pointing posteriorly. There is also a dark streak running up and back from the eye and a light one arching over to the flipper insertion.

Weight

All available weight data, obtained as the carcasses were being processed during commercial whaling operations and amounting to 34 Antarctic and 8 Northern Hemisphere fin whales, were used by Lockyer (1976) to calculate the weight:length formula for this species. She also reviewed the adjustment necessary to allow for blood and other fluid losses during flensing, which account for about 6% of the body weight in baleen whales. The resulting calculated body weight at length is plotted in Fig. 3.

Distribution

The fin whale, as is typical of the large baleen whales, is found in all the major oceans of the world. Its regular seasonal migrations between the temperate waters, where it mates and calves, and the more polar feeding grounds occupied in the summer months are a characteristic feature of its life history (Mackintosh, 1965). Evidence for the seasonal migrations is available from direct recoveries of marked whales, the seasonal pattern of whaling activity in the different parts of the world, and sightings of whales at particular times of the year from survey vessels. Because the seasons are opposite in the two

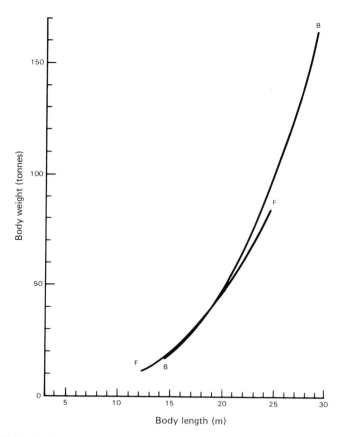

FIG. 3 Calculated total body weight curve for fin whale compared to that of the blue whale. B, Blue whale; F, fin whale. (From Lockyer, 1976.)

hemispheres, the northern and southern populations of whales do not converge towards the equator at the same time; it is possible that occasional interchanges of individuals can occur sufficient to prevent the genetic isolation of the northern and southern populations.

In the North Pacific in summer fin whales are found in the Chukchi Sea, around the Aleutian Islands, the Gulf of Alaska, and down to California in the east. In the west they occur from the Sea of Okhotsk down to the coast of Japan. The winter grounds extend from California southwards in the east and from the Sea of Japan, the East China and Yellow seas, through the Philippine Sea in the west. There appears to be a resident population of fin whales in the Gulf of California.

The North Atlantic fin whales summer from the North American coast to the Arctic, around Greenland, Iceland, north Norway, Jan Meyen, Spitzbergen and the Barents Sea. The wintering areas extend from the ice edge southwards to the Caribbean and the Gulf of Mexico on the west, and from southern Norway, the Bay of Biscay, and Spain in the east. Some fin whales migrate into the Mediterranean, and the species is present in the area throughout the year.

The Southern Hemisphere fin whales are broadly distributed south of 50° S in the summer months, although they do not occur right up to the ice edge as do the blue and minke whales. They migrate into the Atlantic, Indian, and Pacific oceans in winter, both along the coasts of South America as far north as Peru and Brazil, Africa north of South Africa, and the islands north of Australia and New Zealand, as well as in the central ocean areas far from shore.

In the Southern Hemisphere the fin whales tend to enter and leave the Antarctic after the blue whales but before the sei whales. The bigger and older animals generally penetrate farther south than the younger whales. There is also a segregation by sexual class as well as age. Males tend to precede the females, and the pregnant females migrate in advance of the other sexual classes, with the immature whales at the rear of the stream (Laws, 1961). This pattern and succession are not so apparent in the Northern Hemisphere, perhaps due to the constriction by the land masses in the north of the North Pacific and North Atlantic oceans which causes changes in the pattern of environmental conditions as compared with the more open Southern Hemisphere waters.

Stocks

Fin whales are thought to be divided into a number of stocks in each hemisphere, which approximate to separate breeding groups. However, a degree

TABLE 2   A suggested division of the Southern Hemisphere
fin whale populations[a]

Stock	Wintering area	Summer distribution
Chile–Peruvian	west of N. Chile and Peru	100/110° W–60° W
South Georgian	east of Brazil	60° W–25° W
West African	west coast of Africa	25° W–0°
East African	east coast of Africa and Madagascar	0°–40° E
Crozet–Kerguelen	east of Madagascar	40° E–80° E
West Australian	northwest of W. Australia	80° E–110/120° E
East Australian	Coral Sea	140° E–170° E
New Zealand	Fiji Sea and adjacent waters	170° E–145° W

[a] After Ivashin (1969).

of interchange is evident from the recovery of marked whales which indicates that the stocks recognised are not totally independent. The following information is summarised from Gambell (1976).

In the Southern Hemisphere, and based on whale mark recoveries, iodine values of whale oil, the length composition of the catches, and serological studies, it is thought that separate breeding stocks exist broadly on each side of the three oceans with the addition of a central stock in the Indian and Pacific oceans. These proposed stocks overlap and intermingle to a limited extent on the Antarctic feeding grounds. The approximate limits of these stocks are summarised in Table 2 and management areas are shown in Fig. 4A.

There appear to be a number of small independent stocks of fin whales in the North Atlantic (Mitchell, 1973; Sergeant, 1976). The evidence of catch histories and length frequencies has led to the adoption of the stocks shown in Fig. 4b for management purposes, although this subdivision has been questioned and the suggestion made that there is really only a single stock which has become focussed in certain areas as the numbers were reduced by whaling. In addition, Kellogg (1929) suggested that the stock(s) are stratified, the summer feeding grounds of some fin whales being occupied during the winter by whales which had spent the summer farther north.

In the North Pacific there is evidence from whale marking, blood typing, and morphological analyses for three stocks—on the eastern and western sides which intermingle and overlap to varying extents in the Aleutian area and arbitrarily divided at 130°, and in the east China Sea. The latter whales have longer head and shorter tail regions than the others (Ichihara, 1957) and may reach sexual maturity at a smaller size (Fujino, 1960).

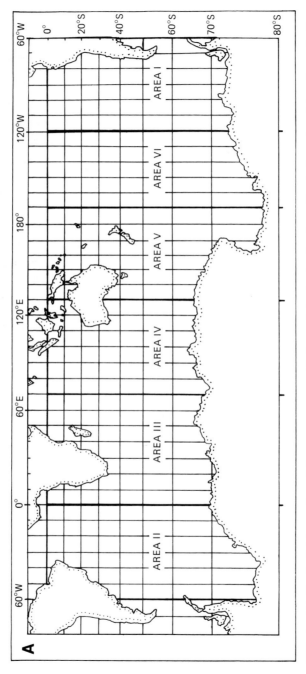

FIG. 4 (A) Stock divisions adopted for management of baleen whales in the Southern Hemisphere, based largely on feeding concentrations of blue whales in the Antarctic; (B) North Atlantic stock divisions for management of fin whales.

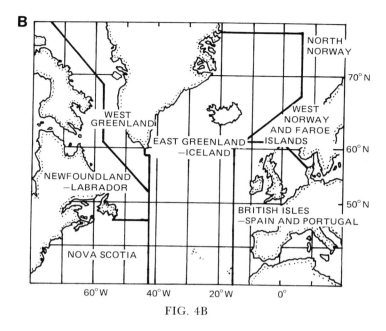

FIG. 4B

Abundance and Life History

Whaling

The fin whale has been a prime target for commercial whaling since the Norwegian development in 1864 of the explosive grenade harpoon fired from a steam-powered vessel. This combination brought the previously unobtainable fast swimming fin whales and other large rorquals within the reach of the whalers, and the history of the fisheries has been well documented by Tonneson and Johnson (1982). The stocks in the eastern North Atlantic were heavily fished, there were episodic fisheries off Newfoundland and Labrador on the west, and similar coastal fisheries on both the Asian and North American sides of the North Pacific. These northern stocks were relatively small and quickly depleted, but the Antarctic held a very large population. From the start of land-based whaling in South Georgia in 1904, and especially after the development of floating factory ships in the 1920s, the fin whale became increasingly important in the catches, especially when the blue whale stocks started to decline in the 1930s. Catches were also taken at the winter end of the migration routes by land stations in South Africa, Chile, and Peru, and the fin whale formed the mainstay of the Antarctic fishery through the 1950s. But these stocks too began to decline, and while pelagic whaling gave a short extension to the industry in the North Pacific, all fin whales are now protected from commercial catching except those off Iceland and Spain.

Population estimates

Because of the need to provide quantitative advice as the basis for management decisions, the Scientific Committee of the International Whaling Commission placed considerable emphasis on estimating the population sizes of the fin whale stocks when they were being harvested. However, as more and more stocks were put under protection or were not fished for other reasons, the analyses were not updated except in the case of the last two fisheries, off Iceland and Spain. As a result, estimates available for fin whale stocks worldwide are of variable reliability, depending on the data available and the analytical techniques used. The original, preexploitation abundance estimate of fin whales in the Southern Hemisphere at the beginning of this century was 490 000, and there are now 103 000 remaining. In the North Pacific the corresponding figures for the original and present stock abundance are 53 000 and 20 000. These estimates are for the total populations of all sizes and ages of whales (Allen, 1980). The North Atlantic populations number a few thousands.

Life history

The life cycle of the fin whale is very closely related to the pattern of seasonal migrations outlined above. The whales mate in the warm waters of lower latitudes in both hemispheres during the winter months and then migrate towards the respective polar feeding grounds where they spend the summer. After 3 to 4 months of intensive feeding they migrate back to the temperate waters once again so that the females give birth in the same waters as those in which they conceived. The newborn calves accompany their mothers on the poleward migration in the spring, living on the rich milk until they are weaned on the feeding grounds some 6 months after birth and can follow the cycle of migrations independently.

The sex ratio at birth and throughout the greater part of life is approximately 1:1, but because of the differential segregation of the sexes and sexual classes at various times in the seasonal migrations, the numbers of males and females present in an area at a particular time may show an imbalance. This is reflected in the catches taken by land stations along the migration routes and by whaling operations at the extremities of the migration range.

Age composition data from catches indicate that natural mortality is 3.5 and 4.5% in southern male and female fin whales, respectively (Allen, 1972). The value is probably higher in the immature animals; although it is difficult to measure directly, calculation suggests a figure of about 12%. Fin whales can live for up to 90 to 100 years. This is based on ages determined by counting the annual growth layers, each made up of one light and one dark lamina, deposited in the horny plug at the proximal end of the ear canal (Roe, 1967).

Food

Fin whales feed mainly on planktonic crustacea but also consume some fish and cephalopods. There is considerable variation by area and season.

In the Southern Hemisphere the main food item in the Antarctic is the krill *Euphausia superba*. Other euphausiids may also occur, particularly in lower latitudes. In both the North Atlantic and North Pacific oceans fish are commonly taken: herring, cod, mackerel, pollock, sardine, and capelin, together with squid and euphausiids and copepods. Just what is consumed is probably determined by availability as much as preference in these northern seas.

Internal Characteristics

The following characteristics of the cranial and skeletal anatomy have been condensed from Tomilin (1957).

Skull

Ventral, dorsal, and lateral views of the skull are shown in Figs. 5–7. The rostrum tapers uniformly to the anterior, without any lateral curves. The basal width of the rostrum is about 30% greater than the median width, while in the blue whale these measures are almost equal. The posterior end of the maxillopalatines protrudes to the rear as a pointed process (Fig. 5). Compared with other balaenopterids, the fin whale skull is characterised by the relatively smaller size of the anteriorly emarginated nasal bones, relatively long frontonasal processes of the maxillaries, and a distinctly expressed posterior expansion of the vomer. Broadening anteriorly, the nasal bones terminate at the anterior end of the suture, forming a small angular prominence. They are usually 2.5 times shorter than the frontonasal processes of the maxillaries. The front of the nasal bones is situated far behind a line connecting the base of the frontonasal processes. The orbital processes of the

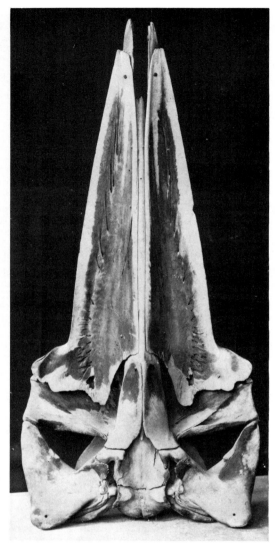

FIG. 5 Ventral view of fin whale skull, ear bones missing. (From True, 1904, courtesy of J. G. Mead, Smithsonian Institution, Washington, D.C.)

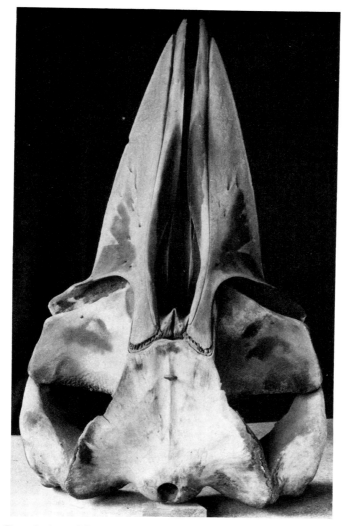

FIG. 6 Dorsal view of fin whale skull. (From True, 1904, courtesy of J. G. Mead, Smithsonian Institution, Washington, D.C.)

frontals are trapeziform viewed from above, with their posterior margin making a right angle with the long axis of the skull (Fig. 6).

The inclined supraoccipital bears a longitudinal median crest. The occipital bone narrows near the top of the skull, with almost parallel margins, and expands laterally in the posteroventral direction, forming the posterior part of the cranium.

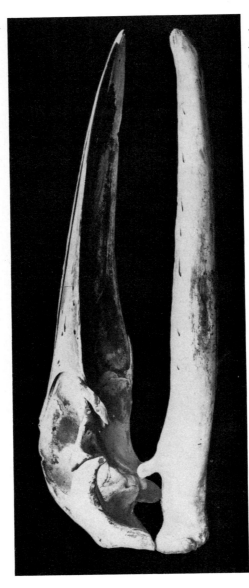

FIG. 7 Lateral view of fin whale skull. (From True, 1904, courtesy of J. G. Mead, Smithsonian Institution, Washington, D.C.)

TABLE 3 Organ weights (tonnes) of the largest and smallest male and female fin whales weighed in the Antarctic[a]

Organ	Female (22.7 m)	Male (20.9 m)	Female (18.5 m)	Male (19.1 m)
Ovaries	2.7	—	0.6	—
Testes	—	26.9	—	10.2
Blubber	13.78	10.78	8.27	9.57
Meat	25.22	21.96	18.39	19.50
Internal organs	6.21	4.54	3.65	3.96
Heart	0.13	0.22	0.13	0.13
Lungs and bronchus	0.54	0.29	0.25	0.29
Stomach	0.20	0.24	0.20	0.12
Intestine	1.04	0.79	0.48	0.49
Kidney	0.23	0.17	0.14	0.22
Liver	0.56	0.48	0.34	0.35
Bone	11.42	7.88	5.91	6.27
Skull	2.62	1.91	1.52	0.98
Vertebrae	4.76	3.82	2.80	3.47
Ribs	1.89	0.97	0.61	0.67
Jaw	1.25	0.76	0.61	0.72
Total weight	57.60	47.58	37.37	40.21

[a]From Nishiwaki (1950).

Skeleton

The number of vertebrae varies, but is usually 60–63, made up of 7 cervical, 15–16 thoracic, 13–16 lumbar, and 24–27 caudal. There are 14–16 ribs, most often 15, with only the first pair joined to the sternum; those in the middle of the series are longest (Ivashin et al., 1972).

The sternum shape varies from rhomboid to cruciate, with wide individual and age-determined variations.

The flipper skeleton has no third digit in the balaenopterids, although rudiments can be found in embryos. The carpal formula is variable: I—3/4, II—5/7, IV—5/7, V—3/4.

Organ weights

The weights of the chief internal organs in the largest and smallest fin whales of each sex recorded in the literature are shown in Table 3.

For a general review of cetacean gross anatomy, see Green (1972). For a review of cetacean microanatomy with some comments on the fin whale, see Simpson and Gardner (1972). These authors show a section of fin whale lung

TABLE 4 Works on the anatomy or physiology of the fin whale

Bone	Ogden et al. (1981), True (1904), Tont et al. (1977), Tomilin (1957)
Baleen, lips, and mouth	Aguilar et al. (1981), Nemoto (1962a,b), Ogawa and Shida (1950), Pivorunas (1977)
Brain and central nervous system	Jacobs and Jensen (1964), Jansen (1952, 1953), Jansen and Jansen (1953), Langworthy (1935), Morgane et al. (1980), Ries and Langworthy (1937), Straus (1935)
Cardiovascular system	Ommanney (1932a), Walmsley (1938)
Endocrines	Slijper (1962), Sverdrup (1952)
Kidney and urogenital system	Ommanney (1932b), Murie (1865), Slijper (1962)
Lungs and respiration	Engle (1966), Laurie (1933), Murie (1865), Simpson and Gardner (1972), Scholander (1940)
Reproductive system	Laws (1958), Lennep (1950), Slijper (1966), Utrecht (1968)
Skin	Giacometti (1967)

and discuss various aspects of the organ histology, comparing large and small whales of different species.

Kulu (1972) displays the chromosome karyotpye of a fin whale, showing 44 chromosomes, in agreement with Arnason (1969). Giacometti (1967) reported that the epidermis of the fin whale measured 3.0 mm in thickness. Lauer and Baker (1969) reported an analysis of fin whale milk as follows: water 57.9%, total solids 42.1%, fat 28.6%, protein 11.49%, lactose 2.58%, and ash 1.59%.

Kanwisher and Senft (1960) recorded temperature, respiration, and electrocardiographic measures from a stranded 13.8 m fin whale. The pulse rate was about 27 beats per min, the whale blew about every 20 sec, and the rectal temperature at a depth of about 46 cm was 33 °C. The authors observed that these were probably abnormal values since the whale died in 36 hr; however, they are the only measures of pulse and body temperature available for a living fin whale.

Additional works on the anatomy or physiology of the fin whale are given in Table 4.

Behaviour

Although fin whales are sometimes found singly or in pairs, they commonly form larger groupings of 3 to 10 or 20 which may in turn coalesce into a broadly spread concentration of a hundred or more individuals, especially on the feeding grounds.

The blow is tall, 4–6 m high. The whale appears to wheel over as it starts to dive, so that after the blowholes have been submerged the back rolls into view to expose the dorsal fin before the animal slips below the surface without showing the tail flukes. A series of shallow dives lasting 10–20 sec may be followed by a longer dive for 15 min or more, down to a depth of 230 m. Breaching does occur, with the animal falling back into the water with a splash.

Fin whales have the reputation of being one of the fastest of the great whales, speeds of over 20 knots being achieved in short bursts. In terms of sustained swimming, it is believed that southern fin whales migrating from the breeding grounds to the Antarctic in spring cover some 2400 miles in about a month (Laws, 1961), or roughly 90 miles a day.

Ray *et al.* (1978) tracked a whale tagged with a radio transmitter at an average speed of just over 9 km/hr. Watkins (1981) followed a radiotagged fin whale between Iceland and Greenland. During a 10-day period, the whale traveled 2095 km and on one of these days traveled 292 km.

Feeding is generally by an engulfing technique: the whale takes in a mouthful of water containing the food, the mouth is closed and the water squeezed out between the baleen plates, leaving the food organisms behind to be swallowed. Fin whales have been observed feeding at the surface by swimming on their right sides and making a lateral scoop with the open mouth and distended throat region (Fig. 1). Such behaviour is in accord with the asymmetrical colour pattern of the head, by placing the head's white right side downwards, to conform with the pale underside of the rest of the body. Reports of aerial observations of feeding behaviour were made by Watkins and Schevill (1979).

Fin whales are known to produce higher frequency sounds (under about 100 Hz, often pulses sweeping downward from 75 to 40 Hz), 20-Hz pulses (both single pulses and patterned sequences), ragged low-frequency pulses, and low-frequency rumbles, as well as nonvocal sharp impulse sounds. The higher frequency sounds appear to be used for communication between nearby fin whales, the 20-Hz single pulses for both local and longer distance communication, and the patterned seasonal 20-Hz pulses are associated with courtship displays, at least in the western North Atlantic (Watkins, 1981).

Reproduction

Although conceptions and births can occur at almost any time of year in the fin whale, most activity is confined to relatively short peak periods in the northern and southern populations. Analysis of foetal length:frequency distributions in the Southern Hemisphere indicates that 77% of fin whales are conceived in the 4 months between April and August (Laws, 1961). The corresponding peak months for breeding in the Northern Hemisphere centre around December and January. The female carries a single foetus for $11\frac{1}{4}$ months and at birth the southern calf averages 6.4 m in length and weighs 1.9 tonnes (Laws, 1959). Lactation extends over 6 or 7 months before the calf is weaned in the higher latitudinal feeding grounds when at a length of about 12 m. The mother then passes through a resting period of 5 or 6 months before mating again in the winter. This resting phase may last another year if the female fails to conceive in the first breeding season, in spite of the possibility of more than one ovulation occurring.

It is characteristic of fin whales, as of many other mammals, that they reach sexual maturity at a given body size, which is most conveniently measured in terms of body length. The southern fin whale becomes sexually mature at a length of 19.9 m in females and 19.2 m in males; the corresponding sizes in the Northern Hemisphere are 18.3 and 17.7 m, respectively.

Because there has been an increase in the growth rates of the fin whale and other rorqual species in the Southern Hemisphere since the major reduction of the populations of blue and fin whales, the fin whales are now reaching the critical sizes at which they become sexually mature at younger ages than in earlier years. For the year classes born up to 1930 the mean age at maturity was a little over 10 years. From the mid-1930s onwards this mean age fell and reached about 6 years for the 1950 year class. These data come from the study of the transition layer formed at sexual maturity in the earplug of the whales (Lockyer, 1972). Direct study of the reproductive status of fin whales sampled in the mid-1960s also indicates an age at maturity of 6 to 7 years for both sexes (Ohsumi, 1972).

The pregnancy rates of fin whales in the Southern Hemisphere also have apparently changed in response to exploitation over the past 50 years, and now some 50–55% of mature nonlactating females are pregnant in the sampled catches (Laws, 1961). This reflects an approximate halving of the average interval between successive births during this period.

Parasites and Disease

The external parasites of fin whales are mostly crustaceans, while the commonest internal parasites are tapeworms and Acanthocephala (Mackintosh and Wheeler, 1929). The copepod *Penella* is particularly common in the Southern Hemisphere, although it tends to be thrown off in the cold waters of the Antarctic. Cirripeds including *Coronulla regina, Conchoderma auritum, C. virgatum,* and *Xenobalanus globicipitis,* and amphipods of the genus *Cyamus* also infect fin whales in the warmer waters but are lost in the cold southern waters. Diatoms such as *Cocconeis*, however, are contracted as surface films in the summer months in the Antarctic.

Granuloma malignum (Hodgkin's disease) has been described in a fin whale (Simpson and Gardner, 1972). On several occasions swordfish swords have been recovered from fin whale carcasses (Jonsgård, 1962; Peers and Karlsson, 1976).

Stranding

There are numerous reports of fin whale strandings. Sergeant (1976) relates the number of strandings to two situations: (1) human causes (whaling, pollution) and (2) mortality due to natural population increase. He suggests that strandings may be a good index of population size (Sergeant, 1976).

References

Aguilar, A., Jover, L., and Grau, E. (1981). Some anomalous dispositions of the Jacobson's organ in the fin whale. *Sci. Rep. Whales Res. Inst.* **33,** 125–126.
Allen, K. R. (1972). Further notes on the assessment of Antarctic fin whale stocks. *Rep. int. Whal. Commn* **22,** 43–53.
Allen, K. R. (1980). "Conservation and Management of Whales". Univ. of Washington Press, Seattle.
Arnason, U. (1969). The karyotype of the fin whale. *Sep. Hered.* **62,** 273–284.
Burmeister, H. (1865). Description of a new species of porpoise. *Proc. Zool. Soc. London*, pp. 190–191.
Engle, S. (1966). The respiratory tissue of the blue whale and fin whale. *Acta. Anat.* **65,** 381–390.
Fischer, (1829). "Synopsis Mammalium". J. G. Cottae, Stuttgart.
Fujino, K. (1960). Immunogenetic and marking approaches to identifying sub-populations of the North Pacific whales. *Sci. Rep. Whales Res. Inst.* **15,** 85–142.
Gambell, R. (1976). Population biology and the management of whales. *Appl. Biol.* **1,** 247–343.
Giacometti, L. (1967). The skin of the whale (*Balaenoptera physalus*). *Anat. Rec.* **159,** 69–76.
Gray, J. E. (1846). "The Zoology of the Voyage of HMS *Erebus* and *Terror*". Janson, London
Green, R. F. (1972). Observations on the anatomy of cetaceans and pinnipeds. *In* "Mammals of the Sea: Biology and Medicine" (Ed. S. H. Ridgway), pp. 247–297. Thomas, Springfield, Illinois.
Ichihara, T. (1957). An application of linear discriminant function to external measurments of fin whale. *Sci. Rep. Whales Res. Inst.* **12,** 127–189.
Ivashin, M. V. (1969). Olokal'nosti nekotorykh promyslovykh vidov kitov v iuzhnom polusharii. *Rybn. Khoz. (Moscow)* **45**(10), 11–13.

Ivashin, M. V., Popov, L. A., and Tsapko, A. S. (1972). "Morskie Mlekopptayuschie". Pischevaya Promychiennost, Moscow.

Jacobs, M. S., and Jensen, A. V. (1964). Gross aspects of the brain and a fiber analysis of cranial nerves in the great whale. *J. Comp. Neurol.* **123,** 55–72.

Jansen, J. (1952). On the whale brain with special reference to the weight of the brain of the fin whale (*Balaenoptera physalus*) *Nor. Hvalfangst-Tid.* **9,** 480–486.

Jansen, J. (1953). Studies on the cetacean brain: the gross anatomy of the rhombencephalon of the fin whale (*Balaenoptera physalus* L.). *Hvalradets Skr.* **37,** 1–35.

Jansen, J., Jr., and Jansen, J. (1953). A note on the amygdaloid complex in the fin whale (*Balaenoptera physalus* L.). *Hvalradets Skr.* **39,** 1–14.

Jonsgård, Å. (1962). Three finds of sword from swordfish (*Xiphais gladius*) in Antarctic fin whales (*Balaenoptera physalus* L.). *Nor. Hvalfangst-Tid.* **51,** 287–291.

Kanwisher, J., and Senft, A. (1960). Physiological measurements on a live whale. *Science* **131,** 1379–1380.

Kellogg, R. (1929). What is known of the migrations of some of the whale bone whales. *Rep. Smithson. Inst. 1928,* 467–494.

Klinowska, M. (1980). "A World Review of the Cetacea". Nature Conservancy Council, London.

Kulu, D. D. (1972). Evolution and cytogenetics. *In* "Mammals of the Sea: Biology and Medicine" (Ed. S. H. Ridgway), pp. 503–527. Thomas, Springfield Illinois.

Lacépède, B. G. (1804). "Histoire Naturelle des Cétacés". Paris.

Langworthy, O. R. (1935). The brain of the whale bone whale (*Balaenoptera physalus* L.). *Bull. Johns Hopkins Hosp.* **57,** 143–147.

Lauer, B. H., and Baker, B. E. (1968). Whale milk I. Fin whale (*Balaenoptera physalus*) and beluga whale (*Delphinapterus leucas*) milk: Gross composition and fatty acid constitution. *Can. J. Zool.* **47,** 95.

Laurie, A. H. (1933). Some aspects of respiration in blue and fin whales. *Disc. Rep.* **7,** 363.

Laws, R. M. (1958). Recent investigations on fin whale ovaries. *Nor. Hvalfangst-Tid.* **47,** 225–254.

Laws, R. M. (1959). The foetal growth rates of whales with special reference to the fin whale, *Balaenoptera physalus* Linn. *Disc. Rep.* **29,** 281–308.

Laws, R. M. (1961). Reproduction, growth and age of southern fin whales. *Disc. Rep.* **31,** 327–486.

van Lennep, E. W. (1950). Histology of the corpora lutea in the blue and fin whale ovaries. *Proc. K. Ned. Akad. Wet.* **53,** 593–599.

Linnaeus, C. (1758). "Systema Naturae", 10th Ed., Vol. 1. Photographic facsimile reprinted, British Museum (Natural History), London.

Lockyer, C. (1972). The age at sexual maturity of the southern fin whale (*Balaenoptera physalus*) using the annual layer counts in the ear plug. *J. Cons. Cons. Int. Explor. Mer* **34,** 276–294.

Lockyer, C. (1976). Body weights of some species of large whales. *J. Cons. Cons. Int. Explor. Mer* **36,** 259–273.

Mackintosh, N. A. (1965). "The Stocks of Whales". Fishing News, London.

Mackintosh, N. A., and Wheeler, J. F. G. (1929). Southern blue and fin whales. *Disc. Rep.* **1,** 257–540.

Mitchell, E. D. (1975). The status of the world's whales. *Nat. Can. (Que.)* **2**(4), 9–25.

Morgane, P. J., Jacobs, M. S., and McFarland, W. L. (1980). The anatomy of the brain of the bottlenosed dolphin (*Tursiops truncatus*). Surface configurations of the telencephalon of the bottlenosed dolphin with comparative anatomical observations on four other species. *Brain Res. Bull.* **5,** Suppl. 3, 1–102.

Murie, J. (1865). On the anatomy of a fin whale (*Physalus antiquorum* Gray) captured near Gravesend. *Proc. Zool. Soc. London*

Nishiwaki, M. (1950). On the body weight of whales. *Sci. Rep. Whales Res. Inst.* **4,** 184–209.

Nemoto, T. (1962a). A secondary sexual character of fin whales. *Sci. Rep. Whales Res. Inst.* **16,** 29–34.

Nemoto, T. (1962b). Food of baleen whales collected in recent Japanese Antarctic whaling expeditions. *Sci. Rep. Whales Res. Inst.* **16,** 89–103.

Ogawa, T., and Shida, T. (1950). On the sensory turbercles of lips and oral cavity in the sei and fin whales. *Sci. Rep. Whales Res. Inst.* **3,** 1–16.

Ogden, J. A., Conloge, G. J., Light, T. R., and Slaon, T. R. (1981). Fractures of the radius and ulna in a skeletally immature fin whale *Balaenoptera physalus*. *J. Wildl. Dis.* **17,** 111–116.

Ohsumi, S. (1972). Examination of the recruitment rate of the Antarctic fin whale stock by use of mathematical models. *Rep. int. Whal. Commn* **22,** 69–90.

Ommanney, F. D. (1932a). The vascular networks (retia mirabilia) of the fin whale (*Balaenoptera physalus*). *Disc. Rep.* **5,** 327–363.

Ommanney, F. D. (1932b). The urogenital system of the fin whale (*Balaenoptera physalus*) with appendix of the dimensions and growth of the kidney of the blue and fin whales. *Disc. Rep.* **5,** 363–466.

Peers, B., and Karlsson, B. (1976). Recovery of a swordfish (*Xiphias gladius*) sword from a fin whale (*Balaenoptera physalus*) killed off the west coast of Iceland. *Can. Field-Nat.* **90,** 492–493.

Pivorunas, A. (1977). The fibrocartilage skeleton and related structures of the ventral pouch of balaenopterid whales. *J. Morphol.* **151,** 299–314.

Racovitza, E. G. (1903). "Resultats du Voyage *S. Y. Belgica* en 1897–1898–1899." Zoologie. Cétacés. (English translation by F. W. True "A Summary of General Observations on the Spouting and Movement of Whales". *Smithson. Inst. Annu. Rep.*, 1903).

Ray, G. C., Mitchell, E. D., Wartzok, D., Kozicki, V. M., and Maiefski, R. (1978). Radio tracking of a fin whale (*Balaenoptera physalus*). *Science* **202,** 521–524.

Ries, F. A., and Langworthy, O. R. (1937). A study of the surface structure of the brain of the whale (*Balaenoptera physalus* and *Physeter catodon*). *J. Comp. Neurol.* **68,** 1–47.

Roe, H. S. J. (1967). Seasonal formation of laminae in the ear plug of the fin whale. *Disc. Rep.* **35,** 1–30.

Scholander, P. F. (1940). Experimental investigations on the respiratory function in diving mammals and birds. *Hvalradets Skr.* **22,** 1–90.

Sergeant, D. E. (1976). Stocks of fin whales *Balaenoptera physalus* L. in the North Atlantic Ocean. *Rep. int. Whal. Commn* **27,** 460–473.

Simpson, J. G., and Gardner, M. (1972). Comparative microscopic anatomy of selected marine mammals. *In* "Mammals of the Sea: Biology and Medicine" (Ed. S. H. Ridgway), pp. 298–418. Thomas, Springfield, Illinois.

Slijper, E. J. (1962). "Whales". Basic Books, New York.

Slijper, E. J. (1966). Functional morphology of the reproductive system in Cetacea. *In* "Whales, Dolphins and Porpoises" (Ed. K. S. Norris), pp. 277–319. Univ. of California Press, Berkeley.

Straus, W. L., Jr. (1935). Note on the spinal cord of the finback whale (*Balaenoptera physalus*). *Bull. Johns Hopkins Hosp.* **57,** 317–229.

Sverdrup, A., and Arnessen, K. (1952). Investigations on the anterior lobe of the hypophysis of the finback whale. *Hvalradets Skr.* **36,** 1–15.

Tomilin, A. G. (1957). "Mammals of the U.S.S.R. and Adjacent Countries. Vol. IX Cetacea" (Ed. V. G. Heptner), Nauk S.S.S.R., Moscow. English Translation, 1967, Israel Program for Scientific Translations, Jerusalem.

Tonnessen, J. N., and Johnson, A. O. (1982). "The History of Modern Whaling". Hurst, London.

Tont, S. A., Pearcy, W. G., and Arnold, J. S. (1977). Bone structure of some marine vertebrates. *Mar. Biol. (Berlin)* **39,** 191–196.

True, F. W. (1904). The whalebone whales of the western North Atlantic. *Smithson. Contrib. Knowl.* pp. 1–317.

van Utrecht, W. L. (1968). Notes on some aspects of the mammary glands in the fin whale, *Balaenoptera physalus* L. with regard to the criterion "lactating". *Nor. Hvalfangst-Tid.* **1,** 1–13.

Walmsley, R. (1938). Some observations on the vascular system of a female fetal finback. *Contrib. Embryol. Carnegie Inst.* **164,** 107–178.

Watkins, W. A. (1981). Activities and underwater sounds of fin whales. *Sci. Rep. Whales Res. Inst.* **33,** 83–117.

Watkins, W. A., and Schevill, W. E. (1979). Aerial observation of feeding behavior in four baleen whales: *Eubalaena glacialis, Balaenoptera borealis, Megaptera novaeangliae* and *Balaenoptera physalus. J. Mammal.* **60,** 155–163.

8

Blue Whale

Balaenoptera musculus (Linnaeus, 1758)

Pamela K. Yochem and Stephen Leatherwood

Genus and Species

Blue whales, *Balaenoptera musculus,* are currently divided into three subspecies (Hershkovitz, 1966; Rice and Scheffer, 1968): *B. m. intermedia* (Burmeister, 1871), found in the Southern Hemisphere; the smaller *B. m. musculus* (Linnaeus, 1758), found in the North Atlantic and North Pacific; and the still smaller pygmy blue whale, *B. m. brevicauda* Ichihara 1966 (sometimes attributed to Zemsky and Boronin, 1964, but see Rice and Scheffer, 1968 and Rice, 1977), at present known mainly from the subantarctic waters of the Indian Ocean and southeast Atlantic. Blue whales in other parts of the world (e.g., northern Indian Ocean, Chile, Peru), though recently assigned to *B. m. brevicauda* on the basis of observations of free-ranging animals, are best regarded as subspecies unknown (Rice, 1977).

TABLE 1 Some common names for blue whales[a]

English	Blue whale, Sibbald's rorqual, sulphur-bottom, blue rorqual, great blue whale, great northern rorqual
Norwegian	Rorqual, heipe-reydur, blaahval, blahval, finghval, myrbjonner (pygmy)
Japanese	Shiro nagasu kujira
Russian	Sinii kit, goluboy kit, bolshoi polosatik, blyuval
German	Schanelwal, Riesenwal, Breitmaulige, Finn Fische, Blauwal
Swedish	Muskel finnfisk, jattenhval, blaval
Icelandic	Steypireydr, hrefna, blahvalur
French	Rorqual de sibbald, rorqual bleu, rorqual à ventre cannele, rorqual, baleine jubarte, baleine d'ostende, baleine bleue
Dutch	Blauwe vinvis
Eskimo (tribe unreported)	Tunnolik, akhvokhrikh
Alaska Eskimo Yupik	Takerrkak
Aleutian	Umgulik, omgolia, abugulikh
Indian, Bangladeshi	Great Indian fin whale
Spanish	Ballena azul
Makah	Kwakwe axtLi

[a]From Slijper (1962), Nature Conservancy Council (1980), and Leatherwood et al. (1982b).

Brinkman (1967) recounts developments in discrimination among the various species of "fin" whales, as all rorquals were known in some areas prior to the late nineteenth or early twentieth centuries. As knowledge of differences among blue whales grows so also should ability to distinguish among the subspecies.

Although validity of the putative pygmy blue whale subspecies has been questioned by some authors (e.g., Ehrenfeld, 1970), it now appears generally accepted (Rice, 1977). Measurements supporting the naming of the new subspecies, not appended to the original descriptions (Ichihara, 1961, 1963, 1966), were published in March 1984 (Omura, 1984). After noting in 1977 that it was "not in a position to reassess the status of a possible sub-species for the purpose of calculating population sizes" and including the pygmy blue whale with the blue whale in its stock assessments (IWC, 1977; p. 45), the International Whaling Commission (IWC) subsequently accepted the separation while calling for evidence to clarify the taxonomic status (IWC,

1979) and the identity of the various populations or stocks (IWC, 1980). In this chapter we follow recent practices in referring to blue whales (*B. m. musculus* and *B. m. intermedia*) and pygmy blue whales (*B. m. brevicauda*).

Although the ranges of *B. m. musculus* and *B. m. intermedia* may overlap, interbreeding is unlikely since migration cycles of the two subspecies are roughly 6 months out of phase (i.e., *B. m. musculus* is migrating poleward to feed as *B. m. intermedia* is moving towards tropical or subtropical calving grounds) (Mackintosh, 1965). Although the degree of potential interbreeding is unknown, the range of *B. m. brevicauda* apparently overlaps that of *B. m. intermedia* in the Indian Ocean, or elsewhere during migration (R. G. Chittleborough, cited in Ichihara, 1966), and in the eastern tropical Pacific (Aguayo, 1974; Berzin, 1978; Wade and Friedrichsen, 1979; Donovan, 1984).

Table 1 gives some other common names for blue whales.

External Characteristics and Morphology

The blue whale (Fig. 1) is the largest of the mysticetes and is the largest animal alive today. The longest specimen measured in the scientifically correct manner (in a straight line from the point of the upper jaw to the notch in the tail) was a 33.58-m female from the Antarctic measured at South Georgia

FIG. 1 Blue whale drawn with human divers to show size relationship. (Drawing by Larry Foster, courtesy of General Whale.)

sometime between 1904 and 1920 (Risting, 1922). The heaviest weight reported was that of a 190-tonne female, 27.6 m long, taken 20 March 1947 at South Georgia (Tomilin, 1957). A 29.5-m female taken at South Georgia in 1931 was computed to have weighed 163.7 tons and believed to have actually weighed 174 tons (Laurie, 1933). Concerning this last specimen, Scheffer (1974, p. 274) wrote "After transposing the figures, correcting an error in the published arithmetic and adding 6.5 percent for loss of body fluids (mostly blood) I arrive at a . . . whole weight of 196 short tons." Lockyer (1976) reviewed published data on body weights for blue whales and presented a formula for weight/length relationships adjusted for blood and fluid lost during piecemeal weighings. Females are larger than males of the same age, and animals in the Southern Hemisphere are larger than those in the Northern Hemisphere. Differences between blue whales and pygmy blue whales have been summarised by Ichihara (1966, 1981) (Table 2). According to those accounts blue whales (Fig. 2) differ from pygmy blue whales (Fig. 3) in colour, body proportion, skull and postcranial skeletal structure, baleen plate shape, length at sexual and physical maturity, and distribution.

Blue whales are fundamentally bluish grey in colour with white undersides to the flippers. Considerable individual variation in coloration has been described, from uniform dark slate blue with little whitish mottling to very light blue with considerable mottling (Mackintosh and Wheeler, 1929; Tomilin, 1957). Pygmy blues are reportedly generally lighter than blue whales (Ichihara, 1966; Zemsky and Boronin, 1964), although 14 presumed pygmy blue whales recently sighted off the coast of Peru were apparently "considerably more blue" than blue whales seen in the North Atlantic (Donovan, 1984), and animals tentatively identified as pygmy blue whales off Sri Lanka were dark blue with only moderate spotting (Leatherwood and Clarke, 1983; Leatherwood *et al.*, 1984). The yellowish or mustard-coloured hue (which prompted the common name sulphur-bottom) sometimes seen on blue whales in colder waters is due to the presence of diatoms (*Cocconeis ceticola*) (Hart, 1935; Omura, 1950; Tomilin, 1957).

The head is wide and flat. Its shape has been likened to that of a Gothic arch, somewhat flattened at the tip (Leatherwood *et al.*, 1976, 1982b). The blowholes are surrounded anteriorly and laterally by fleshy crests like, but larger than, those of other rorquals; a single ridge extends from these blowhole crests towards the tip of the rostrum (True, 1904; Tomilin, 1957; Leatherwood *et al.*, 1982b).

The small (to only about 0.4 m high) dorsal fin is located about 25% of the body length forward of the notch in the tail flukes. Its shape may be distinctly triangular, broadly rounded, smoothly falcate, or little more than a pointed nubbin or step in the dorsal profile (True, 1904; Fig. 14 in Leatherwood *et al.*, 1982b).

TABLE 2 Comparison of selected characteristics of blue whales and pygmy blue whales in the Antarctic

Character	Blue whale	Pygmy blue whale	Source
Distribution in Antarctic waters	Widely distributed, apparently summers south of Antarctic Convergence	Restricted to subantarctic zone north of Antarctic Convergence; major population north of 54° S and from 0° to 80° E but marking results and presence along Western Australian coast suggest wider distribution	Ichihara (1963, 1966)
Body colour	Steel blue	Silvery grey	Ichihara (1966)
		More blue than North Atlantic blue whales (pygmies off Peru)	Donovan (1984)
		Somewhat lighter than blues; no oblong scars at the tail as in blues	Zemsky and Boronin (1964)
Baleen plates	Breadth/length[a] = 1.79 ± 0.04	Breadth/length = 1.48 ± 0.03 (significantly shorter relative to breadth)	Ichihara (1963, 1966)
Length of tail region (notch of flukes to anus)		Significantly shorter than for blue[a] of the same length	Ichihara (1963, 1966), Zemsky and Boronin (1964)

(*continued*)

TABLE 2 (*Continued*)

Character	Blue whale	Pygmy blue whale	Source
Length of trunk region		Significantly longer than for blue whales of the same length [a]	Ichihara (1966)
Body weight	Average = 64 348 kg for 22.4 m whale	Average = 68 916 kg for 22 m whale (significantly heavier than for blue of the same length)	Ichihara (1966)
Internal organ weight	11% of total	15.8% of total	Ichihara (1966)
Length at sexual maturity (male)	22.6 m[a]	<19.2 m	Ichihara (1966)
	22.3 m	<21.3 m	Ichihara (1961)
Length at sexual maturity (female)	23.7 m[a]	19.2 m	Ichihara (1966)
	24.1–24.4 m		Laurie (1937)
	23.7 m	20.7 m	Ichihara (1961)
	23.5–23.8 m		Brinkmann (1948)
		19.6 m	Zemsky and Boronin (1964)
Length at physical maturity (male)	24.1 m		Nishiwaki and Hayashi (1950)
		20.4–20.7 m	Ichichara (1966)
Length at physical maturity (female)		21.6–21.9 m	Ichihara (1966)
	26.2–26.5 m		Brinkmann (1948)
	25.9 m		Nishiwaki and Hayashi (1950)
Breeding season(s)		Two: main season austral winter, a second from November to January	Ichihara (1966)
	Austral winter (peak in July)		Mackintosh and Wheeler (1929)

TABLE 2 (*Continued*)

Character	Blue whale	Pygmy blue whale	Source
Skull (rostrum)	Convex margins; width in middle approximately 29% of skull length		Tomilin (1957)
		Outer margin less curved than in blues; width in middle 25% of skull length	Omura et al. (1970)

^aInformation derived by Ichihara from Mackintosh and Wheeler (1929).

[a]Information derived by Ichihara from Mackintosh and Wheeler (1929).

The tail flukes are broad and uniformly colored bluish grey above and below. The trailing edge is straight or slightly concave and has a median notch. The flippers can be up to 15% of the body length, are bluntly pointed at the tip, and are generally blue–grey dorsally and white below (Tomilin, 1957; Leatherwood et al., 1976, 1982b).

The ventral surface contains 55–88 longitudinally parallel grooves or pleats, the longest of which extend from chin to navel and the shortest of which may be found high along the side of the face (True, 1904; Tomilin, 1957; Leatherwood et al., 1976, 1982b). Though ventral grooves in rorquals have been often speculated to relate to the animal's hydrodynamics, there is much recent evidence from blue whales (Storro-Patterson, 1981; Sears, 1983) and other rorquals (e.g., minke whales, *B. acutorostrata*—see Hoyt, 1984, figure on p. 27; and minke, fin, *B. physalus,* and sei whales, *B. borealis*—see Lambertsen, 1983) that they do at least function in feeding by allowing the throat to distend.

The tongue, palate, and baleen are black, and the largest baleen plates generally do not exceed 1 m in length and are broader than they are long. Specimens contain 270–395 plates per side (Mackintosh and Wheeler, 1929; Tomilin, 1967). In a male described by Tomilin (1957) over 400 "shafts" formed a semicircle in front of the palate, connecting the baleen rows on the right and left sides. The fringe bristles of blue whales are black, ellipsoid in cross section, and thickened proximally. These characteristics distinguish the bristles of blue whales from those of other rorquals (Tomilin, 1957). Fringes of old individuals may be grey (Millais, 1906).

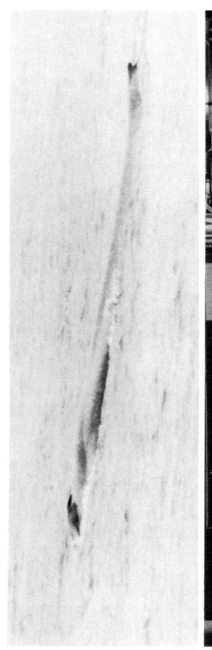

(A)

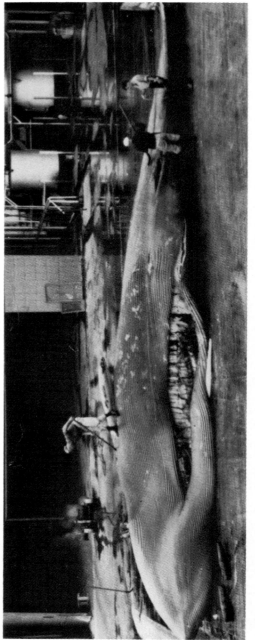

(B)

Distribution and Movements

North Pacific

Most North Pacific blue whales reportedly migrate poleward to feeding grounds in spring and summer after wintering in subtropical and tropical waters off southern California and Baja California in the east, and Taiwan, Japan, and Korea in the west (Tomilin, 1957; Rice, 1978) (Fig. 4). Blue whales have also been sighted in the mid-Pacific from 20° to 35° N, (Rovnin, 1979; Thompson and Friedl, 1982) and 700 nautical miles off Guatemala (Rice, 1978), though the southern and seaward limits of distribution in the North Pacific are not well defined. Some animals spend the winter off both coasts of Baja California and there is evidence of a resident population in the eastern tropical Pacific, between 5° and 10° N and 70° and 90° W (Tomilin, 1957; Berzin, 1978; Rice, 1978; A. A. Berzin, cited in Ivashin, 1984). Veinger (1979) reported that whales sighted north of the Galapagos between 8° and 10° in March and June 1975 were identified as pygmy blue whales on the basis of size, coloration, behaviour, and other characteristics. Blue whales are present in the Gulf of California at least from fall through spring (Rice, 1978; Anonymous, 1979; Leatherwood et al., 1982b; Patten and Soltz, 1980); some may occur there year-round (Leatherwood et al., 1982b).

During the spring and summer months blue whales are found in the Gulf of Alaska, along the Aleutian Islands (especially on the south side), and near the Kurile Islands and the Kamchatka Peninsula (Murie, 1959; Berzin and Rovnin, 1966; Tomilin, 1957). On the western side of the North Pacific and adjacent arctic waters blue whales have been reported from as far north as the Chukotsky Peninsula (northwest Bering and southwest Chukchi seas) (Tomilin, 1957; Sleptsov, 1961; Berzin and Rovnin, 1966), though they do not appear to be regular visitors there (Nikulin, 1946). Blue whales are not commonly seen anywhere in the Bering Sea away from the Aleutian Islands (Omura, 1955; Murie, 1959; Nishiwaki, 1966). Hanna (1920) found a skull with baleen on St. Matthew Island in July 1916. The Japanese vessel *Kinjyo-Maru* sighted a few blue whales in the Bering Sea in 1954 (Omura, 1955), and Eskimos whaling at Gambell on St. Lawrence Island reported that the species was present in the area prior to the 1950s but was absent in the 1950s, 1960s, and early 1970s. A few individuals have been seen there more recently

---

FIG. 2 Blue Whales. (A) An approximately 24-m animal swimming slowly at the surface off San Clemente Island, southern California, August 1971. (B) An animal on the ramp of a Canadian whaling station. [Photos by S. Leatherwood (A) and G. C. Pike, courtesy of I. MacAskie (B).]

FIG. 3 Pygmy blue whales. (A) A dead animal on the deck of a Japanese factory ship in the Antarctic during the 1961–1962 season. (B) and (C) A free-ranging animal off Trincomalee, Sri Lanka, in February 1983, tentatively identified by M.

in late summer (Leatherwood *et al.*, 1982b). Despite extensive survey effort in the Chukchi and Eastern Bering seas in recent years, no blue whales have been sighted there (Leatherwood *et al.*, 1983).

The northward migration apparently begins in April and May. On the east side of the North Pacific blue whales reportedly migrate north along the west coast of North America, then apparently split into two groups, one moving to the Queen Charlotte Islands and the northern Gulf of Alaska and the other heading west towards the Aleutian Islands (Berzin and Rovnin, 1966). On the west side of the North Pacific blue whales reportedly migrate to Kamchatka or the waters off the Kurile Islands, then travel along the

Nishiwaki from photos as a pygmy blue whale. [Photos by T. Ichihara, courtesy of H. Omura (A) and S. Leatherwood (B) and (C).]

islands to the northeast. Whales moving northward in spring apparently are found farther offshore than whales migrating south in the fall, perhaps in response to the seasonal differences in the distribution of their euphausiid prey (Omura, 1950; Nemoto, 1959, 1970).

Tomilin (1967) considered it unlikely that populations on either side of the North Pacific are entirely separate. Tag returns provide the only evidence of exchange between such populations on the feeding grounds. Ivashin and Rovnin (1967) report a blue whale tagged in the Okhotsk Sea (50°13′ N, 153°06′ W) killed 4 years and 1 month later in the Gulf of Alaska east of Kodiak Island (57°42′ N 147°16′ W); a clerical error at the time of marking recorded the tagged animal as a sperm whale. Another blue whale marked by the Soviets on 22 May 1958 moved from Vancouver Island, British Columbia, to the southern end of Kodiak Island (Ivashin and Rovnin, 1967). Twelve of 14 whales marked in the North Pacific by the Japanese between 1963 and 1972 were recaptured in the same areas marked; the other two were marked in the western Gulf of Alaska and recovered on the south side of the eastern Aleutian Islands (Ohsumi and Masaki, 1975).

North Atlantic

North Atlantic blue whales migrate to Arctic waters to feed during spring and summer. In these months they are found on the east side as far as 80° N, near Spitzbergen (Scoresby, 1820; Jonsgård, 1955) and on the west side as far as Davis Strait and southern Greenland (Kapel, 1979), though some animals probably move into Baffin Bay (Allen, 1916). S. Leatherwood (unpublished data) saw them in September 1976 near the ice edge southwest of Jan Mayen Island. At present they apparently do not pass into Hudson Bay (Allen, 1916), although some have entered Hudson Strait in the past (Jonsgård, 1966; Tomilin, 1967; Leatherwood *et al.*, 1976).

The present wintering grounds and southern limits in the North Atlantic are not known. Millais (1907, p. 165) wrote that "many winter to the east of the West Indies" and though he does not provide the source or supporting data, it is reasonable to suspect, by analogy with other stocks, that some should occur there. On the west side of the North Atlantic there are records from Long Island, New York, and Ocean City, New Jersey (Allen, 1916) and as far south as Florida (Baughman, 1946), though Moore (1953) did not include blue whales on his list of Florida marine mammals. Strandings of single whales in Texas in August 1940 (Baughman, 1946) and December 1924 (Schmidly, 1981) and at San Cristobal, Panama, in January 1922 (Harmer, 1923) should not be taken as proof of a tropical Caribbean component of the North Atlantic population(s). On the eastern side, the southernmost records are from the waters between the Cape Verde Islands and the west coast of Africa (Ingebrigsten, 1929; Kirpichnikov, 1950; Tomilin, 1957), and blue whales are known seasonally along the coasts of the British Isles

FIG. 4 Approximate known distribution of the blue whale, *Balaenoptera musculus* (modified from Ruud, 1956). Overlapping grids may indicate differences of published opinions about subspecific identity in those areas rather than sympatric occurrence of two subspecies.

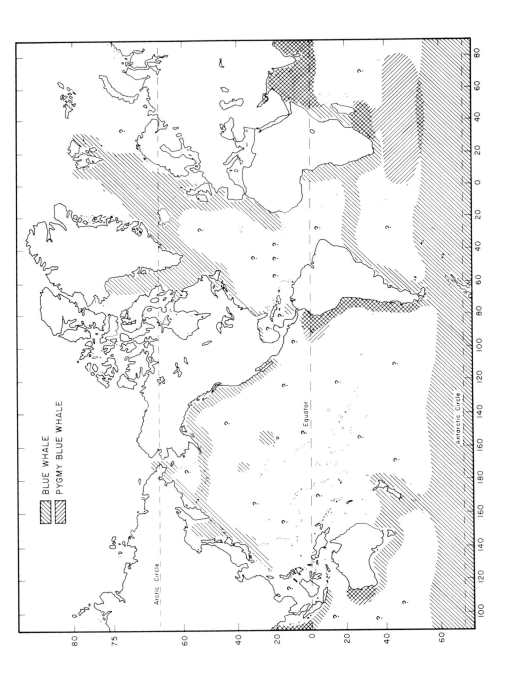

(Harmer, 1927, Fraser, 1974) and mainland Europe (Casinos and Vericad, 1976; Duguy and Robineau, 1982). Earlier reports of their occurrence around the Azores have been discounted by Clarke (1981). Blue whales have been taken in the Strait of Gibraltar (Aloncle 1964) and Collett (1912) reported blue whales in the Mediterranean, but the species is absent from recent lists for that area (Duguy and Robineau, 1982; Cagnalaro et al., 1983), and earlier records are thought to have been of misidentified fin whales (Duguy and Cyrus, 1972).

It has been suggested that blue whales in the North Atlantic, as in the North Pacific, move in and out of the feeding grounds in relation to plankton fronts along the edge of the continental shelf (Evans, 1980).

Southern Hemisphere (including northern Indian Ocean)

The main feeding grounds of Southern Hemisphere blue whales are in the circumpolar belt between the Antarctic pack ice and the Antarctic Convergence (Mackintosh, 1946, 1966; Rice, 1972). Blue whales have been reported as far as 78°38′ S, 116°17′ W (Lillie, 1915). During the austral winter, the population(s) moves north to subtropical and tropical breeding grounds, the exact locations of most of which are unknown; a female blue whale and newborn calf were taken off Saldanha Bay, southwest Africa, in September (Hinton, 1925), Deraniyagala (1948) reported that a female gave birth in Trincomalee Harbor, Sri Lanka, in May 1932, and Leatherwood et al. (1984) observed an adult with accompanying calf off Foul Point, near Trincomalee, in May 1983. Blue whales have been reported in the Northern Hemisphere portions of the Indian Ocean practically year-round (e.g., Gibson-Hill, 1950; Al-Robaae, 1974; Wray and Martin, 1983; Leatherwood et al., 1984). These may well be a separate stock.

No marks have been recovered from whales tagged in the Antarctic that give evidence of the routes of migration (Brown, 1962). The small number of reports of blue whales along normal shipping routes between Australia and South Africa and South Africa and South America has been interpreted to mean that blue whales disperse widely on the wintering/breeding grounds (Harmer, 1931; Mackintosh, 1942; Brown, 1957, 1958). However, Wheeler (1946) suggested that blue whales winter in a more limited region and cited as evidence the number of blue whales seen by him during travels along 20° S latitude between South America and Africa (outside normal shipping lanes). Tomilin (1957) and Mackintosh (1966), while acknowledging that data are scarce, postulated that some degree of concentration is more likely than wide dispersal. May (1979) speculated that the differences in migration times and extents (i.e., southward penetration) of the larger rorquals in the Southern Hemisphere relate to the size and relative proportions of blubber of the various species; the largest species, the blue whale, has the highest proportion of blubber to other tissue and penetrates the farthest south.

In the Southern Hemisphere, blue whales have been reported as far north as Madagascar on the east coast of Africa and Walvis Bay and Angola [and rarely to Gabon (IWS, 1972)] on the west (Kellogg, 1929; Tomilin, 1957). Off the coast of South America, they are reported as far north as Ecuador and Peru (Ingebrigtsen, 1929; Clarke, 1962; Valdivia and Ramirez, 1981; Donovan, 1984), and Hinton (1925) reported a possible sighting of a blue whale north of Rio de Janeiro, Brazil, at roughly 22° S. Blue whales are known from Australian and New Zealand waters (Oliver, 1922; Zenkovich, 1962; Baker, 1972; Nasu, 1973; Chittleborough, 1953; Isles, 1981; Cawthorn 1978, 1984) and from the Indian Ocean including the Bay of Bengal and the Arabian Sea (Blyth, 1859; Fernando, 1912; Pillay, 1926; Deraniyagala, 1932, 1948; Gibson-Hill, 1949, 1950; Daniel 1963; Venkataraman and Girijavallabhan, 1966; Nagabhushanam and Dhulkhed, 1964; Moses, 1947; Leatherwood and Clarke, 1983; Leatherwood et al. 1984). Weber (1923) reported strandings in Java and in 1980 S. J. Holt recounted sightings of blue whales near western Java (IWC 1980).

The discovery of fragments of swordfish bills in blue whales taken in the Antarctic in February 1951 (Ruud, 1952) and 1959 (Jonsgård, 1959) prompted speculation by the latter author that "the blue whale to a more pronounced degree than the fin whale undertakes migrations to tropical and temperate climes, where the swordfish has its habitat" (p. 356).

The pygmy blue whale is found primarily in the waters around Marion Island, Crozet Island, and the Kerguelen Islands, from 80° to 0° E and north of 54° S (Ichihara, 1966). Pygmy blue whales have also been reported from various more northerly areas: one of the 56 blues taken off Durban, South Africa, between 1954 and 1963 was actually a pygmy blue whale though it was not clear whether this was an isolated case or indicative that pygmy blue whales constituted a significant proportion of the catch there (Bannister and Gambell, 1965). There are records from the western coast of Australia (R. G. Chittleborough, personal communication to Ichihara, 1966), from Chilean (Aguayo, 1974; Clarke et al., 1978) and Peruvian (Valdivia and Ramirez, 1981; Donovan, 1984) waters, and near the Galapagos Islands [near 90° N (Berzin, 1978)]. Blue whales seen off northeastern Sri Lanka in spring 1982, 1983, and 1984 (Leatherwood and Clarke, 1983; Leatherwood et al., 1984; Donovan, 1984), off southern Madagascar during summer (Gambell et al., 1975), and in the Gulf of Aden in 1968 (Yukhov, 1969) were thought to be pygmy blue whales.

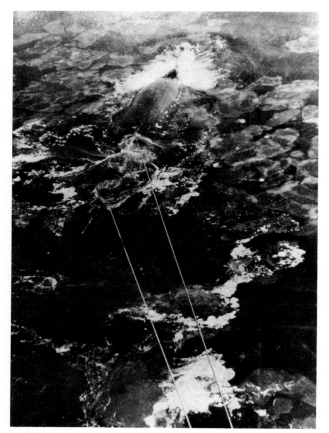

FIG. 5 Two views of a blue whale being killed in the Antarctic during the 1947–1948 season. (Photos by M. Yamada, courtesy of H. Omura.)

Abundance and Life History

Synopsis—history of exploitation

Nineteenth century whalers, operating from small open boats and using hand-held harpoons, generally did not hunt blue whales because these animals were too swift and powerful and sank when killed (Scammon, 1874). Norwegian Svend Føyn's invention and perfection of the harpoon gun, coupled with the arrival on the whaling scene of steam-driven whaling vessels, marked the start of modern whaling, during which it was possible for whalers effectively to hunt the larger, faster rorquals, such as the blue whale.

FIG. 5 (*continued*)

General reviews of modern whaling are given by Burton (1973), Ommanney (1971), Allen (1980), and Tønnessen and Johnsen (1982). In addition, regional whaling is discussed as follows: in the North Pacific by Kellogg (1931), Starks (1922), Omura (1955), Rice (1963, 1974), and Bockstoce (1978); in the North Atlantic by Jonsgård (1955), Jonsson (1965), Mitchell (1974b), Brown (1976), and Holt (1977); and in the Southern Hemisphere by Tillman and Ohsumi (1981), Aguayo (1974), Best (1974), Clarke (1980), and Maturana (1981).

As the largest and most valuable rorquals, the blue whale was the most sought after target of modern whalers and underwent a significant reduction in all areas. Beginning in the late 1800s blue whales were relentlessly pursued and taken in the North Atlantic and North Pacific. In the early 1900s whalers began to hunt them from land stations in the Antarctic. In the 1920s the introduction of the floating factory ship resulted in continued expansion of the blue whale harvest in the Antarctic (e.g., see Omura, 1973). Harvests of blue whales peaked during the 1930–1931 season, when nearly 30 000 were killed worldwide (Chapman, 1974b). Chapman (1974b) states that over 280 000 blue whales (including pygmy blue whales) were caught between 1924–1925 and 1970–1971 (see Fig. 5). The species was not afforded complete protection by the International Whaling Commission (IWC) until 1966 even though evidence of overhunting of blue whale stocks had been presented at least 3 decades earlier (see, for example, Laurie, 1937). The IWC continues to list blue whales as a protected species. The Nature Conservancy Council (1980, p. 73) notes, however, that "in view of the great value of a specimen, the incentive to include any animal encountered on an expedition is great." Clarke *et al.* (1978) report that 676 blue whales were caught off the coast of Chile between 1964 and 1967 and that one whale was captured in 1970 and one in 1971.

Various authors (e.g., Mackintosh, 1942; Laws, 1961; Gambell, 1963, 1976) have suggested that for blue whales, as for other heavily hunted rorquals, there has been an increase in pregnancy rates as a result of exploitation but Mizroch (1981, 1983a, 1983b) and Mizroch and York (1981) reexamined the data and reported that while they observed high variability in pregnancy rates, there was no apparent trend.

Pygmy blue whales identified as such were first caught by Japanese whalers in 1959–1960 near Kerguelen Island. Ichihara (1961) discusses the possibility of earlier unidentified catches of pygmy blue whales in the Antarctic, and Clarke *et al.* (1978) suggest that whales caught off Chile may have been pygmy blue whales.

Current status

Various estimates of abundance of blue whales are given in Table 3. Stocks everywhere have been drastically reduced by over-whaling and are recovering only slowly. Even so, the situation has not yet reached a crisis point for the populations of the magnitude alleged by Small (1971) and widely quoted, that is, a total world population of only 200.

North Pacific. The virgin stock of blue whales in the North Pacific is estimated to have contained from 4500 to 6000 animals (Omura and Ohsumi, 1974; Rice, 1974; Tillman, 1975; Scheffer, 1976; Berzin and Vladimirov, 1981). Current estimates of abundance range from 1400 to 1900 individuals (Nishiwaki, 1966; Omura and Ohsumi, 1974; Tillman, 1975; Scheffer, 1976; Mitchell, 1978; Rice, 1978; Berzin and Vladimirov, 1981). More of those are thought to be distributed on the east side of the North Pacific than on the west side (Omura, 1955; Tomilin, 1957). Blue whales have been completely protected since 1966. Even so, despite contentions that their numbers have apparently increased since the cessation of whaling, their absence or shortage in some areas of former abundance (e.g., Gulf of Alaska) suggests that the population(s) is still depressed (Fiscus *et al.*, 1976; Rice and Wolman, 1982). Farther south, they are being seen with increasing frequency in the Gulf of California and along the Pacific coast of Baja California and

TABLE 3 Some estimates of abundance of blue whales

Area	Estimated abundance		Reference	Comments
	Preexploitation	Present		
North Pacific	5 000	1 400–1 900	Berzin and Vladimirov (1981)	
North Pacific	4 900	1 340	Omura and Ohsumi (1974)	
North Pacific	4 900	1 400–1 900	Gambell (1976)	Various sources
North Pacific	—	1 500	Rice (1978)	
North Pacific	—	1 700	Maser et al (1981)	Sighting data
North Pacific	—	1 700	U.S. Department of Commerce (1983)	
North Pacific	—	920–3 450	Wada (1971)	Indices of abundance from scouting boats, 1965–1975
Eastern North Pacific	6 000	—	Rice (1974)	Catch data
Northwest Atlantic	1 100	Very low hundreds	Mitchell (1974), Allen (1970)	Sighting data, strip censuses
North Atlantic	1 100–1 500	100	Gambell (1976)	Various sources
North Atlantic	—	A few hundred	Rice (1978)	
North Atlantic	—	More than 500	U.S. Department of Commerce (1983)	

(*continued*)

TABLE 3 (*Continued*)

Area	Estimated abundance		Reference	Comments
	Preexploitation	Present		
Northern Norway	~3 500	—	Rørvik and Jonsgård (1981)	Roughly 3 500 were killed from 1868 to 1904
Denmark Strait	Several thousand	—	Rørvik and Jonsgård (1981)	
Antarctic	—	43 925	Masaki (1980)	Sighting data: author states it is an overestimation and "is remote from the true abundance"
Southern Hemisphere	150 000–210 000 including 10 000 pygmy blues	10 000 including 5 000 pygmy blues	Gambell (1976)	Various sources
Antarctic south of 30° S	—	14 450	Masaki (1977)	Sighting data
Antarctic whaling area III and South Africa	—	130	Best (1974)	

Region			Source	Notes
Antarctic whaling area III and South Africa	10 000 pygmies	—	Best (1974)	
Antarctic	—	6 000	Gulland (1972)	Based on various types of estimates
Subantarctic	—	6 000 (pygmies)	Gulland (1972)	Based on various types of estimates
Antarctic	150 000	5 000–10 000	Chapman (1974a)	Based on variety of methods
Antarctic south of 30° S	—	3 110–16 470 ($\bar{x} = 11\,840$)	Masaki and Yamamura (1978)	Sightings data 1965–1977
Antarctic south of 30° S	—	3 542–43 925 ($\bar{x} = 8\,442$)	Masaki (1979)	Sightings data 1965–1978
Southern Hemisphere	—	9 000	U.S. Department of Commerce (1983)	
World	More than 200 000	—	Rice (1978)	
World	200 000	13 000	Maser et al. (1981)	
World	13 000 pygmies	6 500 pygmies	Zemsky and Sazhinov (1982)	
World	—	11 200	U.S. Department of Commerce (1983)	Present estimate based on sighting data

southern California (S. Leatherwood, unpublished data, 1968–1982). Huber et al. (1983) reported sightings off the Farallon Islands in 1981 and 1982 and stated that the 1981 sightings were the first there since their survey had begun in 1970. They also noted that bird- and whale-watching boats in the area have reported a recent increase in sightings.

North Atlantic. The current stock of blue whales in the North Atlantic is estimated to be a few hundred animals (Allen, 1970; Mitchell, 1974), down from an initial minimum population of roughly 3500 in northern Norway [3500 were killed in that area from 1868–1904 (Rørvik and Jonsgård, 1981)], several thousand in Denmark Strait (Rørvik and Jonsgård, 1981), and 1100 in the northwest Atlantic (Allen, 1970; Mitchell, 1974b). Christensen (1980, p. 208) interpreted low sighting rates during cruises in the eastern North Atlantic as evidence that blue whales are "scarce in those waters where minke whaling takes place." Rørvik and Jonsgård (1981) cited data presented by Brown (1972) that imply an average rate of increase for blue whales in the North Atlantic of 15.6% per annum from 1957 to 1971. They state that this is probably too high and that while both Brown's data and their own suggest an increase, if such an increase exists its magnitude is unknown.

Southern Hemisphere (including northern Indian Ocean). Chapman (1974a) estimated the then-current and virgin populations of Antarctic blue whales to have been 5 000–10 000 and 150 000, respectively. Gulland (1972) estimated that as of 1971 the Antarctic blue whale population was at 6 000 but that there were another 6 000 pygmy blue whales present. Gambell (1976) examined five sources (Chapman *et al.*, 1964; Ichihara and Doi, 1964; Gulland, 1972; IWC, 1974; and Ohsumi and Masaki, 1974) and concluded there were 150 000 to 210 000 blue whales (including 10 000 pygmy blue whales) initially but that only 5 000 of each subspecies remained by 1975. Gulland (1981) reviewed methods used to estimate blue whale abundance in the Antarctic and stated that at that time the total population in the southern ocean (blues and pygmy blues combined) was approximately 10 000 and that it was increasing at a rate of 4 to 5% per year.

Marking

Japan began marking whales in the North Pacific in 1949 (Omura and Ohsumi, 1964). The Soviet Union, Canada, and the United States began marking in 1954, 1955, and 1962, respectively (Ivashin and Rovnin, 1967). Marking methods, including types of marks used, have been described by Arsen'ev (1959, 1965) and Omura and Kawakami (1956).

Although most blue whale marks recovered by Japanese researchers have been from the same areas in which the whales were marked (Omura and

Kawakami, 1956; Ohsumi and Masaki, 1975), Omura and Masaki (1975) reported that two whales moved from their area IIIB (140°–160° W, 40° N to the Aleutian Islands) to their area IVB (160°–180° W, 40° N to the Aleutian Islands). As described above (Distribution and Movements, North Pacific), Ivashin and Rovnin (1967) also reported that a blue whale tagged in the eastern Sea of Okhotsk was killed 4 years and 1 month later east of Kodiak Island.

Ohsumi and Masaki (1975) state that the hit rate for marks is low in rorquals, in general, due to their swiftness and that it is lowest in the blue whale. They also state that Japanese researchers shoot few marks into protected species, such as the blue whales, because of the improbability that they will be recovered. Of 64 blue whales marked from 1949 to 1972, only six were tagged after protection was afforded the whales in 1966. Rice (1978) noted that although 76 marks were placed in blue whales off Baja California from 1962 to 1970, the species was protected soon after the marking program began and no marks were recovered. Similarly, although blue whales are occasionally marked in the North Atlantic (Brown, 1977, reported 20 blue whales marked by Canada from 1950 to 1975) and stranded and accidentally killed blue whales are routinely examined for marks, "the reduced state of the stock is such that no scientific sampling has been or will be contemplated that involves the killing of blue whales for examination in these waters" (Mitchell, 1974, p. 158). Variations in skin pigmentation have been used by Sears (1983) in recent years to identify individuals in the Gulf of St. Lawrence. These natural marks have remained unchanged over at least a 2–3-year period. Leatherwood and Clarke (1983) and Leatherwood et al. (1984) have been using distinctive marks on fins and flukes to identify photographically blue whales off Sri Lanka between 1982 and 1984; 12 are known and one was seen in two successive years.

Rayner (1940) and Brown (1954, 1962) reported on blue whale marking in the Southern Hemisphere and analysed information on whale movements obtained from recaptures of marked whales. Most marking was carried out in IWC-defined Antarctic whaling areas II and III, though some whales were marked in each of the six areas. Some whales appeared to return consistently to the same areas on the Antarctic feeding grounds but considerable individual variation was noted. One whale was recaptured 170° (from area III to area V) of longitude from where it had been marked; the elapsed time was a little over 2 years. As mentioned above, there have been no mark recoveries since blue whale harvests ceased, but Brown (e.g., see 1979a,b, 1980, 1981) has provided regular progress reports on current marking programs.

Food and feeding

Blue whales feed almost exclusively on a few species of euphausiids (Nemoto, 1970; Kawamura, 1980). In the North Pacific, their major prey species are

Euphausia pacifica, *Thysanoessa spinifera*, *T. inermis*, *T. longipes*, *T. raschii*, and *Nematoscelis megalops* (Thompson, 1940; Nemoto, 1957, 1959; Nemoto and Kasuya, 1965; Tomilin, 1967; Kawamura, 1980). They also take copepods of the genus *Calanus* (Thompson, 1940; Tomilin, 1957; Nemoto, 1970; Kawamura, 1980) and, less frequently, *Sergestes* spp., amphipods, and squid (Thompson, 1940; Mizue, 1951; Tomilin, 1957; Nemoto, 1970; Kawamura, 1980). Rice (1978) and Leatherwood (in Leatherwood *et al.*, 1982b) have observed blue whales apparently feeding on swarms of "red crabs" (*Pleuroncodes planipes*) off Baja California.

In the North Atlantic blue whales are reported to feed on the euphausiids *T. inermis*, *T. raschii*, *T. longicaudata*, and *Meganyctiphanes norvegica* and on the copepod *Temora longicornis* (Nemoto, 1957; Kawamura, 1980).

Southern Hemisphere blue whales feed almost exclusively on *E. superba* (Mackintosh, 1946; Rice, 1978), although *E. crystallarophias*, copepods, and amphipods are additional prey items (Nemoto, 1970).

The preferred food of the pygmy blue whale is *Euphausia vallentini* (Nemoto, 1962). Gambell (1964) reported *E. recurva* and *E. diomedeae* from the stomach of a pygmy blue captured off Durban, South Africa. Referring to the same specimen, Bannister and Baker (1967) stated that the two euphasiid species were found in a roughly 3:1 proportion, and, from the degree of digestion, were probably eaten together.

Although blue whales have been reported to feed on small fish (Sleptsov, 1955) such as sardines (Mizue, 1951) and capelin (Kawamura, 1980), others have suggested that such fish are probably ingested accidentally (Tomilin, 1957; Nishiwaki, 1972).

Blue whales are generally characterised as "swallowing" (Nemoto, 1959; Mitchell, 1975, 1978) or "gulping" (Rice, 1978; Pivorunas, 1979) feeders, although side- and lunge-feeding behaviours (similar to that of the fin whale), both variations of gulping, have been observed in the northern Gulf of St. Lawrence (Sears, 1979, 1983) and Gulf of California (Storro-Patterson, 1981; Patten and Soltz, 1980). A feeding blue whale swims into an area of heavy prey concentration and scoops up a mouthful of water and prey; the whale then closes its mouth, forcing the water out through the baleen plates and trapping the prey. The collected food is moved back with a motion of the tongue (Mitchell, 1978; Rice, 1978; Pivorunas, 1979; Storro-Patterson, 1981). A description of the internal mechanism of filter feeding in rorquals is given in Lambertsen (1983).

Blue whales feed seasonally in polar waters (Mackintosh, 1965; Tomilin, 1957) and apparently many fast after leaving the feeding grounds; whales taken at subtropical land stations had little or no food in their stomachs (Mackintosh and Wheeler, 1929; Lockyer, 1981). Blue whales feeding off the west coast of Baja California (Rice, 1978), in the Sea of Cortez (Storro-

Patterson, 1981) and off northeastern Sri Lanka (Alling *et al.*, 1983, Leatherwood and Clarke 1983; Leatherwood *et al.*, 1984) are possible exceptions; these latter animals may be nonmigratory. Nemoto (1957) states that movements of blue whales are correlated with the abundance of their euphausiid prey; whales migrate into feeding areas earlier in years when the krill bloom occurs early in the season. Analyses of stomach contents suggest a feeding peak during the evening and early morning hours, apparently coinciding with vertical diurnal migration of prey (Nemoto, 1957; Nishiwaki and Ohe, 1951; Maser *et al.*, 1981).

It has been estimated that in a single day a blue whale may consume from less than 2 (Tomilin, 1967) to 4 (Rice, 1978) tonnes of food.

Mortality

Blue whales have been protected by the IWC from whaling since 1966 although nonmember nations reportedly took a small number in the northeast Atlantic as recently as 1978 (IWC, 1980). Blankley (1980) and Consiglieri and Braham (1982) cite vessel–whale collisions as a cause of blue whale mortality, describing two collisions off the west coast of the United States. Mitchell (1977) noted that entrapment in ice may be a significant source of mortality in the small North Atlantic population. An attack on a blue whale by a pod of killer whales, *Orcinus orca*, was filmed off Cabo San Lucas, Baja California (Tarpy, 1979) (Fig. 6).

Growth and reproduction

In the Northern Hemisphere blue whales calve and mate in late fall and winter (Millais, 1906; Tomilin, 1957). The mating season in the Southern Hemisphere occurs during the austral winter, with the peak in mating in July (Mackintosh and Wheeler, 1929; Nishiwaki, 1952; Tomilin, 1957). Although somewhat protracted, the reproductive seasons of northern and southern whales are roughly 6 months out of phase and it is likely that no interbreeding occurs (Mackintosh, 1965).

After attaining sexual maturity, female blue whales give birth to a single calf every 2 to 3 years (Mackintosh and Wheeler, 1929; Laurie, 1937; Nishiwaki, 1952; Laws, 1961) following a gestation period estimated by most workers to be 10 to 11 months (Mackintosh and Wheeler, 1929; Laws, 1959; Slijper, 1962; Tomilin, 1957). Multiple embryos have been reported (Brinkmann, 1948; Ichihara, 1962; Tomilin, 1957) but are apparently rare. Calves are 6–7 m long at birth (Mackintosh and Wheeler, 1929; Ottestad, 1950; Laws, 1959; Slijper, 1962) and are weaned at 7 months, by which time they are 16 m long (Mackintosh and Wheeler, 1929; Ottestad, 1950; Nishiwaki, 1952; Slijper, 1962). Foetal growth rates are discussed by Frazer and Hug-

FIG. 6 A blue whale under attack by killer whales off Baja California in May 1978. (Photo courtesy of Hubbs Marine Research Institute.)

gett (1959), Brinkmann (1948), Laws (1959) Lockyer (1981), Mizue and Jimbo (1950), and Nishiwaki and Hayashi (1950). The composition of blue whale milk is presented by Clowes (1929a).

Female blue whales attain sexual maturity at lengths of 21 to 23 m in the Northern Hemisphere (Omura, 1955; Tomilin, 1957; Slijper, 1962; Pike and MacAskie, 1969), and 23 to 24 m in the Southern Hemisphere (Mackintosh and Wheeler, 1929; Nishiwaki and Ohe, 1951; Slijper, 1962). Age at sexual maturity is approximately 5* years in both hemispheres (Ruud, et al., 1950; Nishiwaki, 1952; Slijper, 1962). Length at physical maturity is 25 m in the Northern Hemisphere (Pike and MacAskie, 1969) and 26–27 m in the Southern Hemisphere (Slijper, 1962; Nishiwaki and Ohe, 1951). Female pygmy blue whales are sexually mature at about 19 m and are physically mature at about 22 m (Ichihara, 1966).

Male blue whales are 20–21 m in length at sexual maturity and 24 m at physical maturity in the Northern Hemisphere (Omura, 1955; Pike and MacAskie, 1969) and are 22 m at sexual maturity and 24–25 m at physical maturity in the Southern Hemisphere (Mackintosh and Wheeler, 1929; Nishiwaki and Ohe, 1951; Slijper, 1962). Age at sexual maturity is just under 5* years (Ruud et al., 1950; Nishiwaki 1952; Slijper, 1962). Male pygmy blue whales are 21 m at physical maturity and less than 19 m at sexual maturity (Ichihara, 1966).

*Past workers have assumed that laminae in ear plugs are formed at a rate of two per year. If, as some workers currently believe, the rate is one per year, reported estimates should be doubled, that is, age at sexual maturity is 10 rather than 5 years.

The maximum length verified for a blue whale was a 33.58-m female from the Antarctic (Risting, 1922). Pike and MacAskie (1969) reported that the largest blue whale measured from the British Columbia area was 26.2 m in length. The longest pygmy blue whale correctly measured was 24.4 m (Ichihara, 1966).

Two hypotheses have been suggested to explain the size difference between northern and southern blue whales. Mayr (1963) stated that the whales obey Bergmann's (1847) Rule since the temperatures encountered by southern whales on the feeding grounds are lower than those experienced by northern whales. Brodie (1975) argued that the larger size of the Southern Hemisphere blue whales is an adaptation to their relatively longer period of exclusion from the feeding grounds (8 months versus 6 months for northern blues), a larger body size allowing for greater storage of lipid reserves.

Postnatal growth rates have been described by Ottestad (1950) and Lockyer (1981). Lockyer (1981) also calculated metabolic rates and energy budgets of various age classes of blue whales.

Age determination

Methods that have been used to estimate ages of blue whales include counting the number of corpora albicantia found in the ovaries of sexually mature females (Laws, 1957), determining the degree of coloration of the eye lens (Nishiwaki, 1950b), and examining baleen plates for annual growth increments (Ruud *et al.*, 1950; Nishiwaki, 1951, 1952). Age determination using baleen plates becomes impossible, however, after the fifth to seventh (annual) period because the tips of the plates begin to chip or wear off (Nishiwaki, 1950a). The preferred method for age determination of baleen whales is analysis of growth layers in the waxy ear plug (Purves, 1955). Unfortunately, the population of blue whales was so low when this technique was developed [the first extensive collection of plugs was made during the 1955–1956 Antarctic season (Ruud, 1956)] that the data are "too sparse to be useful" (Beddington and May, 1982, p. 65). Chapman (1974b) stated that the lack of adequate age data for blue whales is the most important problem affecting estimates of blue whale population parameters.

Maximum age estimates for blue whales range from 30 years (Slijper, 1962) to 80–90 years (Nishiwaki, 1972; Nature Conservancy Council, 1980).

Internal Anatomical Characteristics

The skull of the blue whale differs from that of other balaenopterids in several ways. In ventral aspect it shows an abrupt termination of palatine

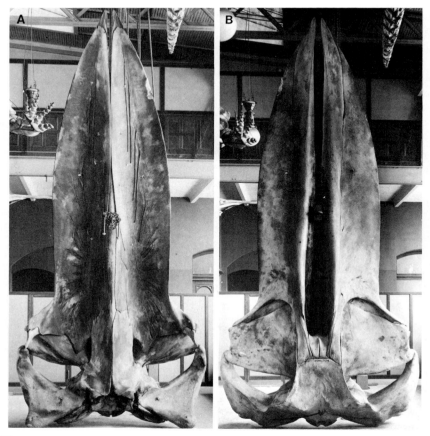

FIG. 7 Ventral (A), dorsal (B), and lateral (C) views of the skull [length 5.79 m "beak to condyles (straight)"] of a blue whale. Dorsal (D) and lateral (E) views of the skull (length 4.86 m) of a pygmy blue whale. For purported differences between the skulls of the blue and the pygmy blue whales, see Table 2; such differences are not

surface of the premaxillaries, a resemblance to the skull of the humpback whale (*Megaptera novaeangliae*); in dorsal aspect, the frontal bone can be seen to occupy a much narrower space than in the other species (Fig. 7). The wide rostrum with its convex margins is the main characteristic feature of the blue whale cranium. Condylo-premaxillary length represents 21.2–23.9% and 23–27% of the body length in females and males, respectively. (Tomilin, 1967). In contrast to the skull of the blue whale, the rostrum of the pygmy blue is less curved on its outer margin and tapers more from the base. While

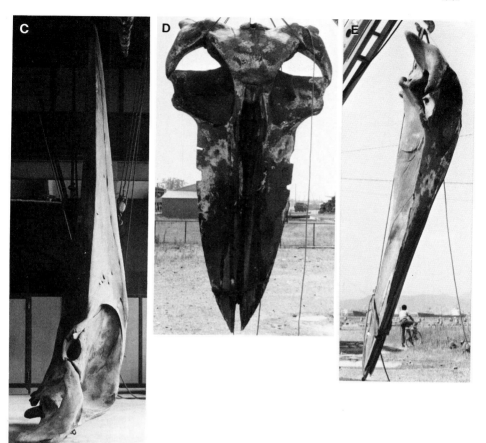

readily apparent to the authors in the above specimens [(A), (B), and (C) from Miller (1924), courtesy of J. G. Mead, U.S. National Museum. Specimen in (D) and (E) taken in the Antarctic, 1966–1967, photographs by T. Ichihara, courtesy of H. Omura].

the width of the rostrum (as percentage of skull length) at the base is similar for blues and pygmy blues, the width in the middle of the rostrum is less in pygmy blue whales (25% of skull length versus about 29% of skull length for blues as described in Tomilin, 1957). Differences in nasal bones and mandibles have also been described (Omura et al., 1970).

Selected references on the organ systems of the blue whale are included in Table 4.

TABLE 4 Selected references on the internal anatomy of blue whales

Organ or organ system	Blue whale, B. musculus	Pygmy blue whale, B. m. brevicauda	Reference
Baleen	x		Nemoto (1959), Slijper (1962), Tomilin (1957)
Blubber	x		Clowes (1929b), Slijper (1962), Tomilin (1957)
Digestive system	x		Ohe (1951), Slijper (1962), Hosokawa and Kamiya (1971)
Reproductive system	x		Mackintosh and Wheeler (1929), Brinkmann (1948), van Lennep (1950), Slijper (1956, 1966), Ohsumi (1964, 1969)
Respiratory system	x		Laurie (1933), Murata (1951), Slijper (1962)
Brain	x		Wilson (1933), Friant (1958), Slijper (1962)
Eye	x		Hosokawa (1951b)
Ear	x		Yamada (1948, 1953), Slijper (1962)
Skeletal system	x	x	Van Beneden and Gervais (1880), Hosokawa (1951a), Omura (1964, 1971), Omura et al. (1970), Tomilin (1957)
Endocrine system	x		Valsø (1938), Jacobson (1941), Slijper (1962)
Circulatory system	x		Slijper (1962)
Histology	x		Murata (1951), Hosokawa and Sekino (1958)
Serology	x		Fujino (1953, 1960)
Blood types		x	Fujino (1962)

Behaviour

Swimming, diving, and grouping

Scammon (1874, p. 72) described the blue whale as the "swiftest whale afloat", citing its speed as the reason why it was rarely pursued by nineteenth century whalers in their small open boats. (Subsequent authors have most often noted the sei as the fastest of the fleet rorquals; see Andrews, 1916; Mackintosh, 1965; Leatherwood et al., 1982b). Current estimates of blue whale swimming speeds range from 2 to 6.5 km/hr while feeding (Lockyer, 1981), 5 to 33 km/hr while crusing or migrating (Slijper, 1962; Zenkovich, 1962; Mackintosh, 1965; Lockyer, 1981) to a maximum of 20 to 48 km/hr when being chased or otherwise harassed (Slijper, 1962; Lockyer, 1981).

Although blowing and diving patterns vary with activity level (Leatherwood et al., 1982a), blue whales generally make 10–20 shallow dives at 12–20-sec intervals (blowing 8–15 times between dives) followed by a deep dive of 10–30 min (Mackintosh, 1965; Leatherwood et al., 1976; Maser et al., 1981). The blow is tall (6–12 m), narrow, and vertical (Slijper, 1962; Leatherwood, et al., 1976). On slow, shallow dives the head and blowhole may still be visible when the dorsal fin breaks the surface and the whale may submerge without exposing the tail flukes. On faster dives and just before a deep dive, the blowhole and head disappear after the blow is seen and are followed by a broad expanse of back sliding by in a low roll. The dorsal fin does not become visible until well after the blow, emerging just before the tail stock and flukes lift above the surface (Leatherwood et al., 1976; Mitchell, 1978). Winning and Minasian (1981) observed blue whales in the Gulf of St. Lawrence lifting their flukes higher than 3 m (10 feet) above the surface on terminal dives and Leatherwood et al. (1984) described frequent pronounced "fluking" in blue whales off northeast Sri Lanka and southwest of the Maldives. The Makah Indians of the northwest coast of North America called this species *kwakwe axtLi* or "noisy tail," referring to the species' habit of raising its tail on a dive (Waterman, 1920). Most observers, however, have noted that blue whale fluke exposure, if it occurs at all, is generally brief and low (Leatherwood et al., 1976, 1982b; Mitchell, 1978).

Blue whales are usually found swimming alone or in groups of two or three (Ruud, 1956; Slijper, 1962; Nemoto, 1964; Mackintosh, 1965; Pike and MacAskie, 1969; Aguayo, 1974) though Zenkovich (1962) reported groups of three to five animals, and only rare encounters with single animals. Larger feeding groups have been observed (Jonsgård, 1955; Slijper, 1962; Rice, 1978) and Nemoto (1964) described mixed schools of blue and fin whales in which the number of blues was always less than or equal to the number of fins.

Sound production

Known vocalisations of blue whales include a variety of sounds described as low frequency moans or long pulses (Cummings and Thompson, 1971, 1977; Edds, 1982; Thompson and Friedl, 1982), buzzes and rasps (Poulter, 1968), and ultrasonic clicks (Beamish and Mitchell, 1971; Beamish, 1974, 1979).

Using a trio of hydrophones to record a blue whale trapped in ice, Beamish (1979, p. 299) calculated that the higher frequency component of the recorded sounds was apparently emanating from the anterior part of the head, an area which contains no probable sound source. He suggested that the long, tapered head of the blue whale "may contain a directional acoustic antenna."

Beamish and Mitchell (1971) speculated that the ultrasonic clicks which they recorded from blue whales were being used to echolocate zooplankton. Beamish (1979, p. 300) reported that the clicks he recorded from an entrapped blue whale were "stronger and more frequent during and directly following the introduction of a lead weight into the water" near the whale, but cautioned that there is still insufficient evidence to support the hypothesis that baleen whales echolocate.

Geographic variations or dialects have been suggested for blue whales (Cummings and Thompson, 1977; Thompson *et al.*, 1979; Thompson and Friedl, 1982), but Edds (1982, p. 346) points out that "we have no knowledge of the roles that context, sex, and season may play in vocalisation variations." Low-frequency moans recorded by Edds (1982) in the western North Atlantic were similar to those reported from an "unidentified source" in the western South Pacific (Kibblewhite *et al.*, 1967). Low-frequency long pulses, attributed to blue whales, recorded by Thompson and Friedl (1982) in Hawaiian waters were different from possible blue whale pulses reported from Chile, California, Midway Island, and the Atlantic Ocean.

Diseases and Parasites

There is little information on diseases in blue whales. Cockrill (1960a,b) reviewed pathology of whales hunted in the Antarctic and present his own findings as veterinary officer of two pelagic whaling expeditions. Parasitic conditions, tumours, abscess formation, other lesions and inflammatory processes, traumata, bacterial conditions, and abnormalities of pregnancy are described.

Rice (1978, p. 34) presents information on endo- and ectoparasites of blue whales, stating that "probably because of its fastidious diet, the blue whale is much less prone to infestations of parasitic worms than are other species of baleen whales." Internal parasites include the thorny-headed worm *Bolbosoma nipponicum* (small intestine) and the giant kidney worm *Crassicauda crassicauda*. Ectoparasites include *Balaenophilus unisetus* and *Odotobius ceti* on the baleen, the skin parasite *Penella*, and the barnacle *Xenobalanus globicipitis*. Blue whales are rarely infested with the amphipod crustacean *Cyamus balaenopterae* (whale lice). Cockrill (1960a) lists the sessile barnacle *Coronula reginae* and diatoms as ectoparasites of Antarctic blue whales. Clarke (1966) reviewed records of occurrence of two species of stalked barnacles, *Conchoderma* spp., on blue whales from three geographical regions: South Africa (*C. virgatum* on a *Penella* sp. and *C. auritum* on baleen plates and on the barnacle *Coronula reginae*), South Georgia (*Conchoderma auritum* on baleen plates and on *Coronula reginae*) and the "South Atlantic" (*Conchoderma auritum* on baleen plates and on *Coronula diadema*). D. W. Rice (personal communication) speculates that *Coronula* spp. are probably only fortuitous epizoites on blue and most other great whales as they are virtually restricted to humpback whales.

Endoparasites include *Tetrabothrius* spp. (Cestoda) and *Crassicauda crassicauda* (Nematoda). Markowski (1955) reported the cestodes *Tetrabothrius affinis*, *T. wilsoni*, *T. schaeferi*, *Priapocephalus grandis*, and *Diplogonoporus balaenopterae* from blue whales examined by members of the staff of the *Discovery* Investigations.

The yellowish brown diatom film acquired by blue whales during stays in cold waters was studied by Omura (1950). He found it to be almost without exception *Cocconeis ceticola* and noted that since it takes the spores about 1 month to develop into the visible diatom film, those animals seen without a film are probably recent arrivals on the polar (Antarctic) grounds. Hart (1935) discussed the use of the diatomaceous deposit as a clue to migratory movements of whales; diatom films are rarely seen in warmer waters (Nishiwaki and Hayashi, 1950; Omura, 1950), and as the deposit increases throughout the season it can be used to estimate the amount of time a whale has been at the Antarctic feeding grounds.

Blue whales frequently are hosts to remoras, *Remora* sp. (Rice and Caldwell, 1961). Individuals off northeastern Sri Lanka have been seen and filmed with several hundred remoras in tow (Leatherwood *et al.*, 1984).

Addison et al. (1972) reported that tissue from a blue whale taken in the Antarctic about 1950 contained significantly lower levels of DDT-complex (Σ DDT = 0.18 ppm of oil) than did some samples from sperm whales (*Physeter macrocephalus*) from the Antarctic, taken 1962–1966 (0.07, 34.98, and 28.55 ppm), and from sperm, fin, bottlenose (*Hyperoodon ampullatus*), pilot (*Globicephala melaena*), and beluga (*Delphinapterus leucas*) whales in the North Atlantic, taken 1962–1970. The significance and effects of pollutant levels in marine mammals are currently disputed.

Acknowledgements

We thank Dale W. Rice, Ray Gambell, and Brent S. Stewart for thorough and thoughtful reviews of various drafts. Hideo Omura showed continuing generosity, expressed this time as a sharing of data and photographs; J. G. Mead provided photographs; Larry Foster, General Whale, granted permission to reprint his fine illustration; Chick Hayashi prepared the distribution map; and Angela Rowley and Steve Karl typed the manuscript. This chapter was submitted to the volume editors 1 April 1983.

References

Addison, R. F., Zinck, M. E., and Ackman, R. G. (1972). Residues of organochlorine pesticides and polychlorinated biphenyls in some commercially produced Canadian marine oils. *J. Fish. Res. Board Can.* **29**, 349–355.

Aguayo, L. A. (1974). Baleen whales off continental Chile. In "The Whale Problem: A Status Report" (Ed. W. E. Schevill), pp. 209–217, Harvard Univ. Press, Cambridge, Massachusetts.

Allen, G. M. (1916). The whalebone whales of New England. *Mem. Boston Soc. Nat. Hist.* **8**, 107–322.

Allen, K. R. (1970). A note on baleen whale stocks of the northwest Atlantic. *Rep. int. Whal. Commn* **20**, 112–113.

Allen, K. R. (1980). "Conservation and Management of Whales". Univ. of Washington Press, Seattle, and Butterworth, London.

Alling, A., Gordon J., Rotton, N., and Whitehead, H. (1982). Indian Ocean sperm whale study, 1981–1982. Interim report. March 1982. World Wildlife Fund–The Netherlands.

Aloncle, H. (1964). Premières observations sur les petits cétaces des côtes marocaines. *Bull. Inst. Pêch. Maritimes Maroc.* **12**, 21–42.

Al-Robaae, K. (1974). *Tursiops aduncus* bottlenosed dolphin: a new record for Arab Gulf, with notes on Cetacea of the region. *Bull. Basrah Nat. Hist. Mus.* **1**, 7–16.

Andrews, R. C. (1916). Monographs of the Pacific Cetacea. II.—The sei whale (*Balaenoptera borealis* Lesson). *Mem. Am. Mus. Nat. Hist.* **1**, Part VI, 291–503.

Anonymous. (1979). United States. Progress report on cetacean research June 1977–May 1978. *Rep. int. Whal. Commn* **29**, 137–141.

Arsen'ev, V. A. (1959). "Migratsiya Zhivotnykh. [Migration of Animals]". Vol. 1. Akad. Nauk. SSSR, Moscow.

Arsen'ev, V. A. (1965). Mechevie kitov na dal'nem vostoka. [Tagging whales in the Far East]. *Rybn. Khoz. (Moscow)* **41**, 25–29.

Baker, A. N. (1972). New Zealand whales and dolphins. *Tuatara (J. Biol. Soc. Victoria Univ. Wellington)* **20** (1), 1–49.

Bannister, J. L., and Baker, A. de C. (1967). Observations of food and feeding of baleen whales at Durban. *Nor. Hvalfangst-Tid.* **56**, 78–82.

Bannister, J. L., and Gambell, R. (1965). The succession and abundance of fin, sei, and other whales off Durban. *Nor. Hvalfangst-Tid.* **54**, 45–60.

Baughman, J. L. (1946). On the occurrence of a rorqual whale on the Texas coast. *J. Mammal.* **27**, 392–393.
Beamish, P. (1974). Whale acoustics. *J. Can. Acoust. Assoc.* **2** (4), 8–12.
Beamish, P. (1979). Behavior and significance of entrapped baleen whales. *In* "Behavior of Marine Animals, Vol. 3. Cetaceans" (Eds. H. E. Winn and B. L. Olla), pp. 291–309. Plenum, New York.
Beamish, P., and Mitchell, E. (1971). Ultrasonic sounds recorded in the presence of a blue whale *Balaenoptera musculus*. *Deep-Sea Res.* **18**, 803–809.
Beddington, J. R., and May, R. M. (1982). The harvesting of interacting species in a natural ecosystem. *Sci. Am.* **247**, 62–69.
Berzin, A. A. (1978). Whale distribution in tropical eastern Pacific waters. *Rep. int. Whal. Commn* **28**, 173–177.
Berzin, A. A., and Rovnin, A. (1966). Raspredelenie i migratsii kitov v serverovostachnoi chasti tikhogo okeana, v Beringovom, Chukotskom moryakh. [Distribution and migration of whales in the northeastern part of the Pacific Ocean, Bering, and Chukchi Seas]. *Izv. Tikhookean. Nauchno-Issled. Inst. Rybn. Khoz. Okeanogr. (TINRO)* **58**, 179–207. (In Russian). Translated by U.S. Dept. Interior, Bureau Commercial Fisheries, Seattle, Washington, 1966, *In* "Soviet Research on Marine Mammals of the East" (Ed. K. I. Panin), pp. 103–136.
Berzin, A. A., and Vladimirov, V. L. (1981). Changes in the abundance of whalebone whales in the Pacific and the Antarctic since the cessation of their exploitation. *Rep. int. Whal. Commn* **31**, 495–499.
Bergmann, C. (1847). Über die Verhältnisse der Wärmeökonomie der Thier zu ihrer Grösse, *Göttinger. Stud. Pt. 1*, 595–708. (Cited in Mayr, 1963; original not seen.)
Best, P. B. (1974). Status of the whale populations off the west coast of South Africa, and current research. *In* "The Whale Problem: A Status Report" (Ed. W. E. Schevill), pp. 53–81, Harvard Univ. Press, Cambridge, Massachusetts.
Blankley, M. A. (1980). The blue whale: the story from San Pedro. *Whalewatcher* **14**(4), 16–17.
Blyth, E. 1859. On the great rorqual of the Indian Ocean. *J. Asiat. Soc. Bengal* **28**(5), 481–498.
Bockstoce, J. (1978). History of commercial whaling in Arctic Alaska. *Alaska Geogr.* **5**(4), 17–25.
Brinkmann, A. (1948). Studies on female fin and blue whales. *Hvalradets Skr.* **31**, 5–38.
Brinkmann, A. (1967). The identification and names of our fin whale species. *Nor. Hvalfangst-Tid.* **56**(3), 49–56.
Brodie, P. F. (1975). Cetacean energetics, an overview of intraspecific size variation. *Ecology* **56**, 152–161.
Brown, S. G. (1954). Dispersal in blue and fin whales. *Disc. Rep.* **26**, 355–384.
Brown, S. G. (1957). Whales observed in the Indian Ocean. *Mar. Observ.* **27**, 157–165.
Brown, S. G. (1958). Whales observed in the Atlantic Ocean. *Mar. Observ.* **28**, 142–146, 209–216.
Brown, S. G. (1962). The movements of fin and blue whales within the Antarctic zone. *Disc. Rep.* **3**, 1–54.

Brown, S. G. (1972). Blue and humpback whales in Icelandic waters. ICES, C.M. 1972/N:4:5p. (Cited in Rørvik and Jönsgård, 1981; original not seen.)

Brown, S. G. (1976). Modern whaling in Britain and the north-east Atlantic Ocean. *Mammal Rev.* **6,** 25–36.

Brown, S. G. (1977). Whale marking in the North Atlantic. *Rep. int. Whal. Commn* **27,** 451–455.

Brown, S. G. (1979a). International cooperation in Antarctic whale marking 1971–1977. *Rep. int. Whal. Commn* **29,** 147–148.

Brown, S. G. (1979b). Whale marking—progress report 1978. *Rep. int. Whal. Commn* **29,** 89–90.

Brown, S. G. (1980). Whale marking—progress report 1979. *Rep. int. Whal. Commn* **30,** 128–130.

Brown, S. G. (1981). Whale marking—progress report 1980. *Rep. int. Whal. Commn* **31,** 154–155.

Burmeister, H. (1871). *Bol. Mus. Pub. Buenos Aires,* p. vii. (As cited by Hershkovitz, 1966, p. 174; original not seen).

Burton, R. (1973). "The Life and Death of Whales". Universe Books, New York.

Cagnalaro, A., di Natale, A., and Notarbartolo di Sciara, G. (1983). "Cetacei. Guide per il Riconseimento delle Specie Animali delle Acque Lagunari e Costiere Italiane". AQ/1/224/#9, Consiglio Nazionale Delle Ricerche, Rome.

Casinos, A., and Vericad, J.-R. (1976). The cetaceans of the Spanish coasts: A survey. *Mammalia* **40**(2), 267–289.

Cawthorn, M. W. (1978). Whale research in New Zealand. *Rep. int. Whal. Commn* **28,** 109–113.

Cawthorn, M. W. (1984). New Zealand progress report on cetacean research, May 1982–May 1983. *Rep. int. Whal. Commn* **34,** 213–215.

Chapman, D. G. (1974a). Status of Antarctic rorqual stocks. *In* "The Whale Problem: A Status Report" (Ed. W. E. Schevill), pp. 218–238. Harvard Univ. Press, Cambridge, Massachusetts.

Chapman, D. G. (1974b). Estimation of population parameters of Antarctic baleen whales. *In* "The Whale Problem: A Status Report" (Ed. W. E. Schevill), pp. 336–351. Harvard Univ. Press, Cambridge, Massachusetts.

Chapman, D. G., Allen, K. R., and Holt, S. J. (1964). Report of the Committee of Three Scientists on the special scientific investigation of the Antarctic whale stocks. *Rep. int. Whal. Commn* **14,** 32–106.

Chittleborough, R. G. (1953). Aerial observations on the humpback whales, *Megaptera nodosa* (Bonnaterre) with notes on other species. *Aust. J. Mar. Freshwater Res.* **4,** 219–226.

Christensen, I. (1980). Observations of large whales (minke not included) in the North Atlantic 1976–78 and markings of fin, sperm and humpback whales in 1978. *Rep. int. Whal. Commn* **30,** 205–208.

Clarke, R. (1962). Whale observations and whale marking off the coast of Chile in 1958 and from Ecuador towards and beyond the Galapagos Islands in 1959. *Nor. Hvalfangst-Tid.* **51,** 265–287.

Clarke, R. (1966). The stalked barnacle, *Conchoderma,* ectoparasitic on whales. *Nor. Hvalfangst-Tid.* **55,** 153–168.

Clarke, R. (1980). Catches of sperm whales and whalebone whales in the southeast Pacific between 1908 and 1975. *Rep. int. Whal. Commn* **30,** 285–288.

Clarke, R. (1981). Whales and dolphins of the Azores and their exploitation. *Rep. int. Whal. Commn* **31,** 607–615.

Clarke, R., Aguayo, A., and del Campo, S. B. (1978). Whale observation and whale marking off the coast of Chile in 1964. *Sci. Rep. Whales Res. Inst.* **30,** 117–177.

Clowes, A. J. (1929a). A note on the composition of whale milk. *Disc. Rep.* **1,** 472–475.

Clowes, A. J. (1929b). A note on the oil content of blubber. *Disc. Rep.* **1,** 476–478.

Cockrill, W. R. (1960a). Pathology of the Cetacea. A veterinary study on whales—Part I. *Br. Vet. J.* **116,** 133–144.

Cockrill, W. R. (1960b). Pathology of the Cetacea. A veterinary study on whales—Part II. *Br. Vet. J.* **116,** 175–190.

Collett, R. (1912). "Norges Pattendyr (Norges Hvirveldyr I)". H. Aschehoug, W. Nygaard, Kristiania (Oslo).

Consiglieri, L. D., and Braham, H. W. (1982). Seasonal distribution and relative abundance of marine mammals in the Gulf of Alaska. Draft Final Report, Contract No. R7120806, Research Unit 68, to the Alaska Outer Continental Shelf Environmental Assessment Program. NOAA, Juneau, Alaska.

Cummings, W. C., and Thompson, P. O. (1971). Underwater sounds from the blue whale, *Balaenoptera musculus. J. Acoust. Soc. Am.* **50,** 1193–1198.

Cummings, W. C., and Thompson, P. O. (1977). Long 20-Hz sounds from blue whales in the northeast Pacific. Abstract. Proceedings (Abstracts) of the Second Conference on the Biology of Marine Mammals, 12–15 December 1977. San Diego, California.

Daniel, J. C. (1963). Stranding of a blue whale *Balaenoptera musculus* (Linn.) near Surat, Gujarat, with notes on earlier literature. *J. Bombay Nat. Hist. Soc.* **60,** 252–254.

Deraniyagala, P. E. P. (1932). A stranded blue whale. *Spolia Zeylan.* **17,** 55–58.

Donovan, G. (1984). Blue whales off Peru, December 1982, with special reference to pygmy blue whales. *Rep. int. Whal. Commn* **34,** 473–476.

Duguy, R., and Cyrus, J.-L. (1972). Note préliminaire à l'étude des cétaces des côtes françaises de Mediterranée. 23rd Congrès-Assemblée plenière. C.I.E.S.M., Paris.

Duguy, R., and Robineau, D. (1982). "Guide des mammifères marins d'Europe". Delachaux and Niestlé, Paris.

Edds, P. L. (1982). Vocalizations of the blue whale *Balaenopera musculus*, in the St. Lawrence River. *J. Mammal.* **63,** 345–347.

Ehrenfeld, D. W. (1970). "Biological Conservation". Holt, Rinehart, and Winston, New York.

Evans, P. G. H. (1980). Cetaceans in British waters. *Mammal. Rev.* **10,** 1–46.

Fernando, H. F. (1912). Whales washed ashore on the coast of Ceylon from 1889 to 1910. *Spolia Zeylan.* **8**(29), 52–54.

Fiscus, C. H., Braham, H. W., Mercer, R. W., Everitt, R. D., Krogman, B. D., McGuire, P. D., Peterson, C. E., Sonntag, R. M., and Withrow, D. E. (1976). Seasonal distribution and relative abundance of marine mammals in the Gulf of

Alaska. NWAFC Processed Rep. NOAA/NMFS. U.S. Dept. of Commerce, Washington, D.C.

Fraser, F. C. (1974). Report on Cetacea stranded on the British coasts from 1948 to 1966. *Rep. Br. Mus. (Nat. Hist.)* **14,** 1–65.

Frazer, J. F. D., and Huggett, A. St. G. (1959). The growth rate of foetal whales. *J. Physiol.* **146,** 21–22.

Friant, M. (1958). Un stade de l'evolution cérébrale du rorqual (*Balaenoptera musculus* L.) *Hvalradets Skr.* **42,** 4–15.

Fujino, K. (1953). On the serological constitution of the sei-, fin-, blue- and humpback-whales (I). *Sci. Rep. Whales Res. Inst.* **8,** 103–125.

Fujino, K. (1960). Immunogenetic and marking approaches to identifying subpopulations of the North Pacific whales. *Sci. Rep. Whales Res. Inst.* **15,** 85–142.

Fujino, K. (1962). Blood types of some species of Antarctic whales. *Am. Nat.* **96,** 205–210.

Gambell, R. (1964). A pygmy blue whale at Durban. *Norsk Hvalfangst-Tid.* **53,** 66–68.

Gambell, R. (1973). Some effects of exploitation on reproduction in whales. *J. Reprod. Fert. Suppl.* **19,** 533–553.

Gambell, R. (1976). World whale stocks. *Mammal. Rev.* **6,** 41–53.

Gambell, R., Best, P. B., and Rice, D. W. (1975). Report on the international Indian Ocean whale marking cruise. *Rep int. Whal. Commn* **26,** 240–252.

Gibson-Hill, C. A. (1949). The whales, porpoises and dolphins known in Malayan waters. *Malay. Nat. J.* **4**(2), 44–61.

Gibson-Hill, C. A. (1950). A note on the rorquals (*Balaenoptera* spp.). *J. Bombay Nat. Hist. Soc.* **49,** 14–19.

Gulland, J. (1972). Future of the blue whale. *New Sci.* **54,** 198–199.

Gulland, J. (1981). A note on the abundance of Antarctic blue whales. *In* "Mammals in the Seas. Volume III. General Papers and Large Cetaceans", FAO Fisheries Series No. 5, pp. 219–228. Food and Agriculture Organization of the United Nations, Rome.

Hanna, G. D. (1920). Mammals of the St. Matthew Islands, Bering Sea. *J. Mammal.* **1,** 118–122.

Harmer, S. F. (1923). Cervical vertebrae of a giant blue whale from Panama. *Proc. Zool. Soc. London 1923*, 1085–1089.

Harmer, S. F. (1927). Report on Cetacea stranded on the British coasts from 1913 to 1926. *Rep. Br. Mus (Nat. Hist.)* **10,** 1–91.

Harmer, S. F. (1931). Southern whaling. *Proc. Linn. Soc. London* **142,** 85–163.

Hart. T. J. (1935). On the diatoms of the skin film of whales, and their possible bearing on problems of whale movements. *Disc. Rep.* **10,** 247–282.

Hershkovitz, P. (1966). Catalog of living whales. *Bull. U.S. Natl Mus.* **246,** 1–259.

Hinton, M. A. C. (1925). "Reports left by the late Maj. G. E. H. Barrett-Hamilton relating to the whales of South Georgia" Crown Agents for the Colonies, London.

Holt, S. J. (1977). Questions about the sex-ratio in catches of rorquals. *Rep. int. Whal. Commn* **27,** 13–140.

Hosokawa, H. (1951a). On the pelvic cartilages of the *Balaenoptera* foetuses, with remarks on the specifical [sic] and sexual difference. *Sci. Rep. Whales Res. Inst.* **5**, 5–15.

Hosokawa, H. (1951b). On the extrinsic eye muscles of the whale, with special remarks upon the innervation and function of the musculus retractor bulbi. *Sci. Rep. Whales Res. Inst.* **6**, 1–34.

Hosokawa, H., and Kamiya, T. (1971). Some observations on the cetacean stomachs, with special considerations on the feeding habits of whales. *Sci. Rep. Whales Res. Inst.* **23**, 91–101.

Hosokawa, H., and Sekino, T. (1958). Comparison of the size of cells and some histological formations between whales and man. *Sci. Rep. Whales Res. Inst.* **13**, 269–301.

Hoyt, E. (1984). "The Whale Watcher's Handbook". A Madison Press Book produced for Doubleday, Garden City, New York.

Huber, H. R., McElroy, T., Boekelheide, R. J., and Henderson, P. (1983). Studies of marine mammals at the Farallon Islands, 1981–82. Final report in fulfillment of NMFS Contract 81-ABC-00129.

Ichihara, T. (1961). Blue whales in the waters around Kerguelen Island. *Nor. Hvalfangst-Tid.* **50**, 1–20.

Ichihara, T. (1962). Prenatal dead foetus of baleen whales. *Sci. Rep. Whales Res. Inst.* **16**, 47–60.

Ichihara, T. (1963). Identification of the pigmy blue whale in the Antarctic. *Nor. Hvalfangst-Tid.* **52**, 128–130.

Ichihara, T. (1966). The pygmy blue whale, *Balaenoptera musculus brevicauda*, a new subspecies from the Antarctic. *In* "Whales, Dolphins and Porpoises" (Ed. K. S. Norris), pp. 79–113. Univ. of California Press, Berkeley.

Ichihara, T. (1981). Review of pygmy blue whale stock in the Antarctic. *In* "Mammals in the Seas. Volume III. General Papers and Large Cetaceans," FAO Fisheries Series No. 5, pp. 211–218. Food and Agriculture Organization of the United Nations, Rome.

Ichihara, T., and Doi T. (1964). Stock assessment of pygmy blue whales in the Antarctic. *Nor. Hvalfangst-Tid.* **53**, 145–167.

Ingebrigtsen, A. (1929). Whales caught in the Northern Atlantic and other seas. *Rapp. P.-V. Reun. Cons. Int. Explor. Mer.* **56**, 1–26.

IWS (Ed. Committee for Whaling Statistics). (1972). "International Whaling Statistics, No. 78". Grondahl and Son, Oslo.

Isles, A. C. (1981). A blue whale *Balaenoptera musculus* stranded near Warrnambool, Victoria. *Victoria Nat.* **98**, 52–53.

Ivashin, M. V. (1978). Soviet investigations of Cetacea June 1976 to May 1977. *Rep. int. Whal. Commn* **28**, 119–122.

Ivashin, M. V. (1984). Progress report on cetacean research June 1982 to May 1983. *Rep. int. Whal. Commn* **34**, 241–247.

Ivashin, M. V., and Rovnin, A. A. (1967). Some results of the Soviet whale marking in the waters of the North Pacific. *Nor. Hvalfangst-Tid.* **56**, 123–135.

IWC. (1974). Report of the Scientific Committee. *Rep. int. Whal. Commn* **24**, 39–54.

IWC. (1980). Report of the sub-committee on other protected species and aboriginal whaling. Blue whales. *Rep. int. Whal. Commn* **30,** 55–56.

IWC. (1981). Report of the sub-committee on other protected species and aboriginal whaling. Blue whales. *Rep. int. Whal. Commn* **31,** 137.

IWC. (1982). Report of the sub-committee on other protected species and aboriginal whaling. Blue whales. *Rep. int. Whal. Commn* **32,** 107.

IWC. (1984). Report of the sub-committee on other protected species and aboriginal/subsistence whaling. Antarctic blue whales. *Rep. int. Whal. Commn* **34,** 130–143.

Jacobson, A. P. (1941). Endocrinological studies in the blue whale. *Hvalradets Skr.* **24,** 1–84.

Jonsgård, Å. (1955). The stocks of blue whales (*Balaenoptera musculus*) in the northern Atlantic Ocean and adjacent arctic waters. *Nor. Hvalfangst-Tid.* **44,** 505–519.

Jonsgård, Å. (1959). New find of sword from swordfish (*Xiphias gladius*) in blue whale (*Balaenoptera musculus*) in Antarctic. *Norsk Hvalfangst-Tid.* **48,** 352–360.

Jonsgård, Å. (1966). The distribution of Balaenopteridae in the North Atlantic Ocean. *In* "Whales, Dolphins, and Porpoises" (Ed. K. S. Norris), pp. 114–124. Univ. of California Press, Berkeley.

Jonsson, J. (1965). Whales and whaling in Icelandic waters. *Nor. Hvalfangst-Tid.* **54**(11), 245–253.

Kapel, F. O. (1979). Exploitation of large whales in west Greenland in the twentieth century. *Rep int. Whal. Commn* **29,** 197–214.

Kawamura, A. (1980). A review of food of balaenopterid whales. *Sci. Rep. Whales Res. Inst.* **32,** 155–197.

Kellogg, R. (1929). What is known of the migrations of some of the whalebone whales. *Annu. Rep. Smithson. Inst. 1928,* 467–494.

Kellogg, R. (1931). Whaling statistics for the Pacific coast of North America. *J. Mammal.* **12,** 73–77.

Kibblewhite, A. C., Denham, R. N., and Barnes, D. J. (1967). Unusual low frequency signals observed in New Zealand waters. *J. Acoust. Soc. Am.* **41,** 644–655.

Kirpichnikov, A. A. (1950). Nablyadeniya nad raspredeleniyem kruphykh Kitoobraznakh v Atlanti cheskom okeane. [Observations on the distribution of large cetaceans in the Atlantic Ocean]. *Priroda (Leningrad)* **10,** 63–64. (Cited in Rørvik and Jonsgård, 1981; original not seen.)

Lambertsen, R. H. (1983). Internal mechanism of rorqual feeding. *J. Mammal.* **64,** 76–88.

Laurie, A. H. (1933). Some aspects of respiration in blue and fin whales. *Disc. Rep.* **8,** 363.

Laurie, A. H. (1937). The age of female blue whales and the effect of whaling on the stock. *Disc. Rep.* **15,** 223–284.

Laws, R. M. (1957). Polarity of whale ovaries. *Nature (London)* **179,** 1011–1012.

Laws, R. M. (1959). The foetal growth rates of whales with special reference to the fin whale, *Balaenoptera physalus* (Linn.). *Disc. Rep.* **29,** 281–308.

Laws. R. M. (1961). Reproduction, growth, and age of southern fin whales. *Disc. Rep.* **21,** 327–486.

Laws, R. M. (1962). Some effects of whaling on the southern stocks of baleen whales. *In* "The Exploitation of Natural Animal Populations" (Eds E. D. le Cren and M. W. Holdgate), pp. 137–158. Blackwell, Oxford.

Laws, R. M. (1977). The significance of vertebrates in the Antarctic marine ecosystem. *In* "Adaptations within Antarctic Ecosystems. Proceedings of the Third SCAR Symposium on Antarctic Biology" (Ed. G. A. Llano), pp. 411–438. Gulf Publishing, Houston, Texas.

Leatherwood, S., Caldwell, D. K., and Winn, H. E. (1976). Whales, dolphins, and porpoises of the western North Atlantic. A guide to their identification. *NOAA Tech. Rep. NMFS Circ.* **396.**

Leatherwood, S., Goodrich, K., Kinter, A. L., and Truppo, R. M. (1982a). Respiration patterns and 'sightability' of whales. *Rep. int. Whal. Commn* **32,** 601–613.

Leatherwood, S., Reeves, R. R., Perrin, W. F., and Evans, W. E. (1982b). Whales, dolphins, and porpoises of the eastern North Pacific and adjacent Arctic waters. A guide to their identification. *NOAA Tech. Rep. NMFS Circ.* **444.**

Leatherwood, S., Bowles, A. E., and Reeves, R. R. (1983). Endangered whales of the eastern Bering Sea and Shelikof Strait, Alaska: Results of aerial surveys, April 1982 through April 1983 with notes on other marine mammals seen. HSWRI Tech. Rep. No. 83-159 1 December 1983. Hubbs–Sea World Research Institute, San Diego.

Leatherwood, S., and Clarke, J. T. (1983). Cetaceans in the Strait of Malacca, Andaman Sea and Bay of Bengal, April 1982; with a preliminary review of marine mammal records from those regions (abstract). *Rep. int. Whal. Commn* **33,** 778.

Leatherwood, S., Peters, C. B., Santerre, R., Santerre, M., and Clarke, J. T. (1984). Observations of cetaceans in the Northern Indian Ocean Sanctuary, November 1980 through May 1983. *Rep. int. Whal. Commn* **34,** 509–520.

van Lennep, E. W. (1950). Histology of the corpora lutea in blue and fin whale ovaries. *Proc. K. Ned. Akad. Wet.* **53,** 593–599.

Lillie, D. G. (1915). Cetacea. British Antarctic ("Terra Nova") Expedition, 1910. *Nat. Hist. Rep. Zool.* **1**(3), 85–124.

Linnaeus, C. (1758). "Systema Naturae", 10th Ed., Vol. 1. Photographic facsimile reprinted, British Museum (Natural History), London.

Lockyer, C. (1981). Growth and energy budgets of large baleen whales from the Southern Hemisphere. *In* "Mammals in the Seas. Volume III. General Papers and Large Cetaceans", FAO Fisheries Series No. 5, pp. 379–487. Food and Agriculture Organization of the United Nations, Rome.

Mackintosh, N. A. (1942). The southern stocks of whalebone whales. *Disc. Rep.* **22,** 197–300.

Mackintosh, N. A. (1946). The natural history of whalebone whales. *Biol. Rev.* **21,** 60–74.

Mackintosh, N. A. (1965). "The Stocks of Whales". Fishing News, London.

Mackintosh, N. A. (1966). The distribution of southern blue and fin whales. *In* "Whales, Dolphins, and Porpoises" (Ed. K. S. Norris), pp. 125–144. Univ. of California Press, Berkeley.

Mackintosh, N. A., and Wheeler, J. F. G. (1929). Southern blue and fin whales. *Disc. Rep.* **1,** 257–540.

Markowski, S. (1955). Cestodes of whales and dolphins from the *Discovery* collections. *Disc. Rep.* **27,** 377–395.

Masaki, Y. (1977). Japanese pelagic whaling and whale sighting in the Antarctic, 1975–76. *Rep. int. Whal. Commn* **27,** 148–155.

Masaki, Y. (1979). Japanese pelagic whaling and whale sightings in the 1977/78 Antarctic season. *Rep int. Whal. Commn* **29,** 225–251.

Masaki, Y. (1980). Additional comments on the result of whale sighting by the scouting boats in the Antarctic whaling season from 1965/66 to 1977/78. *Rep. int. Whal. Commn* **30,** 339–357.

Masaki, Y., and Yamamura, K. (1978). Japanese pelagic whaling and whale sighting in the 1976/77 Antarctic season. *Rep. int. Whal. Commn* **28,** 251–261.

Maser, C., Mate, B. R., Franklin, J. F., and Dyrness, C. T. (1981). Natural history of Oregon coast mammals. *U.S. Dept. Agric. For. Serv. Gen. Tech. Rep.* PNW-133 (Pac. Northwest For. Range Exp. Stn, Portland, Oregon).

Maturana, C. R. (1981). Chile. Progress report on cetacean research 1966–May 1980. *Rep. int. Whal. Commn* **31,** 181–183.

May, R. M. (1979). Ecological interactions in the Southern Ocean. *Nature (London)* **277,** 86–89.

Mayr, E. (1963). "Animal Species and Evolution." Belknap, Harvard Univ. Press, Cambridge, Massachusetts.

Millais, J. G. (1906). "The Mammals of Great Britain and Ireland", Vol. III. Longmans, London.

Millais, J. G. (1907). "Newfoundland and Its Untrodden Ways". Longmans, London.

Miller, G. S., Jr. (1924). Some hitherto unpublished photographs and measurements of the blue whale. *Proc. U.S. Natl Mus.* **66,** 1–4.

Mitchell, E. (1974). Present status of northwest Atlantic fin and other whale stocks. *In* "The Whale Problem: A Status Report" (Ed. W. E. Schevill), pp. 108–169, Harvard Univ. Press, Cambridge, Massachusetts.

Mitchell, E. (1975). Trophic relationships and competition for food in northwest Atlantic whales. *Proc. Can. Soc. Zool. Annu. Meet.* pp. 123–133.

Mitchell, E. (1978). Finner whales. *In* "Marine Mammals of Eastern North Pacific and Arctic Waters" (Ed. D. Haley), pp. 36–45. Pacific Search Press, Seattle, Washington.

Mizroch, S. A. (1981). Further notes on Southern Hemisphere baleen whale pregnancy rates. *Rep. int. Whal. Commn* **31,** 629–633.

Mizroch, S. A. (1983a). Reproductive rates in Southern Hemisphere baleen whales. Unpublished Masters Thesis. Univ. of Washington, Seattle.

Mizroch, S. A. (1983b). Baleen whale life history strategies and population models. Abstract. Fifth Biennial Conference on the Biology of Marine Mammals. 27 Nov.–1 Dec. 1983, Boston.

Mizroch, S. A., and A. E. York (1981). Have Southern Hemisphere baleen whale pregnancy rates increased? Abstract. Cetacean Reproduction Conference. 28 Nov.–7 Dec. 1981, La Jolla, California.

Mizue, K. (1951). Food of whales in the adjacent waters of Japan. *Sci. Rep. Whales Res. Inst.* **5,** 81–90.

Mizue, K., and Jimbo, H. (1950). Statistic study of foetuses of whales. *Sci. Rep. Whales Res. Inst.* **3,** 119–131.

Moore, J. C. (1953). Distribution of marine mammals to Florida waters. *Am. Midl. Nat.* **49,** 117–118.

Morgan, L. (1978). Modern shore-based whaling. *Alaska Geogr.* **5,** 35–43.

Moses, S. T. (1947). Stranding of whales on the coasts of India. *J. Bombay Nat. Hist. Soc.* **47,** 377–379.

Murata, T. (1951). Histological studies on the respiratory portions of the lungs of Cetacea. *Sci. Rep. Whales Res. Inst.* **6,** 35–48.

Murie, O. J. (1959). Fauna of the Aleutian Islands and Alaska Peninsula. *U.S. Fish Wildl. Serv. N. Am Fauna* **61,** 1–364.

Nagabhushanam, A. K., and Dhulkhed, M. H. (1964). On a stranded whale on the south Kanara coast. *J. Mar. Biol. Assoc. India* **6,** 323–325.

Nasu, K. (1973). Results of whale sighting by *Chiyoda Maru* No. 5 in the Pacific sector of the Antarctic and Tasman Sea in the 1966/67 season. *Sci. Rep. Whales Res. Inst.* **25,** 205–217.

Nature Conservancy Council. (1980). "A World Review of the Cetacea". Nature Conservancy Council, London.

Nemoto, T. (1957). Foods of baleen whales in the northern Pacific. *Sci. Rep. Whales Res. Inst.* **12,** 33–89.

Nemoto, T. (1959). Food of baleen whales with reference to whale movements. *Sci. Rep. Whales Res. Inst.* **14,** 149–290.

Nemoto, T. (1962). Food of baleen whales collected in recent Japanese Antarctic whaling expeditions. *Sci. Rep. Whales Res. Inst.* **16,** 89–103.

Nemoto, T. (1964). School of baleen whales in the feeding areas. *Sci. Rep. Whales Res. Inst.* **18,** 89–110.

Nemoto, T. (1970). Feeding pattern of baleen whales in the ocean. *In* "Marine Food Chains" (Ed. J. H. Steel), pp. 241–252. Univ. of California Press, Berkeley.

Nemoto, T., and Kasuya, T. (1965). Foods of baleen whales in the Gulf of Alaska of the North Pacific. *Sci. Rep. Whales Res. Inst.* **19,** 45–51.

Nikulin, P. G. (1946). Distribution of cetaceans in seas surrounding the Chukchi Peninsula. *Izv. Tikhookean. Nauchno-Issled Inst. Rybn. Khoz. Okeanogr. [TINRO]* **22,** 255–257. Translated by U.S. Naval Oceanographic Office, Washington, D.C., 1969 (Transl. 428).

Nishiwaki, M. (1950a). Age characteristics in baleen plates. *Sci. Rep. Whales Res. Inst.* **4,** 162–182.

Nishiwaki, M. (1950b). Determination of the age of Antarctic blue and fin whales by the color changes in crystalline lens. *Sci. Rep. Whales Res. Inst.* **4,** 115–161.

Nishiwaki, M. (1951). On the periodic mark on the baleen plates as the sign of annual growth. *Sci. Rep. Whales Res. Inst.* **6,** 133–152.

Nishiwaki, M. (1952). On the age determination of Mystacoceti, chiefly blue and fin whales. *Sci. Rep. Whales Res. Inst.* **7,** 87–119.

Nishiwaki, M. (1966). Distribution and migration of the larger cetaceans in the North

Pacific as shown by Japanese whaling results. *In* "Whales, Dolphins, and Porpoises" (Ed K. S. Norris), pp. 171–191. Univ. of California Press, Berkeley.

Nishiwaki, M. (1972). General biology. *In* "Mammals of the Sea. Biology and Medicine" (Ed. S. H. Ridgway), pp. 3–204. Thomas, Springfield, Illinois.

Nishiwaki, M., and Hayashi, M. (1950). Biological survey of fin and blue whales taken in the Antarctic season 1947–48 by the Japanese fleet. *Sci. Rep. Whales Res. Inst.* **3,** 132–190.

Nishiwaki, M., and Ohe, T. (1951). Biological investigation on blue whales (*Balaenoptera musculus*) and fin whales (*Balaenoptera physalus*) caught by the Japanese Antarctic whaling fleets. *Sci. Rep. Whales Res. Inst.* **5,** 91–167.

Ohe, T. (1951). Iconography on the abdominal cavity and viscera of the *Balaenoptera*, with special remarks upon the peritoneal coverings. *Sci. Rep. Whales Res. Inst.* **5,** 17–40.

Ohsumi, S. (1964). Comparison of maturity and accumulation rate of corpora albicantia between the left and right ovaries in Cetacea. *Sci. Rep. Whales Res. Inst.* **18,** 123–148.

Ohsumi, S. (1969). Occurrence and rupture of vaginal band in fin, sei, and blue whales. *Sci. Rep. Whales Res. Inst.* **21,** 85–94.

Ohsumi, S., and Masaki, Y. (1974). Status of whale stocks in the Antarctic, 1972/73. *Rep. int. Whal. Commn* **24,** 102–113.

Ohsumi, S., and Masaki, Y. (1975). Japanese whale marking in the North Pacific, 1963–1972. *Bull. Far Seas Fish. Res. Lab.* **12,** 171–219.

Oliver, W. R. B. (1922). A review of the Cetacea of the New Zealand seas. *Proc. Zool. Soc. London* pp. 557–585.

Ommanney, F. D. (1971). "Lost Leviathan". Dodd, Mead, New York.

Omura, H. (1950). Diatom infection on blue and fin whales in the Antarctic whaling area V (the Ross Sea area). *Sci. Rep. Whales Res. Inst.* **4,** 14–26.

Omura, H. (1955). Whales in the northern part of the North Pacific. *Nor. Hvalfangst-Tid.* **44**(6), 323–345, **44**(7), 395–405.

Omura, H. (1964). A systematic study of the hyoid bones in the baleen whales. *Sci. Rep. Whales Res. Inst.* **18,** 149–170.

Omura, H. (1971). A comparison of the size of vertebrae among some species of the baleen whales with special reference to whale movements. *Sci. Rep. Whales Res. Inst.* **23,** 61–69.

Omura, H. (1973). A review of pelagic whaling operations in the Antarctic based on the effort and catch data in 10° squares of latitude and longitude. *Sci. Rep. Whales Res. Inst.* **25,** 105–203.

Omura, H. (1984). Measurements of body proportions of the pygmy blue whale, left by the late Dr. Tadayoshi Ichihara. *Sci. Rep. Whales Res. Inst.* **35,** 199–203.

Omura, H., and Kawakami, T. (1956). Japanese whale marking in the North Pacific. *Nor. Hvalfangst-Tid.* **45,** 555–563.

Omura, H., and Ohsumi, S. (1964). A review of Japanese whale marking in the North Pacific to the end of 1962, with some information on marking in the Antarctic. *Nor. Hvalfangst-Tid.* **53,** 90–112.

Omura, H., and Ohsumi, S. (1974). Research on whale biology of Japan with special reference to the North Pacific stocks. *In* "The Whale Problem: A Status Report" (Ed. W. E. Schevill), pp. 196–208. Harvard Univ. Press, Cambridge, Massachusetts.

Omura, H., Ichihara, T., and Kasuya, T. (1970). Osteology of pygmy blue whale with additional information on external and other characteristics. *Sci. Rep. Whales Res. Inst.* **22**, 1–27.

Ottestad, P. (1950). On age and growth of blue whales. *Hvalradets Skr.* **33**, 67–72.

Patten, D. R., and Soltz, D. L. (1980). Blue whales in the Sea of Cortez. *Whalewatcher* **14**(2), 10–11.

Pike, G. C., and MacAskie, I. B. (1969). Marine mammals of British Columbia. *Fish. Res. Board Can. Bull.* **171**, 1–54.

Pillay, R. S. N. (1926). List of cetaceans taken in Travancore from 1902–1925. *J. Bombay Nat. Hist. Soc.* **31**, 815–817.

Pivorunas, A. (1979). The feeding mechanisms of baleen whales. *Am. Sci.* **67**, 432–440.

Poulter, T. C. (1968). Marine mammals. *In* "Animal Communication" (Ed. T. Sebeok), pp. 405–465. Indiana Univ. Press, Bloomington.

Purves, P. E. (1955). The wax plug in the external auditory meatus of the Mysticeti. *Disc. Rep.* **27**, 239–302.

Rayner, G. W. (1940). Whale marking, progress and results to December 1939. *Disc. Rep.* **19**, 245–284.

Rice, D. W. (1963). Pacific coast whaling and whale research. *Trans. N. Am. Wildl. Nat. Res. Conf.* **28**, 327–335.

Rice, D. W. (1972). The great blue whale. *Pac. Search* **6**, 8–10.

Rice, D. W. (1974). Whales and whale research in the eastern North Pacific. *In* "The Whale Problem: A Status Report" (Ed. W. E. Schevill), pp. 170–195. Harvard Univ. Press, Cambridge, Massachusetts.

Rice, D. W. (1977). A list of marine mammals of the world. NOAA Tech. Rep. NMFS SSRF-711.

Rice, D. W. (1978). Blue whale. *In* "Marine Mammals of Eastern North Pacific and Arctic Waters" (Ed. D. Haley), pp. 30–35. Pacific Search Press, Seattle.

Rice, D. W., and Caldwell, D. K. (1961). Observations on the habits of the whale-sucker (*Remilegia australis*). *Nor. Hvalfangst-Tid.* **5**, 189–193.

Rice, D. W., and Scheffer, V. B. (1968). A list of the marine mammals of the world. *U.S. Fish Wildl. Ser. Spec. Sci. Rep. Fish.* **579**, 1–16.

Rice, D. W., and Wolman, A. W. (1982). Whale census in the Gulf of Alaska, June to August 1980. *Rep. int. Whal. Commn* **32**, 491–497.

Risting, S. (1922). "Av Hvalfangstens Historie". Kommandør Chr. Christensens Hvalfangst Museum, Sandefjord, Norway.

Rørvik, C. J., and Jonsgård, Å. (1981). Review of balaenopterids in the North Atlantic Ocean. *In* "Mammals of the Seas. Volume III. General Papers and Large Cetaceans", FAO Fisheries Series No. 5, pp. 379–487. Food and Agriculture Organization of the United Nations, Rome.

Rovnin, A. A. (1969). Distribution of the large cetaceans in the tropical part of the Pacific Ocean. *In* "Morskie Mlekopitayuschie" (Eds. V. A. Arsen'ev, B. A. Zenkovich, and K. K. Chapskiy). Third All-Union Conf. Mar. Mamm. Fisheries Research Board of Canada Transl. No. 1510.

Ruud, J. T. (1952). Do swordfish attack the large baleen whales? *Nor. Hvalfangst-Tid.* **4,** 191–193.
Ruud, J. T. (1956). The blue whale. *Sci. Am.* **195,** 46–50.
Ruud, J. T., Jonsgård, Å., and Ottestad, P. (1950). Age studies on blue whales. *Hvalradets Skr.* **33,** 5–66.
Scammon, C. M. (1874). "The Marine Mammals of the Northwestern Coast of North America". Carmany and Co., San Francisco.
Scheffer, V. B. (1974). The largest whale. *Defend. Wildl. Int.* **49**(4), 272–274.
Scheffer, V. B. (1976). The status of whales. *Pac. Disc.* **29**(1), 2–8.
Scheffer, V. B., and Slipp, J. W. (1948). The whales and dolphins of Washington state. *Am. Midl. Nat.* **39,** 257–337.
Schmidly, D. J. (1981). Marine mammals of the southeastern United States coasts and the Gulf of Mexico. U.S. Dept. of Interior, Bureau of Land Management, Fish and Wildlife Service. Biological Services Program FWS/OBS-80/41.
Scoresby, W. (1820). "An Account of the Arctic Regions, with a History and Description of the Northern Whale-fishery". Constable, Edinburgh.
Sears, R. (1979). Observations of cetaceans along the north shore of the Gulf of St. Lawrence, August 1979. East Falmouth, Massachusetts, USA. Mingan Island Cetacean Study, mimeo report. (Cited in Gaskin, 1982; original not seen.)
Sears, R. (1983). A glimpse of blue whales feeding in the Gulf of St. Lawrence. *Whalewatcher* **17**(3), 12–14.
Sleptsov, M. M. (1961). O kolebanii chislennosti kitov v Chukotskom more v raznye gody. [Fluctuations in the number of whales of the Chukchi Sea in various years]. *Tr. Inst. Morfol. Zhivotn.* **34,** 54–64. Translated by U.S. Naval Oceanographic Office, Washington, D.C., 1970 (Transl. 478).
Slijper, E. J. (1956). Some remarks on gestation and birth in Cetacea and other aquatic mammals. *Hvalradets Skr.* **41,** 5–62.
Slijper, E. J. (1962). "Whales". Hutchinson, London.
Slijper, E. J. (1966). Functional morphology of the reproductive system in Cetacea. *In* "Whales, Dolphins, and Porpoises" (Ed. K. S. Norris), pp. 278–319. Univ. of California Press, Berkeley.
Small, G. L. (1971). "The Blue Whale". Columbia Univ. Press, New York.
Starks, E. C. (1922). A history of California shore whaling. *Fish. Bull. (California State Fish and Game Commission)* **6,** 1–38.
Storro-Patterson, R. (1981). Great gulping blue whales. *Oceans* **14,** 16.
Tarpy, C. (1979). Killer whale attack. *Natl Geogr.* **155**(4), 542–545.
Thompson, R. J. (1940). Analysis of stomach contents of whales taken during the years 1937 and 1938 from the North Pacific. Unpublished masters thesis. Univ. of Washington, Seattle.
Thompson, P. O., and Friedl, W. A. (1982). A long term study of low frequency sounds from several species of whales off Oahu, Hawaii. *Cetology* **45,** 1–19.
Thompson, T. J., Winn, H. E., and Perkins, P. J. (1979). Mysticete sounds. *In* "Behavior of Marine Animals. Cetaceans" (Eds H. E. Winn and B. L. Olla), Vol. 3, pp. 403–431. Plenum, New York.
Tillman, M. (1975). Assessment of North Pacific stocks of whales. *Mar. Fish. Rev.* **37,** 1–4.

Tillman, M., and Ohsumi, S. (1981). Japanese Antarctic pelagic whaling prior to World War II: Review of catch data. *Rep. int. Whal. Commn* **31,** 625–627.

Tønnessen, J. N., and Johnsen, A. O. (1982). "The History of Modern Whaling". Univ. of California Press, Berkeley. (Translation by R. I. Christopherson and abridged version of "Den Moderne Hvalfangst Historie: Opprinnelse og Utvikling", Vols. 1–4, 1959–1970, Norwegian Whaling Association, Sandefjord, Norway.)

Tomilin, A. G. (1967). "Mammals of the USSR and Adjacent Countries. Vol. IX. Cetacea". Israel Program for Scientific Translations, Jerusalem.

True, E. W. (1904). The whalebone whales of the western North Atlantic compared with those occurring in European waters, with some observations on the species of the North Pacific. *Smithson. Contrib. Knowl.* **33,** 1–332.

U.S. Department of Commerce. (1983). Marine Mammal Protection Act of 1972. Annual Report 1982/1983. NOAA/NMFS, Washington, D.C.

Valdivia, J., and Ramirez, P. (1981). Peru. Progress report on cetacean research June 1979–May 1980. *Rep. int. Whal. Commn* **31,** 211–214.

Valsø, J. (1938). The hypophysis of the blue whale (*Balaenoptera musculus* L.). *Hvalradets Skr.* **16,** 5–30.

van Beneden, P. J., and Gervais, P. (1880). "Osteographie des Cétacés Vivants et Fossiles". Bertand, Paris. (Cited in Omura *et al.,* 1970; original not seen.)

Veinger, G. M. (1979). Distribution of whales in the North Pacific on expeditional data of 1965/66, 1969/70 and 1975. *Rep. int. Whal. Commn* **29,** 341.

Venkataraman, G., and Girijavallabhan, K. G. (1966). On a whale washed ashore at Calicut. *J. Mar. Biol. Assoc. India* **8,** 373–374.

Wada, S. (1977). Indices of abundance of large-sized whales in the North Pacific in the 1975 whaling season. *Rep. int. Whal. Commn* **27,** 189–194.

Wada, S. (1980). Japanese whaling and whale sighting in the North Pacific 1978 season. *Rep. int. Whal. Commn* **30,** 415–424.

Wade, L. W., and Friedrichsen, G. L. (1979). Recent sightings of the blue whale, *Balaenoptera musculus*, in the northeastern tropical Pacific. *Fish. Bull.* **76,** 915–919.

Waterman, T. T. (1920). The whaling equipment of the Makah Indians. *Univ. Washington Pub. Political Social Sci.* **1** (1), 1–67.

Weber, M. (1923). "Die Cetaceen der Siboga Expedition". E. J. Brill, Leiden.

Wheeler, J. F. G. (1946). Observations on whales in the South Atlantic Ocean in 1943. *Proc. Zool. Soc. London.* **116,** 221–224.

Wilson, R. B. (1933). The anatomy of the brain of the whale (*Balaenoptera sulfurea*) *J. Comp. Neurol.* **58,** 419–480.

Winning, B., and Minasian, S. M. (1981). La Baleine Bleu. *Pac. Disc.* **34,** 1–8.

Wray, P., and Martin, K. R. (1983). Historical whaling records from the western Indian Ocean. *Rep. int. Whal. Commn (Spec. Issue)* **5,** 213–242.

Yamada, M. (1948). Auditory organ of the whalebone whales. *Sci. Rep. Whales Res. Inst.* **2,** 21–30.

Yamada, M. (1953). Contribution to the anatomy of the organ of hearing of whales. *Sci. Rep. Whales Res. Inst.* **8,** 1–79.

Yukov, V. L. (1969). Observations of cetaceans in the Gulf of Aden and the northwestern part of the Arabic Sea. *In* "Morskie Mlekopitayushchie" (Eds. V. A. Arsen'ev, B. A. Zenkovich, and K. K. Chapskiy). Third All-Union Conf. Mar. Mamm. Fisheries Research Board of Canada, Transl. No. 1510.

Zemsky, V. A., and Boronin, V. A. (1964). On the question of the pygmy blue whale taxonomic position. *Nor. Hvalfangst-Tid.* **53,** 306–311.

Zemsky, V. A., and Sazhinov, E. G. (1982). The distribution and current abundance of pygmy blue whales. *In* "Morskie Mlekopitayushchie" (Ed. V. A. Arsen'ev). All-Union Research Institute of Marine Fisheries and Oceanography. VNIRO. Marine Mammals, Collected Papers. Moscow. (In Russian with English summary).

Zenkovich, B. A. (1962). Sea mammals as observed by the round-the-world expedition of the Academy of Sciences of the USSR in 1957/58. *Nor. Hvalfangst-Tid.* **51,** 198–210.

9

Humpback Whale
Megaptera novaeangliae (Borowski, 1781)

Howard E. Winn and Nancy E. Reichley

Genus and Species

Megaptera novaeangliae, meaning "great wing of New England," combines Gray's (1846) generic classification with Borowski's (1781) specific distinction. Although there is considerable variation in colour pattern, size, and other morphological features, a single species is accepted. While the humpback is similar to the rorquals in the shape of its dorsal fin and in the existence of ventral grooves, there are several anatomical differences that place it in a separate genus. Some authors do consider *Megaptera* a rorqual (see Tomilin, 1957); however, most today concur that the number and spacing of ventral grooves and the overall appearance of the humpback separate it from the other members of the Balaenopteridae.

Common names

The tendency to round its back when diving most likely led to the name "humpback." Other common names include gorback (Russian), Buckelwal, Pflockfish, Knurrwhal (German), baleine à bosse (French), knolhval (Norwegian), jorobada (Spanish), and zatokuzira (Japanese).

External Characteristics and Morphology

The humpback whale's body is short and stouter than most other members of the Balaenopteridae (Fig. 1). At birth, humpbacks average between 4 and 5 m in length (Matthews, 1937; Nishiwaki, 1959; Chittleborough, 1965; Tomilin, 1957). Sexual maturity is reached around 2 to 5 years, physical maturity 10 years later (Matthews, 1937; Chittleborough, 1959, 1965; Nishiwaki, 1959). Mean length at sexual maturity is 11.58 m (males) and 12.09 m (females). At physical maturity, average length is 13.35 m (males) and 13.72 m (females) (Nishiwaki, 1959; Dawbin, 1960; Chittleborough, 1965). The largest animal documented was 18 m (sex unlisted; Tomilin, 1957); the oldest, 48 years (Chittleborough, 1965). In any given adult age class, females are larger than males by 40 to 70 cm (Tomilin, 1957).

Weights differ considerably due to varying blubber thicknesses throughout the year. For example, an 11.67-m animal weighed 34.0 tonnes in February,

FIG. 1 Underwater lateral view of humpback mother and calf showing typical body shape. The characteristic knobs on the flipper can be seen. (Photograph courtesy of David Woodward.)

and a 12.89 m whale weighed 31.5 tonnes in April (Ash, 1953). Estimates calculated from whale meat and fillings give average values of 24.1–29.8 tonnes per animal (12.80–12.98 m in length) (Ash, 1957). A newborn calf averages 2 tonnes (Goodal, 1913, in Tomilin, 1957).

The humpback's head is rounder than that of the other rorquals, and constitutes 28–30% of the total body length (Tomilin, 1957). Rounded, subcutaneous knobs, or dermal tubercles, are found on the dorsal surface of the snout, chin, and mandibles, with variation in number and location. On the snout, the knobs are arranged in three rows (one median and two lateral), the lateral ones forming a double row on each margin of the upper jaw. There are usually 5–8 median knobs and 5–15 in each of the double lateral rows. Three or four knobs also lie on each side of the blowhole. The chin carries a group of tubercles at the tip of the jaw. The lower jaw has 10–15 (Tomilin, 1957) or 2–11 (Matthews, 1937) tubercles per side. Each tubercle contains a sensory hair which is 1–3 cm long (Tomilin, 1957).

Body coloration ranges from completely black to black with white markings on the throat, abdomen, and sides. In some animals the belly may be completely white. [See Lillie's (1915) colour patterns.] There may also be white marks on the head, behind the eyes, and on the back (Tomilin, 1957).

Baleen plates are primarily blackish brown or grey and may have some lighter fibres. Outer portions of the mouth usually match baleen colour. Tomilin (1957) notes that humpback baleen bristles are coarser than those of the finback whale and are relatively short. In the adult, one side may contain 270–400 plates and reach 85–104 cm in length. In a 12.1-m male, length of a baleen side was 240 cm (Tomilin, 1957). Matthews (1937) and Tomilin (1957) found that baleen growth increases dramatically during weaning as the baleen then becomes functional. Baleen spacing increases with body length in both sexes (Matthews, 1937).

The dorsal fin is low (less than 30 cm; Tomilin, 1957) and varies in shape from falcate to only slightly rounded. It may be the same colour as the back or have speckled areas.

The most distinctive external features of the humpback whale are the number of ventral pleats, flipper size and form, and fluke coloration and shape. The humpback has fewer and more widely spaced ventral grooves than other members of the family Balaenopteridae. True (1904), Matthews (1937), and Tomilin (1957) found 12–36 pleats per animal; Matthews (1937) gives an average of 28. The pleats run from the chin to the navel. Some appear at the corners of the mouth and on the base and proximal end of the flippers. Between the navel and penis pouch, pleats become one deep-set groove up to 1 m long in males. In females the groove is shorter and ends before the urogenital slit (Tomilin, 1957).

The flippers are long and narrow and may equal 23–31% of the total body

FIG. 2 Aerial photograph of two humpbacks, showing long white flippers. (Photograph courtesy of the University of Rhode Island, under sponsorship of the Minerals Management Service.)

FIG. 3 Humpback flukes (ventral view). Note distinctive colour pattern and fringed edge. (Photograph courtesy of the University of Rhode Island, under sponsorship of the Minerals Management Service.)

length (Tomilin, 1957). The dorsal surface may range from black through intermediate stages to white, while the ventral surface is usually all white (but may have speckles or a greyish hue). There is some indication that the majority of Southern Hemisphere humpbacks are darkly coloured on the dorsal side of their flippers, while those of the Northern Hemisphere animals are largely white on top (see Matthews, 1937). There is so much variability, however, that the degree of differentiation has not yet been determined. The anterior margin of all humpback flippers is scalloped (Fig. 1), and each knob usually contains barnacles. The posterior edge of the flipper is smoother and thinner than the anterior.

Humpback flukes are also highly characteristic in both shape and colour (Fig. 2). Their trailing edge is usually serrated and may be deeply notched in the center. Barnacles may be found clustered on the tips. The dorsal surface is basically black, while the ventral side is a combination of white and black (Fig. 3). The ventral coloration is distinctive and may be used to identify individual animals (Katona et al., 1979). Flukes of some animals contain numerous scars, some of which are believed to be killer whale teeth marks (Katona, 1980). Tomilin (1957) states that the fluke spread is 27–33.5% of the total body length.

Internal Anatomy

Skull and skeleton

Several anatomical features distinguish the humpback from the other rorquals. The humpback's rostrum is shorter and broader than the finback's, and the zygomatic process is relatively slender in the posterior part of the skull, expanding laterally towards the anterior. The width of the humpback's zygomatic process is the largest in the Balaenopteridae, constituting 57–67% of the condylomaxillary length of the skull. This percentage decreases with age as the skull grows more rapidly in length than width. The orbital process of the humpback's maxilla is more developed than that of the other rorquals. The dorsal portions of the maxilla and premaxilla are more curved, and the vomer narrows towards the posterior of the skull (Tomilin, 1957).

The humpback, like the right whale, has a broad temporal opening laterally, and prominent frontals near the median line of the cranium. The frontals are wider along the axis of the cranium than near the orbits. The nasal bones are narrow and anteriorly pointed. The lower jaw, wide at the base, has a relatively low coronoid process. When compared with the finback's supraoccipital, that of the humpback is narrower than the condyles and has little or no median ridge (Allen, 1916; Tomilin, 1957) (Figs. 4–6).

FIG. 4 Lateral view of cranium of *Megaptera novaeangliae*. Scale, 1 m. (From True, 1904; U.S. National Museum No. 21492. Scale provided by J. G. Mead.)

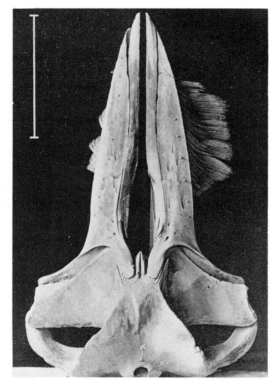

FIG. 5 Dorsal view of cranium of *Megaptera novaeangliae*. Scale, 1 m. (From True, 1904; U.S. National Museum No. 21492. Scale provided by J. G. Mead.)

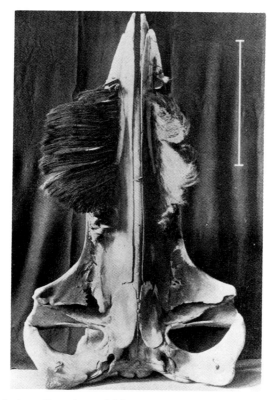

FIG. 6 Ventral view of cranium of *Megaptera novaeangliae*. Scale, 1 m. (From True, 1904; U.S. National Museum No. 21492. Scale provided by J. G. Mead.)

The vertebral count is C7, T14, L11(10), Ca21: total 52–53. In most humpbacks, the cervical vertebrae are separate, have reduced spinous processes, and are less expanded at the tips than the finback's (Allen, 1916; Tomilin, 1957). The humpback's thoracic vertebral spines are narrower laterally and also less expanded at the tips. Transverse processes are reduced and less flattened (Allen, 1916). Spinous processes disappear at the fortieth to forty-second vertebrae, transverse processes at the thirty-seventh to thirty-ninth (True, 1904; Allen, 1916).

In the humpback (and most rorquals), ankylosis, or fusion of the epiphyses and vertebral centra, begins at the ends of the vertebral column and stops in the posterior section of the thoracic and lumbar regions. The degree of fusion is an indication of degree of physical maturity (Chittleborough, 1955a). Tomilin (1957) found various stages of ankylosis in animals of different lengths, indicating that some large (approximately 14 m) males may still be physically immature because of the existence of cartilage layers between the epiphyses and centra, indicating incomplete fusion.

Like all mysticetes, the humpback's sternum is in one piece. Distinctive to the humpback, however, is a triangular or V-shaped sternum (Klima, 1978). Thick and broad, it has two rounded lateral wings and a shortened posterior section. There are 14 pairs of ribs, and only the first pair articulates with the sternum. Unlike the finback, which has ligaments binding the two ribs to the sternum, the humpback has a cartilaginous intercalary layer attached to each rib (Tomilin, 1957). The rest of the ribs are floating with free ends.

The scapula is usually fan-shaped, although its form is variable. It lacks coracoid and acromial processes, although they may exist as rudimentary knobs. The humpback, like all mysticetes, has no clavicle (True, 1904; Tomilin, 1957).

Typical phalangeal formula in the flippers is I3, II7, IV7, V2, but may vary within one or two elements in each digit. The ulna has a reduced olecranon and is smaller than the radius. There are approximately five cartilaginous elements in the carpus as well as a large pisiform on the ulnar side (Tomilin, 1957). The humpback has a rudimentary pelvic girdle approximately 23 cm long, which may have vestigal femurs 9.3–12.5 cm long (Struthers, 1889, in Tomilin, 1957). Rudimentary limbs in the form of skinless bony and cartilaginous knobs have been found in some animals (Tomilin, 1957).

Blubber thickness

Blubber thicknesses vary at different times of the year and with different ages and physiological conditions. Some younger animals have thicker layers of blubber than adults, and pregnant females have more blubber than other adults. The humpback has the greatest relative blubber thickness for its size of any rorqual, and is usually second only to the blue whale in absolute thickness. Tomilin (1957) found that maximum blubber thicknesses in females varied from 14 cm (two pregnant females, 13.1 and 14.4 m long) to 19 cm (a 15.4 m female). Average thicknesses ranged from 11.5 to 12 cm. Matthews (1937) gives similar figures: a 9-m male had blubber 14 cm thick; a 9-m female, 10.9 cm; a 14-m male, 9.75 cm; and a 14-m female, 12.2 cm.

Ventral pouch

Pivorunas (1977) has discussed the ventral pouch and mandibular areas which he describes as characterised by a Y-shaped fibrocartilage structure that lies in the substance of the muscular ventral pouch. Lambertsen (1983) has discussed the internal mechanism of feeding.

Brain

Breathnach (1955) described a brain of 4030 g from a female of unknown size. The cerebellum constituted 18% of the total brain weight. The brain did not differ markedly from those of other mysticete whales. Pilleri (1966) studied eight brains taken from humpback whales caught in the Indian Ocean. Their average weight was 4675 g. Various structures were measured for comparison with other species. Humpback whales were found to have a level of cerebralisation between right and sei whales.

Gonads

At puberty, the paired testes weigh about 4.0 kg (Chittleborough, 1965). Nishiwaki (1959) found similar results and noted that the testes are heavier in breeding than in feeding areas. Ovaries of sexually immature females weigh between 150 and 500 g (Tomilin, 1957), or as high as 680 g (Matthews, 1937). Sexually mature females have ovaries weighing from 600 to 3000 g (Tomilin, 1957), although Mitchell (1973) gives a lower limit of 450 g. See Table 1 for other organ weights.

Distribution

Humpbacks regularly leave colder waters where they feed during spring, summer, and fall travelling to a winter range over shallow tropical banks where they calve and do not feed. Mating also presumably occurs primarily in the winter. Generally speaking, the migration routes are oriented north and south (Fig. 7). The annual cycle is 6 months out of phase between the Northern and Southern hemispheres but is in phase with the climatic cycle. Marked animals have been caught between the tropical calving and polar feeding grounds in the Southern Hemisphere, establishing the connections between the two areas.

It is often stated that the majority of humpbacks are coastal migrators. Although they are coastal on the northern feeding grounds and occur in dense aggregations on shallow banks in the tropics, it seems as if the majority are deep oceanic migrators (beyond the 200-m line) between the feeding and calving grounds. This is true in the western North Atlantic and off the west coast of South America (see also Fig. 7). Humpbacks disperse more widely in deep water than when in shallow water. They move through coastal waters during migrations when a land mass (such as New Zealand) is in their direct route or when the 100 fathom line is near shore. Occasionally young stray individuals can be seen inshore between New York and Florida in the west-

TABLE 1 Humpback organ weights (in kg)

Organ	13.7-m male[a]	12.5-m male[a]	12.9-m male[b]	13.9-m female[b]
Body weight	40 823	37 195	—	—
Adrenal	1.210	—	—	—
Thyroid	2.960	3.237	—	—
Brain	7.229	5.288	—	—
Kidney	—	192	—	162
Liver	635	476	327	315
Heart	193	214	125	243
Spleen	—	6.041	—	—
Eyes	0.980	—	—	—
Lungs	—	—	362	372
Tongue	—	—	792	—
Stomach	—	—	105	148
Ovaries	0.6–3 kg in sexually mature females[b]			
Testes	4 kg in sexually mature males[c]			

[a] From Quiring (1943).
[b] From Tomilin (1957).
[c] From Chittleborough (1965).

ern North Atlantic. The vast majority do not seem to come into coastal waters until they reach the latitudes between Long Island, New York, and Cape Cod, Massachusetts, which bring them onto the feeding grounds. It is believed that the above pattern is general around the world. Further study and the use of radio tags to track animals should clarify this issue.

Tracking

Since 1926, when the first Discovery tags were used to mark individual whales, researchers have been developing more sophisticated equipment to gather information relating the animals to their environment. Discovery tags were stainless steel tubes fired from shotguns into the whale's blubber. Each whale received 2–10 tags in selected places. They were retrieved usually when the animals were flensed and in the boilers at whaling stations, enabling researchers to locate individual animals. Information obtained from

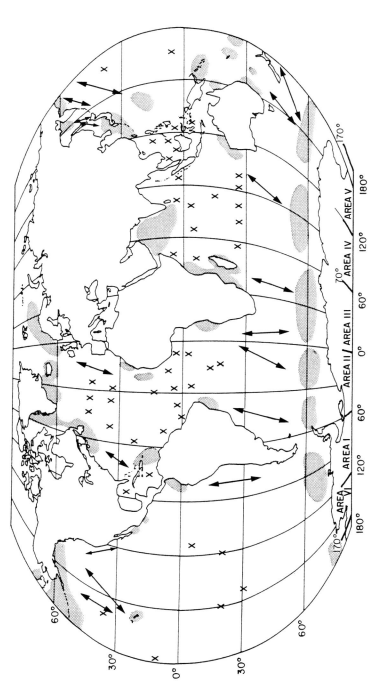

FIG. 7 Migratory cycle and distribution of humpback whales. X, humpback sightings; stippled areas between 40° N and S are tropical calving areas, those nearer the poles are feeding areas; arrows indicate movements between regions. (From unpublished manuscript by H. E. Winn and L. K. Winn.)

the tags was limited, and the tags had a tendency to injure or kill the animal (Ruud et al., 1953). Data on humpbacks were recorded with Discovery tags for many years before new and more detailed methods were devised.

Today, tagging and identification procedures for humpbacks and other whales are much more elaborate and cause minimal, if any, physical damage to the animal (see Watkins, 1981). Natural markings such as fluke patterns, dorsal fin shape and colour, and body scars and coloration are used to track and identify animals (Katona et al., 1979). Winn et al. (1975) and Levenson and Leapley (1976) suggest acoustic techniques combined with visual observations to census humpbacks in breeding areas. Because silent animals would not be recorded, this technique would only provide the lower limit to a population by counting the number of "singing" whales.

Watkins et al. (1978, 1981) monitored daily schedules of tagged humpbacks in Prince William Sound, Alaska. Tags remained on the animals for around 16 days. Recorded activities included fast swimming, milling, changing directions, and resting. In some cases, humpbacks were found to roam as much as 100 km/day. Visual observations coupled with radio tag data provided detailed observations on humpback behaviour including feeding, rolling, rubbing, penis erection, and breaching.

J. D. Goodyear (personal communication) is currently working on a "remora" tag, which can continuously record depth and temperature changes and is affixed to the animal by suction cups and shot from a crossbow. The tag is recoverable and can be made to release after a predetermined time. Whales may be tracked from shore stations or boats at distances up to 80 km. In Newfoundland, humpback movements were monitored over a 24-hr day–night cycle for 80 hr. Maximum dive depth was measured, and the sex of the tagged humpbacks was determined from skin samples obtained via a remote cytological sexing technique.

More sophisticated types of radiotelemetry are currently being designed or tested which will provide information on the animal's environment (water temperature, salinity, dissolved oxygen), behaviour (diving depth, swimming speed, sound production), and physiological state (heart rate, body temperature), all as a function of time and location (Leatherwood and Evans, 1979). The tagged whale might be tracked via aircraft, satellite, shore stations, or hand-held and ship's receivers.

Population Structure and Dynamics

The three oceanic populations have been thought to be divided into various stocks that are for the most part isolated, but with a little interchange in some cases. It has been thought that there are two stocks each in the North Pacific and North Atlantic, and perhaps seven in the Southern Hemisphere, totalling 11 stocks in all. This is, however, an oversimplified view. There are probably a series of stocks around the world with the amount of isolation decreasing as one goes to smaller units. Gaskin (1976) presented a scheme of stocks in the South Pacific (Fig. 8). These are western Australia, eastern Australia, and a series of stocks which pass by New Zealand on migration to a series of islands, such as Tonga, Fiji, and Cook Islands, to the east of Australia. Over 3000 humpbacks have been marked in the South Pacific over a number of years. Most recaptures indicate that separate stocks occur in western Australia, eastern Australia and New Zealand, including islands to the north, and that they go to different but slightly overlapping areas of the Antarctic. One hundred and twenty-three animals showed the segregation, whereas three crossed from eastern to western Australia and 12 showed interchange between the eastern Australia and New Zealand stocks. A large number of animals were caught in the same area as marked in subsequent years.

Recent studies have shown an interchange between Hawaii and the eastern Mexican population, with mixing in the feeding grounds off Alaska. The degree of segregation, if any, is unknown. The Hawaiian humpback popula-

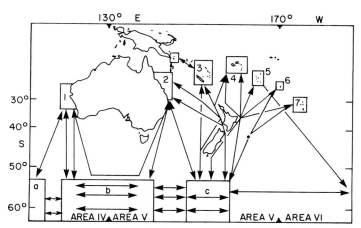

FIG. 8 Southern Pacific stocks of humpbacks and interchange between areas. (a,b,c), Antarctic feeding grounds. Breeding and stock areas are (1) subtropical coast of western Australia; (2) subtropical–tropical coast of Queensland; (3) New Caledonia–Loyalty Islands–New Hebrides; (4) Fiji and Lau Islands; (5) Tonga; (6) Niue; (7) Cook Islands. Arrows indicate apparent migration routes and inferred movements shown by whale mark recoveries. (Redrawn from Gaskin, 1976, with permission.)

TABLE 2 Historical and current sizes of various stocks of humpback whales around the world

High latitude stocks (feeding areas to tropical calving grounds)	Historical population size	Current population size
Southern Hemisphere	102 000+	2 710+
Antarctic area		
I. Go to west coast of South America, and Brazil	25 000	900
II. Go to west coast of Africa (Congo, etc.)	25 000–35 000	170
III. Go to east Africa, Madagascar	30 000+	340
IV. Go to western Australia	12 000–17 000	800
V. Go by New Zealand to east Australia, Tonga, Fiji, Hebrides, Chesterfield Island, and other islands	10 000	500
VI. None present		
Arabian Sea	?	500+
North Pacific		1 000
Eastern Aleutians, Alaska—go to Hawaii, near mouth of Gulf of California and west coast mainland Mexico	15 000+	850
Western Aleutians—go to Northern Marianas, Ryukyu Islands, Bonin Island, Taiwan, Korea, and other islands		150
North Atlantic		5 100
Northeast Atlantic—go to Cape Verde Islands and adjacent coast of Africa		100
Northwest Atlantic—go to southern Bahamas to Venezuela	10 000+	5 000+
TOTAL ALL AREAS		9 310+

tion apparently has an interesting recent history (Herman, 1979). The whales have invaded Hawaii only in the last 200 years. Whether they came to the islands due to harassment in other areas or due to climatic changes is unknown. Since the Hawaiian population appears to have been of recent origin, it may have arisen from the Mexican population and we thus see less segregation than is found with other populations. Five animals marked near the Aleutians were recovered in the Ryukyu Islands of Japan which represents a longitudinal displacement of 70°. Most of the smaller stocks have not been studied.

Abundance and Life History

Exploitation and population sizes

The history of exploitation of the humpback is primarily a story of the early part of the twentieth century, beginning with the modern whaling era. Humpbacks were particularly vulnerable due to their tendency to aggregate on the tropical breeding grounds and to come close inshore on the northern feeding grounds. Well over 60 000 humpbacks were killed between 1910 and 1916 in the Southern Hemisphere (Fig. 9). Other peaks of exploitation occurred in the 1930s and 1950s. However, heavy exploitation had occurred much earlier in the North Atlantic, resulting in low catches after 1900 and giving rise to the notion that there were not many humpbacks in this ocean. Exploitation has occurred throughout the twentieth century in the North Pacific with peak catches of over 3000 in 1962 and 1963.

Catching of humpbacks was prohibited in the Antarctic in 1939 but the seasons were reopened in 1949. Finally, after serious depletion of the stocks, the hunting of humpbacks in the whole Southern Hemisphere was banned in 1963 (Mackintosh, 1965). All hunting was prohibited in 1956 in the North Atlantic and in 1966 in the North Pacific. In all cases, prohibition occurred after the stocks had been seriously depleted.

The total population of humpback whales today numbers at least 6000 individuals, with the likelihood of another 1000 to 3000 stragglers here and there (Table 2). The original world population was at least 150 000 individuals; the historical population sizes given in Table 2 are only estimates for various times earlier in this century. The healthiest population in terms of total numbers (5000 or more) is in the western North Atlantic (Mitchell and Reeves, 1983). It is hoped that this represents an increase due to protection since 1956. Perhaps other populations will increase with time. There is little evidence, however, for any significant increase in the eastern North Atlantic. With levels at about 6% of the original population, it will take many years

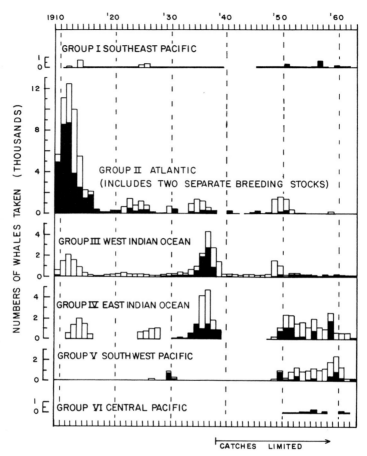

FIG. 9 Number of humpbacks taken in the world's oceans between 1910 and 1963. Group II includes the two breeding stocks from the east and west coasts of the South Atlantic. See Table 2 for explanation of other groups. The number of humpbacks taken from 1910 to 1916 (especially in group II) are estimates from total catch records. In group V, 1961, the actual catch may have been larger than the recorded catch. (□) Warm latitudes—winter; (■) Antarctic—summer. (Redrawn from Mackintosh, 1965, with permission.)

before we will see the results. Estimates for various geographic areas are presented in Table 2.

There are only a few areas where humpbacks are now taken. One or two every other year are killed in Bequia and the Cape Verde Islands. Ten to 12 a year are taken in Greenland. Up to 10 a year were taken in Tonga until hunting was stopped in 1978. These are only subsistance fisheries like the United States Eskimo bowhead fishery. However, it is possible that subsistance fisheries can maintain local stocks at a low level. In addition to subsistence whaling, at least one humpback has been taken by pirate whalers.

Some of the stocks around the world are still seriously depleted. Dr Miyazaki, of the University of the Ryukyus reports no confirmed sightings around Okinawa. Several possible sightings by fishermen suggest that a few animals still come to the area. There are no recent confirmed sightings in the Northern Marianas. We sighted and recorded only one singing humpback in the Cape Verde Islands off Africa in 1979, although a mother and calf had been killed the year before. Certainly the stocks all around the Antarctic still appear to be at low population levels.

As shown above, humpbacks are found throughout the oceans in small numbers. They are presumably in position to exploit new habitats, particularly large shallow banks with temperatures of 25°C which are appropriate calving areas.

Behaviour

The humpback is one of the most acrobatic whales, exhibiting a variety of conspicuous surface and underwater behaviours. The animals typically throw their flukes preceding longer dives, showing the characteristic black and white ventral colour pattern (see Fig. 3). Humpbacks may leap almost completely out of the water (breaching), or vertically lift only their head to the surface so that their eyes are above water (spyhopping; see Fig. 10). Along with lobtailing, or tail slapping, breaching may serve as some type of communication or "acoustic contact," as the sound may carry several miles (Herman and Tavolga, 1980). Payne (1978) has noticed that other humpbacks may join a breaching individual and speculates that breaching may be a spacing mechanism. Tomilin (1957) states that breaching is a reaction to any type of excitation.

The most obvious behavioural differences occur between the summer feeding and winter breeding grounds. During the breeding season several behaviours occur that may be involved with courtship and mating. Scammon (1874) and Herman and Forestell (1977) reported on animals rubbing and patting each other with their pectorals while lying side by side. In some cases one animal would swim between the outstretched pectorals of another and be stroked over its entire length. Headrubbing was also seen. Tyack (1981) found four types of behaviour that may have been associated with courtship or mating from animals off Hawaii. An animal may slap the water with one of its pectoral flippers (flippering), or lie on its back and alternately slap the

FIG. 10 Humpbacks off Cape Cod. The animal on the left is spyhopping (ventral surface is facing photographer). (Photograph courtesy of the University of Rhode Island, under sponsorship of the Minerals Management Service.)

water with one flipper at a time (belly flippering). The whale may raise the dorsal portion of its head horizontally to the surface, and then sink back down underwater without travelling forward (head-up). Rolling may also occur in a courtship context, and corollaries of this include a raised flipper (no slapping) or flukes lifted vertically rather than horizontally. These behaviours have also been reported from animals in feeding areas and are therefore not limited to breeding grounds.

Social groups of up to seven animals may form (Payne, 1978), and "escort" whales within a group may accompany cow–calf pairs. These escorts become aggressive towards other humpbacks approaching the group, and sometimes blow bubbles from their blowhole or mouth as an apparent "screen." There is some indication that the escorts may be males (Herman and Tavolga, 1980; Tyack, 1981; Whitehead, 1982). Agonistic encounters between animals may take the form of rear body thrashing (the whale throws the rear third of its body out of the water and then slams it down sideways against the surface), horizontal tail lashing (the animal lashes its flukes sideways through the water), and lobtailing. Many of these movements are believed to be threat displays, as contact is rarely made (Tyack, 1981).

Agonistic behaviour in the form of threat displays may also be directed towards boats, primarily on the breeding grounds. Groups of whales are more aggressive than single animals. An individual or a small group may charge a boat, "scream," and then veer off. Animals have also begun breaching or lobtailing when approached by boats, or have directed lateral tail swipes at boats or other whales, although the blow usually misses. Head-nodding appears to be another type of threat display commonly seen (Payne, 1978).

Feeding

On summer feeding grounds humpbacks feed on a variety of prey species. In the northern North Pacific and Bering Sea they feed on euphausiids (or krill) and schools of mackerel, sand lance, capelin, and herring. In the Antarctic they eat primarily euphausiids (Nemoto, 1959; Jurasz and Jurasz, 1979). Australian animals eat both euphausiids and fish (Nemoto, 1959). Sand lance, along with some herring and pollock, appear to be the predominant food types of humpbacks in Cape Cod and Greenland waters (Perkins *et al.*, 1982; CETAP, 1982; Hain *et al.*, 1982; Watkins and Schevill, 1979; Overholtz and Nicolas, 1979). In addition to euphausiids and fish, some animals have been found with bottom organisms in their stomachs (Tomilin, 1957; Kawamura, 1980) (Table 3).

Regardless of prey type, humpbacks feed on plankton or fish in large patches or schools, and for this reason are classified as "swallowers" rather than "skimmers" (Nemoto, 1970). When feeding, the humpback typically takes in food and water through its mouth. Ventral grooves in the throat allow it to expand and accommodate large quantities of food and water. The mouth is then closed, water is pressed out, and the food is caught in the baleen plates and subsequently swallowed (Figs. 11 and 12). The internal mechanism of rorqual feeding (particularly the use of the tongue) is described by Lambertsen (1983), and may be applied to the humpback as well.

Ingebrigtsen (1929) was one of the first to describe the elaborate humpback feeding behaviour. An animal would lie on its side on the surface, swim in a circle, and strike the water with its flukes to form a "ring of foam", which would surround the krill. The humpback would then dive under the ring and surface in the center, mouth open. Bubble feeding was also mentioned by Ingebrigtsen and, along with other feeding behaviours, has been studied in greater detail by several researchers.

In Alaskan and Cape Cod waters, the dominant feeding methods are lunging and bubble behaviours. An animal lunge-feeds by swimming vertically or obliquely up through aggregations of plankton or fish. This behaviour occurs particularly when food is abundant. Variations exist, such as

FIG. 11 Humpbacks lunge-feeding. The mouths are open with the top jaw plainly visible. The lower jaws are discernable at water level. Baleen plates and the characteristic knobs on top of the head can also be seen. (Photograph courtesy of the University of Rhode Island, under sponsorship of the Minerals Management Service.)

lateral and inverted lunging. Groups of two or more whales have a tendency to lunge-feed laterally. In a group of whales lunge-feeding, breathing and lunging tend to be synchronous, individual lunging speed is slowed down, and the animals' movements are orchestrated so that they usually avoid each other (Jurasz and Jurasz, 1979; Hain et al., 1982).

Bubble behaviours utilise different types of underwater exhalations called bubble clouds and bubble columns. A bubble cloud is formed by one underwater exhalation which concentrates or herds a mass of prey. Feeding is presumed to occur underwater, after which the whale rises slowly to the surface within the bubble cloud. After several blows and some shallow diving the manoeuvre is repeated. Several feeding variations exist with the use of bubble clouds. The whale may lunge-feed off to one side of the cloud, or swim above it and then horizontally feed between the still rising cloud and the surface of the water. More than one bubble cloud may be used at once (Jurasz and Jurasz, 1979; Hain et al., 1982).

Bubble columns are formed as a humpback swims underwater in a broad circle while exhaling. The individual bubble columns form rows, semicircles, or complete circles (bubble nets). Acting very much like a seine net, the configurations of these columns presumably serve the same function as the bubble clouds: to concentrate or herd the prey. The whale then lunge-feeds at an acute angle to the bubble columns, or below the surface (Jurasz and Jurasz, 1979; Hain *et al.*, 1982).

Two feeding methods observed less often from animals in the Cape Cod region are tail slashing and "inside loop behavior" (Hain *et al.*, 1982). In the first, the animal swims in a large circle while slashing its tail through the water. Feeding is then presumed to take place in the centre of the turbulence. In the inside loop behavior, the whale makes a shallow dive, hitting the water with its flukes as it submerges. A 180° roll is rapidly executed as the animal makes a sharp U-turn ("inside loop") and lunge-feeds slowly through the turbulent area created by its flukes. In some cases a whale feeds beside the area of turbulence, or another humpback would lunge-feed through the disturbance alone or with the original animal. "Flick feeding," a similar behaviour, has been described by Jurasz and Jurasz (1979) from animals in Alaskan waters. They noted that the behaviour occurred only when the whales were feeding on euphausiids.

FIG. 12 Lunge-feeding humpback with widely distended throat. The ventral grooves allowing the throat to expand can be seen. Water is being pressed out of the mouth as food is being caught in the baleen plates. (Photograph courtesy of the University of Rhode Island, under sponsorship of the Minerals Management Service.)

TABLE 3 Food types of humpback whales in different feeding areas

Antarctic[a]	Northern North Pacific, Bering Sea[a,b,c]	Cape Cod, Newfoundland, Greenland[d,e]	Australia and New Zealand[c]
Euphausia superba	*Euphausia pacifica*	Sand lance (*Ammodytes* sp.)	*Euphausia spinifera*
Euphausia crystarollophyas	*Thysanoessa longipes*	Herring (*Clupea harengus*)	*Euphausia hemigibba*
Thysanoessa macrura	*Thysanoessa raschii*	Capelin (*Mallotus villosus*)	*Nyctiphanes australis*
Thysanoessa vicina	*Thysanoessa inermis*		*Munida gregaria* (decapod larvae)
Calanus propinquus	Atka mackerel (family Scombridae)		Some herringlike fishes [*Clupea fimbriata* (?)]
Parathemisto gaudichaudi	Sand lance (*Ammodytes* sp.)		
	Herring (*Clupea harengus*)		
	Capelin (*Mallotus villosus*)		
	Saffron cod (*Eleginus glacilis*)		
	Arctic cod (*Boreogadus saida*)		

[a]From Nemoto (1959).
[b]From Jurasz and Jurasz (1979).
[c]From Kawamura (1980).
[d]From Whitehead (1982); CETAP Annual Report (1982); Hain et al. (1982).
[e]From Mitchell (1973); Whitehead et al. (1980).

Along with specific feeding behaviours, the large amount of white on dorsal flipper surfaces of Northern Hemisphere humpbacks may aid in concentrating prey (Howell, 1930; Brodie, 1977). Different orientations of the flippers may also assist in trapping prey during bubblenetting procedures by helping form the bubble curtain itself (Hain *et al.*, 1982).

Competition may occur among humpbacks feeding on large concentrations of plankton or fish. In one observation the animals rushed to the surface together while pushing and shoving each other with mouths full and throats distended (Watkins and Schevill, 1979).

Swimming and diving

Swimming speeds may reach as high as 27 km/hr from wounded animals (Zenkovich, 1937, in Tomilin, 1957) to a more common range of 3.8–14.3 km/hr (average 8.0 km/hr) from migratory animals (Chittleborough, 1953). Cows with calves swim the slowest, and lone humpbacks tend to travel faster than those in groups (Chittleborough, 1953; Payne, 1978).

Most humpbacks stay submerged for under 6 or 7 min, although longer dives of 15 to 30 min have been recorded (Liouville, 1913; Tomilin, 1957). Levenson (1969) divides underwater durations of animals off Bermuda into short and long dives. Short dives usually last 2–4 min with surface times of approximately 1 min. Blows are not regular, and flukes are not lifted as the whale submerges. This behaviour usually typifies shallow dives. In longer, presumably deeper dives of 8 to 15 min, the flukes are lifted and the animal surfaces between dives for about 4 min, blowing regularly. Groups of two or three animals were found to dive and surface in unison.

Recent studies in Cape Cod waters (CETAP, 1982) have shown typical dive times from 1 sec to 10.1 min, with a median surface activity time between dives of approximately 41.1 sec. Time spent at the surface varied among different animals and activities: surface feeding humpbacks had the shortest median surface duration of 26.4 sec, while calves spent the longest time at the surface (2.04 min). Lone animals spent a mean of 40.2 sec at the surface; groups of two to three whales, 46.2 sec; and cows, 48.6 sec. In terms of percentage of a total time budget, lone and feeding humpbacks spend the least amount of time at the surface (24–27%); groups, an intermediate amount (30%); and cows and calves, the longest (41% and 65%, respectively).

Respiration rates of humpbacks are similar to those recorded for finback and right whales, with a median of 68 breaths/hr. Lone animals have respiration rates of 63 breaths/hr, surface feeding humpbacks 66 breaths/hr, groups of two or three whales 68 breaths/hr, cows 81 breaths/hr, and calves 83 breaths/hr (CETAP, 1982).

Sound production

Of all baleen whale sounds, those of the humpback whale have been the most studied. The humpback produces many different types of sounds, including moans, groans, cries, squeals, chirps, and clicks.

On tropical breeding grounds, many of these sounds are arranged in complex sequences called songs (Payne and McVay, 1971). The songs comprise sounds arranged into themes, phrases, and syllables (or units) which are repeated monotonously for long periods of time. Distinct dialects exist in different ocean basins, with subtle differences among populations, subpopulations, and individual animals. Songs change annually in each population. While several theories exist concerning song functions, the predominant thought is that they are sung by sexually mature, isolated males in a breeding context (Anonymous, 1969; Winn and Winn, 1980; Winn *et al.*, 1981; Thompson, 1981; Tyack, 1981). In Bermuda and the West Indies, most song sounds contain fundamental frequencies from 40 to 5000 Hz, with harmonic energy generally less than or slightly beyond 8000 Hz (Levenson, 1969; Levenson and Leapley, 1976; Winn *et al.*, 1971; Thompson *et al.*, 1979; Winn *et al.*, 1981).

Sounds that are not organised in complex patterns may be heard throughout the humpback's migratory range. These vocalisations have been called "social sounds" by several researchers (see Payne, 1978), although their functions are as yet unclear. The sounds produced in the humpback's colder, summer feeding areas are typically sporadic and few in number when compared with the continuous vocalisations recorded in the tropics. Isolated northern latitude vocalisations similar to song-type sounds were recorded by Winn *et al.* (1979) off Newfoundland. Scattered occurrences of higher frequency sounds (clicks and clacks) with energy from 2 to 14 kHz have been recorded in Cape Cod, Massachusetts, and Newfoundland waters (Winn *et al.*, 1971; Winn *et al.*, 1979). No evidence exists to date that humpbacks or any mysticetes use these higher frequency sounds for echolocation. Lower frequencies, however, may be important in orientation and navigation, or in locating large objects such as masses of prey (Winn, 1972; Norris, 1969; Ayrapetyants and Konstantinov, 1973; Beamish, 1974; Yablokov *et al.*, 1974; Kinne, 1975; Winn and Winn, 1978; Thompson *et al.*, 1979; Herman and Tavolga, 1980).

Watkins (1967) recorded loud wheezing blows from humpbacks off Cape Cod, Massachusetts. Frequencies reached as high as 2 kHz, and duration was 1.5–2 sec. In Alaskan waters, Thompson *et al.* (1977) recorded trains of 20 to 85 Hz broadband pulses from feeding humpbacks, along with scattered grunts, yelps, moans, shrieks, and trumpeting through the animals' blowholes. Rare occurrences of songs have been recorded on Alaskan and Cape Cod feeding grounds (Jurasz and Jurasz, 1979; Winn and Winn, 1980).

Relationships with other species

Reports on the humpback's interrelationships with other species are few. Mitchell (1974) stated that humpbacks may compete with other rorquals, particularly the finback, for food. According to Mitchell, the humpback and fin have a wide range of preferred prey, and by virtue of this generalist feeding strategy, may avoid direct competition. Humpbacks and fins have been observed feeding close to each other with no apparent interaction other than avoidance (Watkins and Schevill, 1979). White-sided dophins (*Lagenorhynchus acutus*) are commonly seen in association with humpbacks, particularly when the whales are feeding (CETAP, 1982). Watkins *et al.* (1981) noted humpbacks apparently feeding on small fish with Dall's porpoises (*Phocoenoides dalli*) in Prince William Sound, Alaska. Many seabirds may also be in the vicinity of feeding humpbacks, presumably preying on the same food (usually fish). Minke whales (*Balaenoptera acutorostrata*) have been seen in close proximity with humpbacks; however, no detailed studies of their interactions exist (CETAP, 1982).

Territoriality

While there is no direct evidence of territoriality in humpbacks, several observations suggest some type of preferred area by individual animals or groups. The very nature of their seasonal return to the same feeding and breeding grounds shows some degree of range preference. Schevill and Backus (1960) reported on one humpback's "residency" in the same area near Portland, Maine, at specific times of the day for 6 days. Mayo (1982) observed what he termed "site tenacity" by an animal in shallow water off Cape Cod. The whale may have been feeding in a strong rip current in the area. One possible function of the well-known song of the humpback may be the transmission of territorial information by males on the breeding grounds (Winn and Winn, 1978; Tyack, 1981).

Reproduction

Humpback calves are born in the warm tropical and subtropical waters of each hemisphere. The gestation period lasts from 11 to 11.5 months. The embryo grows on the average of 17 to 35 cm a month, although this growth may not be uniform throughout pregnancy (Tomilin, 1957). Newborn calves are between 4 and 5 m long and are suckled for approximately 5 months (Matthews, 1937; Nishiwaki, 1959; Chittleborough, 1965; Tomilin, 1957). The females' milk is highly nutritive, containing 20.4–41.3% fat, 10.7–

13.6% protein, 0.2–1.7% lactose, and 40.6–65.4% water (Yablokov *et al.*, 1974). A minimum of 43 kg of milk may be consumed by the calf daily. Growth is quite rapid during this time, and at the end of the lactation period calves are between 7.5 and 9 m long (Matthews, 1937; Tomilin, 1957).

Sexual maturity is reached usually between 4 and 5 years of age (Chittleborough, 1959, 1965; Nishiwaki, 1959), and possibly in as little as 2 years (Matthews, 1937; Tomilin, 1957). In some cases puberty (when reproduction is first possible) may preceed sexual maturity (when full sexual capacity is attained) by 1 year (Chittleborough, 1955a). In sexually mature males the weight of the testes (2.0 kg each at puberty) and rate of spermatogenesis increase during the breeding season, coinciding with ovulation in the female (Chittleborough, 1955a,b). Although highly variable, penis length can also be an indication of sexual maturity: mean length at puberty is approximately 107 cm, with a range of 97 to 250 cm in adults (Chittleborough, 1955a; Tomilin, 1957).

Ovaries of sexually mature females weigh between 600 and 3000 g (Tomilin, 1957). After sexual maturity is reached, ovary weight remains fairly constant with age. The size of the ovaries does not decrease in older females, as it does in blue and finback whales (Chittleborough, 1954). As in other balaenopterids, "resting" Graafian follicles on the ovaries' surface enlarge as the time of ovulation approaches. In the humpback, follicles around 30 mm in diameter signify imminent ovulation. There is generally one ovulation per season; multiple ovulations are not common. Soon after ovulation, whether or not fertilisation has occurred, the follicle wall folds and forms a corpus luteum; this slowly degenerates into a corpus albicans. The corpora albicantia persist throughout the animal's life, and their number may be used as an index of relative age, although not with complete accuracy (Matthews, 1937; Chittleborough, 1954).

Breeding commonly occurs once every 2 years but may take place twice every 3 years. In the latter case, lactation may last longer than 5 months, continuing through the first part of the next ovulatory period. If the female is impregnated shortly after parturition, pregnancy and lactation (from the previous pregnancy) may exist simultaneously (Matthews, 1937; Chittleborough, 1954). Population density may play an important role in postpartum oestrus.

Reports of actual copulation are few and not necessarily reliable. Lillie (1910, in Nishiwaki and Hayashi, 1950) and Nishiwaki and Hayashi (1950) make direct reference to copulation, yet both sources refer only to body positioning; copulation was most likely inferred. In Antarctic waters (which are feeding grounds), two humpbacks swim in single file, and then engage in rolling, flippering, and tail fluk123. Both dive and then surface vertically with ventral surfaces "in close contact," appearing above the water to just below their flippers. They then fall back together or separately. These behaviours were repeated for several hours, with pauses between vertical surfacings (Nishiwaki and Hayashi, 1950).

Diseases, Parasites, and Toxins

Little is known about diseases of the humpback whale. Yablokov *et al.* (1974) state that in the true rorquals cirrhosis of the liver and mastitis may occur, but it is unclear whether humpbacks also contract the diseases. Tomilin (1957) found the humpback to be the most parasitized of the Balaenopteridae, carrying a wide variety of ecto- and endoparasites. The quantity of ectoparasites may be related to swimming speed: the humpback's relatively slow pace enables more ectoparasites to accumulate. Patches of parasites are found primarily in areas protected from strong water flow (Yablokov *et al.*, 1974). Different types of whale lice (order Amphipoda) inhabit scars, scratches, chin, throat, and urogenital slit. Barnacles (*Coronula* and *Conchoderma* spp.) are commonly found in the throat, chin, and urogenital slit. They are less common on the flipper protuberances, the edge of the caudal peduncle, and the posterior ventral pleats. Lillie (1915) noted that barnacles are usually attached to pigmented sections of skin, and rarely on white portions. Some diatoms (genera *Cocconeis* and *Licmophora*) and flagellates (*Hematophagus* sp.) are also common ectoparasites (Matthews, 1937; Tomilin, 1957).

The humpback plays host to a number of endoparasites including various trematodes, cestodes, nematodes, and acanthocephalans (Dailey and Brownell, 1972). Many helminths live in portions of the blubber, liver, mesentery, and intestine. *Ogmogaster ceti*, a commensal nematode specific to the Balaenopteridae, is found in humpback baleen plates (Yablokov *et al.*, 1974). Other species live in the penis, ureter, and kidneys (Yablokov *et al.*, 1974; Tomilin, 1957).

Taruski *et al.* (1975) found significant residue levels of DDT, PCB's, chlordane, and dieldrin in humpback blubber. Differences in DDT levels may have been related to the animals' annual migratory cycle. The highest DDT levels were found on feeding grounds, the lowest in breeding areas. This may be due to fasting in the breeding areas or to sampling of different subpopulations. Reported levels are high enough in some cases to suggest a continued monitoring effort for chlorinated hydrocarbons in several cetacean species, including the humpback.

Predation

The humpback, like most large whales, has few, if any predators other than man. Some animals do have killer whale teeth scars on their bodies (Katona,

1980), and killer whales have been seen harassing humpbacks apparently without attempting to kill them (Whitehead et al., 1982). Sharks usually attack only dead or weakened animals (Yablokov et al., 1974), although crescentic scars have been found on seemingly healthy animals that may have been caused by the teeth of *Isistius brasiliensis*, a squaloid shark (Matthews, 1937; Jones, 1971). Similarly, whitish, oval-shaped scars have been found on many humpbacks and are believed to be the marks of parasitic sea lampreys (*Lampetra* or *Petromyzon*) (Pike, 1957; Nemoto, 1955).

Entrapment

Considerable numbers of humpbacks are entrapped each summer in fishing gear, especially in Newfoundland waters (Beamish, 1979; Lien, 1980; Mitchell, 1982). Some are released alive but others are found dead.

References

Allen, G. M. (1916). The whalebone whales of New England. *Mem. Boston Soc. Nat. Hist.* **8**(2), 1–322.
Anonymous. (1969). Singing whales. *Nature (London)* **244**, 217.
Ash, C. E. (1953). Weights of Antarctic humpback whales. *Nor. Hvalfangst-Tid.* **42**, 279–283.
Ash, C. E. (1957). Weights and oil yields of Antarctic humpback whales. *Nor. Hvalfangst-Tid.* **46**, 569–573.
Ayrapetyants, E. Sh., and A. I. Konstantinov. (1973). "Echolocation in Animals". Acad. Sci. U.S.S.R., Joint Sci. Council Physiol. Man and Animals. Israel Program for Scientific Translations, Jeruselem.
Beamish, P. (1974). Whale acoustics. *Can. Acoust. Assoc. J.* **2**(4), 8–12.
Beamish, P. (1979). Behavior and significance of entrapped baleen whales. In "Behavior of Marine Animals. Volume 3. Cetaceans" (Eds. H. E. Winn and B. L. Olla), pp. 291–309. Plenum, New York.
Borowski, G. H. (1781). "Gemeinnützige Naturgeschichte des Thierreichs", Vol. 2. Lange, Berlin.
Breathnach, A. S. (1955). The surface features of the brain of the humpback whale (*Megaptera novaeangliae*). *J. Anat.* **89**, 343–354.
Brodie, P. F. (1977). Form, function and energetics of Cetacea: A discussion. In "Functional Anatomy of Marine Mammals" (Ed. R. J. Harrison), Vol. 3, pp. 45–58. Academic Press, New York.
CETAP. (1982). A Characterization of Marine Mammals and Turtles in the Mid- and North Atlantic Areas of the US Outer Continental Shelf, Final Report. Ref. No. AA551-CT8-48. Bureau of Land Management, Washington, D.C.
Chittleborough, R. G. (1953). Aerial observations on the humpback whales, *Megaptera nodosa* (Bonnaterre), with notes on other species. *Aust. J. Mar. Freshwater Res.* **4**, 219–226.

Chittleborough, R. G. (1954). Studies on the ovaries of the humpback whale on the western Australian coast. *Aust. J. Mar. Freshwater Res.* **5,** 35–63.

Chittleborough, R. G. (1955a). Puberty, physical maturity, and relative growth of the female humpback whale, *Megaptera nodosa* (Bonnaterre), on the western Australian coast. *Aust. J. Mar. Freshwater Res.* **6,** 315–327.

Chittleborough, R. G. (1955b). Aspects of reproduction in the male humpback whale. *Aust. J. Mar. Freshwater Res.* **6,** 1–29.

Chittleborough, R. G. (1959). Determination of age in the humpback whale, *Megaptera nodosa* (Bonnaterre). *Aust. J. Mar. Freshwater Res.* **10,** 125–143.

Chittleborough, R. G. (1965). Dynamics of two populations of the humpback whale, *Megaptera novaeangliae* (Borowski). *Aust. J. Mar. Freshwater Res.* **16,** 33–128.

Dailey, M. D., and Brownell, R. L., Jr. (1972). A checklist of marine mammal parasites. *In* "Mammals of the Sea: Biology and Medicine" (Ed. S. H. Ridgway), pp. 528–589. Thomas, Springfield, Illinois.

Dawbin, W. H. (1960). An analysis of the New Zealand catches of humpback whales from 1947–1958. *Nor. Hvalfangst-Tid.* **49,** 61–75.

Gaskin, D. E. (1976). The evolution, zoogeography, and ecology of Cetacea. *Oceanogr. Mar. Biol.* **14,** 247–346.

Gray, J. E. (1846). Mammalia: On the cetaceous animals. *In* "The Zoology of the Voyage of HMS *Erebus* and *Terror*" (Eds. J. Richardson and J. E. Gray), Vol. 1, p. 16. Janson, London.

Hain, J. H. W., Carter, G. R., Kraus, S. D., Mayo, C. A., and Winn, H. E. (1982). Feeding behavior of the humpback whale, *Megaptera novaeangliae*, in the western North Atlantic. *Fish. Bull.* **80**(2), 99–108.

Herman, L. M. (1979). Humpback whales in Hawaiian waters: A study in historical ecology. *Pac. Sci.* **33**(1), 1–15.

Herman, L. M., and Forestell P. (1977). The Hawaiian humpback whale: Behaviors. *Proc. Conf. Biol. Mar. Mammals 2nd.*, p. 29.

Herman, L. M., and Tavolga, W. N. (1980). The communication systems of cetaceans. *In* "Cetacean Behavior: Mechanisms and Functions" (Ed. L. M. Herman), pp. 149–210. Wiley, New York.

Howell, A. B. (1930). "Aquatic Mammals: Their Adaptations to Life in the Water". Thomas, Springfield, Illinois. (Reprinted, 1970, Dover, New York.)

Ingebrigtsen, A. (1929). Whales caught in the North Atlantic and other seas. *Rapp. P.-V. Reun. Cons. Int. Explor. Mer.* **56,** 1–26.

Jones, E. (1971). *Isistius brasiliensis*, a squaloid shark, the probable cause of crater wounds on fishes and small cetaceans. *Fish. Bull.* **69,** 791–798.

Jurasz, C. M., and Jurasz, V. P. (1979). Feeding modes of the humpback whale, *Megaptera novaeangliae*, in southeast Alaska. *Sci. Rep. Whales Res. Inst.* **31,** 69–83.

Katona, S. (1980). Paper presented at Humpback Whales of the Western N. Atlantic Workshop, 17–21 Nov., New England Aquarium, Boston, Massachusetts.

Katona, S., Baxter, B., Brazier, O., Kraus, S., Perkins, J., and Whitehead, H. (1979). Identification of humpback whales by fluke photographs. *In* "Behavior of Marine Animals. Volume 3. Cetaceans" (Eds. H. E. Winn and B. L. Olla), pp. 33–44. Plenum, New York.

Kawamura, A. (1980). A review of food of balaenopterid whales. *Sci. Rep. Whales Res. Inst.* **32,** 155–197.
Kinne, O. (1975). Orientation in space: Animals. Mammals. *In* "Marine Ecology. Vol. II: Physiological Mechanisms" (Ed. O. Kinne), pp. 709–916. Wiley, New York.
Klima, M. (1978). Comparison of early development of sternum and clavicle in striped dolphin and in humpback whales. *Sci. Rep. Whales Res. Inst.* **30,** 253–269.
Lambertsen, R. H. (1983). Internal mechanism of rorqual feeding. *J. Mammal.* **64**(1), 76–88.
Leatherwood, S., and Evans, W. E. (1979). Some recent uses and potentials of radio telemetry in field studies of cetaceans. "Behavior of Marine Animals. Volume 3. Cetaceans" (Eds H. E. Winn and B. L. Olla), pp. 1–31. Plenum, New York.
Levenson, C. (1969). Behavioral, physical, and acoustic characteristics of humpback whales(*Megaptera novaeangliae*) at Argus Island. Informal Rep. No. 69–54. U.S. Naval Oceanogr. Office.
Levenson, C., and Leapley, W. T. (1976). Humpback whale distribution in the eastern Caribbean determined acoustically from an oceanographic aircraft. Tech. Note 3700-46-76. U.S. Naval Oceanogr. Office, Washington, D.C.
Lien, J. (1980). "Humpback whale entrapment in inshore fishing gear in Newfoundland." Memorial Univ. of Newfoundland, St. Johns.
Lillie, D. G. (1915). Cetacea. British Antarctic (*"Terra Nova"*) Expedition, 1910. *Nat. Hist. Rep. Zool.* **1,** 85–124.
Liouville, J. (1913). "Cétacés de l'Antarctique (Baleinoptères, Ziphiidés, Delphinidés). Deuxieme Expedition Antarctique Française (1908–1910)". Masson, Paris.
Mackintosh, N. A. (1965). "The Stocks of Whales". Fishing News, London.
Matthews, L. H. (1937). The humpback whale—*Megaptera nodosa*. *Disc. Rep.* **17,** 7–92.
Mayo, C. (1982). Occurrence of humpback whales, *Megaptera novaeangliae*, on Stellwagon Bank, Mass.: A summary paper presented at the Western North Atlantic Whale Research Assoc., 15 Oct., Boston, Massachusetts.
Mitchell, E. D. (1973). Draft report on humpback whales taken under special scientific permit by eastern Canadian land stations, 1969–1971. *Rep. int. Whal. Commn* **23,** 138–154.
Mitchell, E. D. (1974). Trophic relationships and competition for food in northwest Atlantic whales. *Proc. Can. Soc. Zool. Ann. Meet.*, pp. 123–133.
Mitchell, E. (1982). Canada: progress report on cetacean research June 1980 to May 1981. *Rep. int. Whal. Commn* **32,** 161–169.
Mitchell, E., and Reeves, R. (1983). Catch history, abundance, and present status of northwest Atlantic humpback whales. *Rep. int. Whal. Commn (Spec. Issue 5)*, 153–212.
Nemoto, T. (1955). White scars on whales. *Sci. Rep. Whales Res. Inst.* **10,** 69–77.
Nemoto, T. (1959). Food of baleen whales with reference to whale movements. *Sci. Rep. Whales Res. Inst.* **14,** 149–290.

Nemoto, T. (1970). Feeding pattern of baleen whales in the ocean. *In* "Marine Food Chains" (Ed. J. H. Steele), pp. 241–252. Univ. of California Press, Berkeley.

Nishiwaki, M. (1959). Humpback whales in Ryukyuan waters. *Sci. Rep. Whales Res. Inst.* **14**, 49–88.

Nishiwaki, M., and Hayashi, K. (1950). Copulation of humpback whales. *Sci. Rep. Whales Res. Inst.* **3**, 183–185.

Norris, K. S. (1969). The echolocation of marine animals. *In* "The Biology of Marine Mammals" (Ed. H. T. Andersen). pp. 183–227. Academic Press, New York.

Overholtz, W. J., and Nicolas, J. R. (1979). Apparent feeding by the fin whale, *Balaenoptera physalus*, and humpback whale, *Megaptera novaeangliae*, on the American sand lance, *Ammodytes americanus*, in the Northwest Atlantic. *Fish. Bull.* **77**(1), 285–287.

Payne, R. (1978). Report on a workshop on problems related to humpback whales (*Megaptera novaeangliae*) in Hawaii. Marine Mammal Commission Rept. No. MMC-77/03. NTIS PB 280 794.

Payne, R., and McVay, S. (1971). Songs of humpback whales. *Science* **173**, 585–597.

Perkins, J. S., Bryant, P. J., Nichols, G. and Patten, D. R. (1982). Humpback whales (*Megaptera novaeangliae*) off the west coast of Greenland. *Can. J. Zool.* **60**(11), 2921–2930.

Pike, G. C. (1957). Lamprey marks on whales. *J. Fish. Res. Board Can.* **8**, 275–280.

Pilleri, G. (1966). Note on the anatomy of the brain of the humpback whale, *Megaptera novaeangliae*. *Rev. Suisse Zool.* **73**, 161–165.

Pivorunas, R. (1977). The fibrocartilage skeleton and related structures of the ventral pouch of balaenopterid whales. *J. Morphol.* **151**, 299–314.

Quiring, D. P. (1943). Weight data on five whales. *J. Mammal.* **24**(1), 39–45.

Ruud, J. T., Clarke, R., and Jonsgård, A. (1953). Whale marking trials at Steinshamn, Norway. *Nor. Hvalfangst-Tid.* **42**(8), 429–441.

Scammon, C. M. (1874). "The Marine Mammals of the North-western Coast of North America". Carmany, San Francisco.

Schevill, W. E., and Backus, R. H. (1960). Daily patrol of a *Megaptera*. *J. Mammal.* **41**, 279–281.

Taruski, A. G., Olney, C. E., and Winn, H. E. (1975). Chlorinated hydrocarbons in cetaceans. *J. Fish. Res. Board Can.* **32**, 2205–2209.

Thompson, T. J. (1981). Temporal characteristics of humpback whale (*Megaptera novaeangliae*) songs. Unpublished Ph. D. dissertation, Univ. of Rhode Island, Kingston.

Thompson, P. O., Cummings, W. C., and Kennison, S. J. (1977). Sound production of humpback whales, *Megaptera novaeangliae*, in Alaskan waters. *J. Acoust. Soc. Am.* **62**(Suppl. 1), S89 (abstract).

Thompson, T. J., Winn, H. E., and Perkins, P. J. (1979). Mysticete sounds. *In* "Behavior of Marine Animals. Volume 3: Cetaceans" (Eds H. E. Winn and B. L. Olla), pp. 403–431. Plenum, New York.

Tomilin, A. G. (1957). "Mammals of the U.S.S.R. and Adjacent Countries. Vol. IX: Cetacea" (Ed. V. G. Heptner), Nauk S.S.S.R., Moscow. (English Translation, 1967, Israel Program for Scientific Translations, Jerusalem).

True, F. W. (1904). The whalebone whales of the western North Atlantic. *Smithson. Contrib. Knowl.* **33**, 1–332.

Tyack, P. (1981). Interactions between singing Hawaiian humpbacks and conspecifics nearby. *Behav. Ecol. Sociobiol.* **8**, 105–116.

Watkins, W. A. (1967). Air-borne sounds of the humpback whale (*Megaptera novaeangliae*). *J. Mammal.* **48**, 573–578.

Watkins, W. A. (1981). Reaction of three species of whales, *Balaenoptera physalus, Megaptera novaeangliae,* and *Balaenoptera edeni* to implanted radio tags. *Deep-Sea Res.* **28A**, 589–599.

Watkins, W. A., and Schevill, W. E. (1979). Aerial observation of feeding behavior in four baleen whales: *Eubalaena glacialis, Balaenoptera borealis, Megaptera novaeangliae,* and *Balaenoptera physalus. J. Mammal.* **60**(1), 155–163.

Watkins, W. A., Johnson, J. H., and Wartzok, D. (1978). Radio tagging report of finback and humpback whales. *WHO Tech. Rep.* 78–51.

Watkins, W. A., Moore, K. E., Wartzok, D., and Johnson, J. H. (1981). Radio tracking of finback (*Balaenoptera physalus*) and humpback (*Megaptera novaeangliae*) whales in Prince William Sound, Alaska. *Deep-Sea Res.* **28A**, 577–588.

Whitehead, H. (1982). Group structure and stability of humpback whales in the northwestern Atlantic. Paper presented at the western North Atlantic Whale Res. Assoc., 15 Oct. 1982, Boston, Massachusetts.

Whitehead, H., Harcourt, P., Ingham, K., and Clarke, H. (1980). The migration of humpback whales past the Bay of Verde Peninsula, Newfoundland, during June and July, 1978. *Can. J. Zool.* **58**(5), 687–692.

Whitehead, H., Glass, C., and Harcourt, P. (1982). Humpback whales on the Southeast Shoal—Summer 1982. Paper presented at the western North Atlantic Whale Research Assoc., 15 Oct. 1982, Boston, Massachusetts.

Winn, H. E. (1972). A comparison of mysticete and odontocete cetacean acoustic signals and their behavioral consequences. 84th Meeting of the Acoust. Soc. Am., Miami Beach, Florida, 18 Nov.–1 Dec.

Winn, H. E., and Winn, L. K. (1978). The song of the humpback whale (*Megaptera novaeangliae*) in the West Indies. *Mar. Biol. (Berlin)* **47**, 97–114.

Winn, H. E., and Winn, L. K. (1980). Sound production during feeding of humpback whales. Paper presented at the humpback whales of the western North Atlantic workshop., 17–20 Nov., New England Aquarium, Boston, Massachusetts.

Winn, H. E., Perkins, P. J., and Poulter, T. C. (1971). Sounds of the humpback whale. Proc. of the 7th Ann. Conf. on Biol. Sonar and Diving Mammals, pp. 39–52. Stanford Res. Inst., Menlo Park, California.

Winn, H. E., Bischoff, W. C., and Taruski, A. G. (1973). Cytological sexing of Cetacea. *Mar. Biol. (Berlin)* **23**, 343–346.

Winn, H. E., Edel, R. K., and Taruski, A. G. (1975). Population estimate of the humpback whale (*Megaptera novaeangliae*) in the West Indies by visual and acoustic techniques. *J. Fish. Res. Board Can.* **32**, 499–506.

Winn, H. E., Beamish, P., and Perkins, P. J. (1979). Sounds of two entrapped humpback whales (*Megaptera novaeangliae*) in Newfoundland. *Mar. Biol. (Berlin)* **55**, 151–155.

Winn, H. E., Thompson, T. J., Cummings, W. C., Hain, J., Hudnall, J., Hays, H., and Steiner, W. W. (1981). Song of the humpback whale—population comparisons. *Behav. Ecol. Sociobiol.* **8,** 41–46.

Yablokov, A. V., Bel'kovich, V. M., and Borisov, V. I. (1974). "Whales and Dolphins", Parts 1 and 2. Natl. Tech. Info. Serv., Springield, Virginia.

10

Right Whales

Eubalaena glacialis (Müller, 1776) and *Eubalaena australis* (Desmoulins, 1822)

William C. Cummings

Genus and Species

Right whales were so named by early whalers because they were the "right" whales to pursue. Under certain circumstances they were relatively easy to catch, floated when killed, and produced an exceedingly valuable yield of oil and baleen. The term black right whale has been used to distinguish these animals from the bowhead (in the past called Greenland right), which does not have a mass of excrescences on the head. "Black" is misleading in this case, because were it not for the white-tipped lower jaw, the bowhead would appear as black as southern and northern right whales.

Several authors have considered the problems of naming the right whales, a practice sometimes based on geographic separation and not on significant anatomical differences. Eschricht and Reinhardt (1861) established the distinction between the Arctic bowhead and the right whales of more temperate waters. Some authorities have put the right whales and the bowhead in a single genus, *Balaena*. However, right whales are treated here as members of a separate genus, *Eubalaena*, in one chapter with no attempt to judge the validity of the various scientific names of recent usage (see Omura, 1958; Tomilin, 1962; Hershkovitz, 1966; Rice and Scheffer, 1968). Even for the taxonomist, such an effort is hardly possible at present in view of inadequate specimens from all regions. Moreover, the sorely needed protection now afforded these species would make it difficult to obtain the required study material.

There are three major geographic regions of which corresponding trinomials of right whales have been proposed: southern oceans,—southern right whale, *Eubalaena glacialis australis* Desmoulins: North Atlantic—North Atlantic right whale (Nordkaper, Biscayan right), *Eubalaena glacialis glacialis* Müller; and the North Pacific—North Pacific right whale, *Eubalaena glacialis japonica* Lacépède (Hershkovitz, 1966). All of these are clearly distinct from the bowhead (treated in Chapter 11). However, just how the right whales of the genus *Eubalaena* are related is not known; they may even be conspecific. Omura (1958) and Omura *et al.* (1969) compared data from a number of specimens from the North Atlantic and North Pacific. Finding insignificant differences, they concluded that the name *Eubalaena glacialis* should be applied to right whales from both areas.

The trinomials are not used in this chapter. Instead, Omura's findings on northern right whales are accepted, and I follow the practice of using the name *Eubalaena glacialis* (Müller) for all North Pacific and North Atlantic right whales. *Eubalaena australis* (Desmoulins) is used here for all right whales below the equator (southern right whales), thus restricting the generic name *Balaena* for the bowhead, *Balaena mysticetus* Linnaeus, which is not called a right whale for the sake of avoiding confusion.

The student must be careful to recognise that certain anecdotal accounts in the earliest literature may be confusing, because the early whalers did not distinguish in their records between bowheads and right whales of the Northern Hemisphere, or between northern and southern right whales.

External Characteristics and Morphology

Colour

Eubalaena generally is uniformly dark in color (Fig. 1) except for scars, belly patches of varying size, parasites, and head excrescences, most of which are light (Fig. 2). Piebalding on the belly varies between individuals of both

FIG. 1 Northern right whale at a land-based whaling station. (Photo by G. C. Pike, courtesy of I. MacAskie and S. Leatherwood.)

FIG. 2 Swimming northern right whale. (Courtesy of F. Kasamatsu.)

species—from very little or none to huge white blotches that nearly cover the abdomen. Scammon (1874) had seen some northern right whales that were piebald all over. Although white calves of southern right whales have been noted, Best (1970) has reported that it is unlikely that there is any significant difference in the amount of white on northern and southern right whales, as some statistics indicate.

Excrescences

The heads of right whales are decorated with varying amounts of the yellowish, reddish, or whitish excrescences (often called callosities), the largest of which appears as a "bonnet" near the dorsal extremity (Figs. 2 and 3). Other excrescences frequently occur between the bonnet and the region just back of the blowholes, along the lower jaw, and above the eye (Fig. 3). Southern right whales may have excrescences on the upper edge of the lower jaw more frequently than northern right whales (Best, 1970). A comparison of my observations on large numbers of southern right whales, my sightings of right whales off Cape Cod, and published illustrations of northern right whales supports Best's contention. Payne (1972; see also Payne *et al.*, 1981) was able to identify certain individual southern right whales according to their pattern of excrescences. Whitish spots of varying extent also may appear on the flukes and flippers, but *Eubalaena* does not have the conspicuous white chin so characteristic of the bowhead.

FIG. 3   Southern right whale's head showing bonnet on the snout and other excrescences on the head, upper edge of lower jaw, and fore and aft of the blowholes. (After Cummings *et al.*, 1974.)

A number of ideas has been presented concerning the origin of the bonnet and other excrescences (reviewed by Matthews, 1938), for example, excrescences formed from the adhesion of parasitic barnacles, growths caused by irritation of whale lice, formations from natural development, disease-induced growths, pathological corn or callus induced by rubbing off parasites (callosity), and aftermath of hypertrophied hair follicles. Since they do occur on baby southern right whales and even on the foetuses of northern right whales, excrescences probably are not caused by wear or parasites, but their presence does offer a surface of interstices which would facilitate parasitic attachment. Although we have by convention used the term callosity in connection with excrescences of southern right whales (Cummings *et al.*, 1974), W. E. Schevill (personal communication), Woods Hole Oceanographic Institution (who colloquially refers to the "rock garden") told me that "callosity" may not be the best term since it suggests a questionable origin.

Body form

Compared with other mysticetes, especially the balaenopterids, right whales are very large in girth relative to their length. This ratio gives them a noticeably rotund appearance (Figs. 1 and 4). At a length of 16 to 17 m, northern right whales may weigh more than 100 tonnes (Klumov, 1962). The jaws are greatly arched, a suitable receptacle for the extraordinarily long baleen. The head is enormous, close to one-third the body length. Right whales have no dorsal fin, nor do they have the grooved throat that is characteristic of the balaenopterids. The flippers are short and very broad. Hair appears on the tips of the chin and upper jaw on both southern and northern right whales. It is also associated with excrescences.

Baleen

The baleen plates and their bristles are generally dark although some white has been noted on a few individuals. There is a tendency for northern right whale baleen to darken with age of the animal and of the specimen at hand. Plates are very long in relation to those of other whales (Fig. 5), though not as long as those of the bowhead. Hinton (1925) reported a maximum length of 2.2 m for a plate from a southern right whale, and True (1904) reported 2.6 m as maximum for North Pacific right whales. More recently, Omura *et al.* (1969) found a 2.8-m plate in a 16.4-m male northern right whale. Data from numerous independent observations indicate that a series of plates typically numbers from 205 to 270 on each side, with no apparent difference in count between northern and southern right whales. As an example of the

FIG. 4 Southern right whale, posterior view. (After Cummings et al., 1972.)

value of right whale baleen, that from a single whale at one time brought as much as £3375 and, together with the oil, was enough to pay the ship's expenses for a whaling expedition (Norman and Fraser, 1949).

Body length

Based on records of the whale fishery, Omura (1958) thought that North Pacific *Eubalaena* generally attained greater body lengths than those of the North Atlantic. Although this also was the view of Allen (1908), some doubt is cast in Omura's later paper (Omura et al., 1969). Accordingly, the longest right whale on record from the North Pacific was a 17.8-m female. The largest from the North Atlantic, however, was an 18-m whale noted in the Scottish whaling records of 1908–1914 (Norman and Fraser, 1949), and Wang (1978) reported on an 18-m female caught near Hai Yang Island in the Yellow Sea. An 11.5-m female North Pacific right whale weighed 22.8

tons (Omura, 1958). Newborn *E. glacialis* are 5–6 m long and often lighter in colour than adults (Nishiwaki, 1972). Best (1970) has an account of four southern right whale foetuses, of which the largest was 6 m, and Matthews (1938) measured a southern right whale calf at 6.5 m. I measured a beached southern right whale calf at 4.8 m in length and 2.8 m in greatest girth. A Captain Day (Scoresby, 1820) reported that southern right whales captured in the fishery attained maximum lengths of 35 to 40 feet (10.7–12.2 m), and were seldom as large as the bowhead of Greenland and the Davis Straits. This reported figure is considerably smaller than the maximum noted for northern right whales, but it appears to have been only an estimate and quite

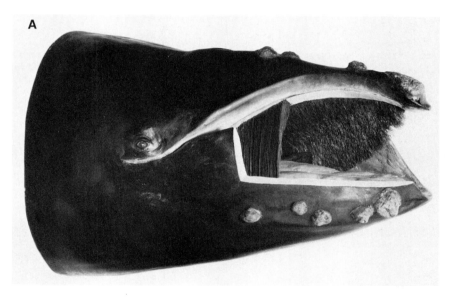

FIG. 5 (A) Model of the head of a Right Whale with the lower right lip and part of the baleen on the right side cut away to show the arrangement of baleen plates in the mouth (Fraser, 1952). (B) Long baleen plate of a northern right whale (top) compared with a plate from a rorqual (bottom) referenced to a 1-meter scale (Fraser, 1952).

possibly the average length was shortened as a consequence of high fishing pressure.

Blowholes

The blowholes of right whales are well partitioned on the exterior surface, a characteristic that results in a nearly vertical V-shaped blow which often makes them readily identifiable (Fig. 6). Contrary to popular notion, shape of the blow alone is not a reliable characteristic for the identity of distant right whales, because the two branches may appear as one when intermingled by the wind or when viewed from the side.

Genital area

As in the other whales, the genital aperture of female southern right whales appears in the same groove containing the anus. Just in front of the aperture is a large single-lobed clitoris. A mammary gland with nipple is situated on each side of the groove. The gland of a 15.4 m, nonlactating female northern right whale was spindle shaped, 1 m in length, and weighed 10 kg (Omura *et al.*, 1969).

FIG. 6 Characteristic V-shaped blow (exhalation) of right (southern) whales. (After Cummings *et al.*, 1972.)

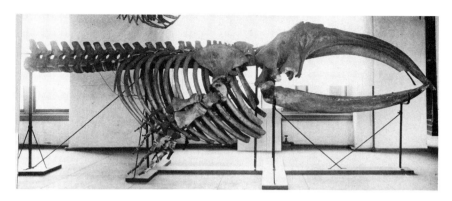

FIG. 7   Anterior skeleton of northern right whale. (Courtesy of J. G. Mead, Smithsonian Institution.)

The penis of southern right whales is pigmented and ends with a "blunt flagellum," according to Matthews' description (1938). That of northern right whales was described by Omura as being up to 2.7 m long, 1.1 m in girth at the base, and pigmented with bluish black. The male genital groove of a large southern right whale was 83 cm long, starting as a shallow recess and deepening towards the posterior. The male's anus is located well behind this groove.

Internal Anatomical Characteristics

Skeleton

The skeleton (Fig. 7) of *E. glacialis* was described in detail by Omura *et al.* (1969), True (1904), and Allen (1908). The skull (Figs 7–10) may be as long as 5.2 m and may weigh up to 1000 kg. Skull length increases proportionately with age. The rostrum and mandibles are greatly arched and the supraoccipital bears a bony shield which appears to be peculiar to *Eubalaena*.

The seven cervical vertebrae (Fig. 7) are fused so that the neck region may be only 2.4% of the entire skeletal length. Howell (1930) advanced the idea that this fusion was necessary to form a firm attachment for the massive head, and he considered the neck region to function as a single thoracic vertebra without ribs. There are 14–15 thoracic, 10–11 lumbar, and 25 caudal vertebrae, and 14–15 pairs of ribs in *Eubalaena*. A total of 55–57 vertebrae is found in the two species. Five digits appear in the pectoral limbs, each having 2–6 phalanges. Humerus, radius, and ulna are short as is typical

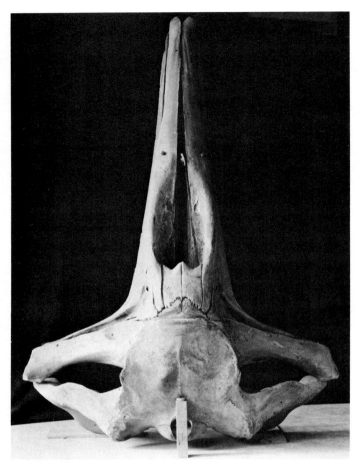

FIG. 8 Dorsal view of northern right whale skull. (Courtesy of J. G. Mead, Smithsonian Institution.)

for whales. The pelvic region is reduced to elongated pelvic bones only 25 cm long which are associated with the small, light vestiges of femurs and tibias. According to Omura's data the bones and the viscera in the largest northern right whales weigh about the same, each amounting to 14–15% of the total weight of the whale.

Auditory bullae of right whales are large and pointed on one end, a feature which distinguishes them from balaenopterids.

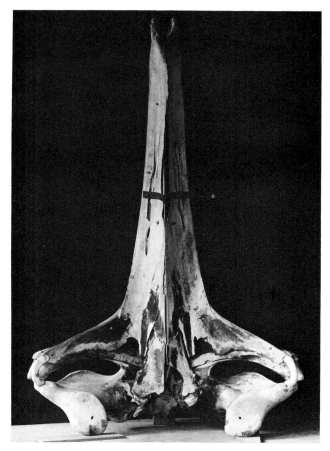

FIG. 9 Ventral view of northern right whale skull. (Courtesy of J. G. Mead, Smithsonian Institution.)

Mouth, baleen, and tongue

Yablokov and Andreyeva (1965) examined cross sections of baleen plates from *E. glacialis*. They found a certain amount of positive correlation between the number of layers in the small tubes of the medullar portion and age of the whale, and they suggested that the baleen might be useful to determine the age of the animals. Omura *et al.* (1969) described transverse growth marks on

FIG. 10 Lateral view of northern right whale skull. (Courtesy of J. G. Mead, Smithsonian Institution.)

the baleen and reasoned that the growth of baleen is fastest between the time an animal is weaned (11 m body length) and the time it reaches 13.5 m. He was not able to establish clearly the relationships between growth marks and age but assumed that the marks were annuli which erupted out of the gum in the winter or spring.

The tongue of southern right whales varies from white to bluish grey. Relatively large in cross section, the tongue is short, replete with oil, muscular, and overlaid by a narrow ridged palate situated between the baleen plates (Matthews, 1938).

Gonads

The ovary of a sexually mature northern right whale weighed 2 kg (Klumov, 1962), and Omura *et al.* (1969) found that the right ovary of a pregnant female was 6.3 kg. Testes of mature northern right whales are very large, up to 2 m long with a combined weight of nearly 1000 kg. For comparison, the combined maximum weight of both testes in the grey whale at the height of the breeding season was only 68 kg (Rice and Wolman, 1971). For some unexplained reason there is a considerable difference between shape of the spermatozoon as shown by Omura and associates (1969) and by Klumov (1962).

Central nervous system

The spinal cord of the northern right whale is only about 15% of the length of the whale (Seki, 1958). The epidural space is well-laced with blood vessels and fatty tissue. There is little difference between humans and right whales in size of the cells at the anterior part of the column, but the whale has much larger cells in the lumbar region. The central canal of the spinal cord of right whales is obscure throughout its length. Rootlets associated with the caudal extension of the spinal accessory nerve are restricted to the first cervical segment in the northern right whale as in the sperm whale. (For a description of the cetacean nervous system see Jansen and Jansen, 1969, and Pilleri and Gihr, 1970.)

The data of Omura *et al.* (1969) indicate that the northern right whale's brain is relatively small—in four specimens it ranged from 2.4 to 3.1 kg, compared to a maximum of 10 kg in the sperm whale. The brain weight of a 13.1 m male southern right whale was 2.8 kg, only 0.02% of the total body weight (Pilleri, 1964). Even though the absolute brain weight of mysticete whales is higher than that of most mammals, the brain-to-body weight ratio is one of the lowest. Of five mysticete species investigated by Pilleri, *Eubalaena* was found to have the lowest degree of encephalization, but the essential features of the brain are similar to those of fin and humpback whales. Data from the southern right whale indicate that the brain is unique among mysticetes by virtue of an unusually long cerebrum and an auditory nerve of very small diameter.

Kidneys

Cetacean kidneys apparently have the highest degree of lobulation noted in mammals. Kamiya (1958) likened this organ to a bunch of grapes represented by numerous reniculi. He determined that a kidney weighing 32.4 kg from a 11.7-m female *E. glacialis* had 5377 reniculi. Many of the reniculi were fused. When compared individually with nine odontocetes and two mysticetes, the right whale had up to five times the number of reniculi.

Skin

The structure of skin layers among cetaceans varies depending upon the species. Contrasted with the balaenopterids, which virtually lack a dermis, the right whale is unique in that it has a well-developed dermis without fat (Sokolov, 1975).

Whaling and Abundance

The population of *E. glacialis* in the North Pacific was estimated to be 200–250, (Wada, 1972; Tillman, 1975) but no estimates were available for the North Atlantic. Hunting of northern right whales began with Basque exploitations in the Bay of Biscay as early as the eleventh (possibly the tenth) century. Large groups of up to 100 right whales were once seen in the North Atlantic as far south as the Azores. In fact, *E. glacialis* formed the basis of the world's whale fishery from the thirteenth to the seventeenth centuries (Allen, 1908). By the end of the sixteenth century large numbers of Basque ships frequented the west side of the Atlantic; 30 were seen near Newfoundland as early as the mid-1500s.

Americans began right whaling in the North Atlantic during the seventeenth century, and by the mid-eighteenth century the scarcity of right whales forced the New Englanders towards the bowhead. Referring to several of his own comments on the New England right whale fishery, Allen (1908) said, "These extracts could be greatly extended, but those already given are sufficient to show how important a role this species of whale played in the early maritime history of New England and Middle States; also its former abundance, and how a century of pursuit for its commercial products reduced it [*E. glacialis* in the northwest Atlantic] to commercial extinction" (p. 281).

Americans were engaged in intensive whaling for *E. glacialis* in the North Pacific, a venture which extended from the Oregon coast to Vancouver and the Aleutian Islands. This Pacific fishery became exhausted to the point of economic failure in much shorter time than in the North Atlantic. The United States, Japan, Russia, Great Britain, Norway, and Holland had joined in the widespread pursuit of northern right whales until they became so scarce that during the 1934–1935 whaling season only two were taken, off Alaska (Norman and Fraser, 1949).

Some people thought *E. glacialis* had become extinct, but fortunately remnant populations survived in the North Pacific and North Atlantic, and there appears to be a slow increase. A total of 31 sightings of right whales was reported by Gilmore (1956) and by Rice and Fiscus (1968) in the southeastern North Pacific from 1955–1967. Klumov (1962) concluded that there are three major stocks of right whales in the North Pacific, two on the Asiatic side and one off North America. Watkins and Schevill (1976; Schevill *et al.*, 1981) see *E. glacialis* in the waters off Cape Cod, Massachusetts (North Atlantic) from December through early summer, with as many as 30 animals once seen in a 1- to 2-square-mile area (1976). The Bay of Fundy is also an area of concentration (Kraus *et al.*, 1982). Japanese and Russian whalers see right whales in the North Pacific, the Kurils, the Bering and Okhotsk seas, the Gulf of Alaska, and near the Aleutians. The distribution of northern right whales and a history of their capture by whalers was reviewed by Reeves *et al.* (1978).

Masaki (1972) estimated a total population of about 4300 *E. australis* in the southern oceans, higher than Best's (1974) figure of 1030 for 1970 as extrapolated from South African data. Best calculated a population of about 200 *E. australis* for the coastal waters of South Africa in 1970. Southern right whales were found in a belt from 60° W to 150° E by Japanese whalers during two successive seasons, 1965–1967, including only the months from November to February (Omura *et al.*, 1969). About 50 southern right whales were sighted off Argentina (Cummings *et al.*, 1974) and some have been observed off Chile (Clark, 1965). Gaskin (1964) encountered southern right whales off New Zealand, Chittleborough (1956) reported their presence off Australia, and Zenkovich (1962) found them in the Indian Ocean and the South Pacific. Southern right whales have reappeared off Tristen da Cunha, mid-South Atlantic, and Parker (1978) records the sighting of two near Macquarie Island (54.5°S, 159°E). As a result of overexploitation of northern right whales, the industry turned its attention to southern right whales towards the end of the eighteenth century. Fishing pressure was so severe over a period of 65 years that by 1835 it was hardly profitable to pursue the southern right whale (Best, 1970).

Distribution

Eubalaena australis has been seen from about 20 to 50° S, and the species occurs virtually around the world. *E. glacialis* has been seen in the North Pacific from about 25 to 60° N, and in the North Atlantic from about 30° to 75° N (Fig. 11). Except for a very few individuals taken by aboriginals, northern right whales have been protected from man's exploitation since 1937 (Anonymous, 1973).

E. glacialis and *E. australis* apparently move from subpolar regions with the onset of winter to lower latitudes, staying more or less near continents or island masses. After mating and calving they return to the subpolar regions in the respective spring months. For unknown reasons they avoid the warm equatorial regions and, since the seasons are opposite, these two species apparently remain separated by 5000 miles or more. Marcuzzi and Pilleri (1971) hypothesised that the more primitive of the two species originated in southern waters on the basis of Pleistocene remains found in South America, and that perhaps during a glacial period the original species was able to migrate across the equator, an event which presumably cannot now be repeated because of warm water.

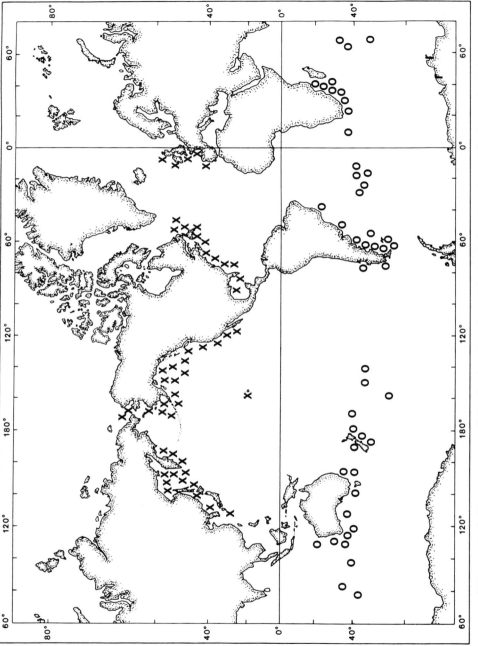

FIG. 11 Probable present general distribution of southern right (circles) and northern right (crosses) whales.

Ellis (1980) provided a good review of the history of whaling, including a summary of the biology and mangificent colour plates.

Behaviour

Sound production

A number of underwater sounds have been recorded from southern right whales in Patagonian waters (Cummings et al., 1971, 1972, 1974; Payne, 1972). Sonagrams of several kinds appear in Fig. 12. The most common sound is a belchlike utterance that averaged 1.4 sec in duration with most of the energy appearing below 500 Hz. Source levels of this powerful sound ranged from 172 to 187 dB, re 1 µPa at 1 m. Southern right whales also produce two kinds of low frequency moans: simple moans in a narrow band of frequencies (centered at 160 Hz) without appreciable frequency shifts, and complex, wide-band moans (centered at about 235 Hz) that exhibit extensive frequency shifts and overtones. Southern right whales produce pulses, 0.06-sec bursts with frequency components extending from 30 to 2100 Hz, and miscellaneous sounds composed of numerous phonations below 1900 Hz that vary in length from 0.3 to 1.3 sec. Their blow sounds are clearly audible in air and from a hydrophone. *E. australis* evidently can control the frequency of lesser components independently of the main components of the blow sound. The range of blowing southern right whales was determined using the sound arrival time differences between waterborne and airborne sound without reliance on visual methods. Payne and Payne (1971) described the sounds of southern right whales as being in the frequency range of 50 to 500 Hz with some energy as high as 1500 Hz. Payne (1972) has recorded sounds from identifiable individual southern right whales using the sound arrival time differences on an array of sonobuoys. At the time of this writing, it is my understanding that Payne and co-workers are summarising their lengthy studies on southern right whales; their monograph is presently unavailable, but some information is available in Payne (1983).

Mating whales sometimes make loud slapping sounds as they roll over and repeatedly beat the water's surface. These slaps can be heard from air and water at the same time. One of the most spectacular sounds of southern right whales, heard in the Golfo San José, Argentina is a very strong pulse resembling a gun shot. This sound, followed by as much as 5 sec of reverberation, ranges from 50 to above 2200 Hz. Based upon very similar sounds that we have recorded from humpback whales off Alaska, the "gun shot" pulses from southern right whales are probably the result of tail slapping, a common

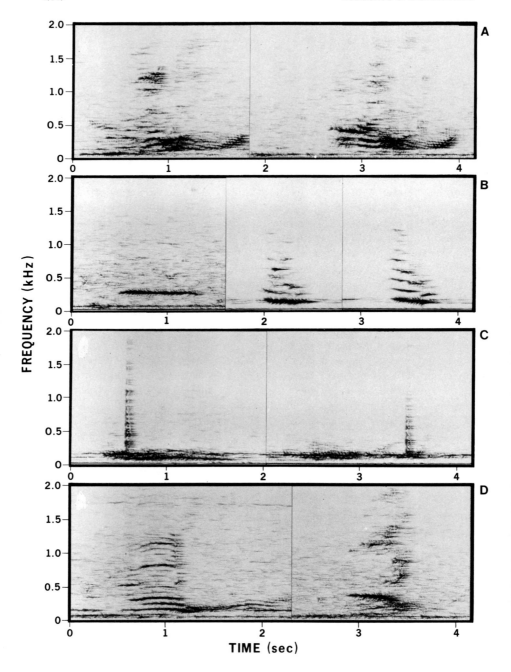

behaviour among large and small whales. This contrasts with the normally smooth sequence of fluking associated with swimming (Fig. 13).

Bellowing sounds and moans of rising pitch are common among mating southern right whales and appear to be part of the mating behaviour. The scarcity of sounds at times other than during overt breeding activities suggests that southern right whale phonations are significant in bringing and keeping the whales together or perhaps as a signal used as a social releaser for mating. Other mammals, insects, frogs, toads, fishes, and birds include sound production as part of their reproductive behaviour. Expecting that southern right whales would respond in some overt fashion to the presence of their underwater sounds, Cummings et al. (1972, 1974) projected a number of different sounds recorded in the same location from other right whales. Regardless of the behavioural state at the time of playback and numerous playback repetitions with calibrated equipment of high fidelity at natural source levels, there were no changes in their movements or vocal behaviours. These results are very different from those of subsequent work by Clark and Clark (1980), who reported both vocal and orientation responses to right whale sound playbacks in the same location as our work.

Numerous sounds have been recorded from *E. glacialis* in the northwest Atlantic (Schevill and Watkins, 1962; Watkins and Schevill, 1976), many of which are of a similar nature to those produced by *E. australis* off Patagonia. Cummings and Philippi (1970) analysed a long series of recordings made off Newfoundland in late December of 1965. Certain characteristics of the highly filtered individual sounds and the winter date of the recording tentatively identified the sounds as being from *E. glacialis*. These sounds were found to be in repeated stanzas that lasted 11–14 min at 8–10-min intervals. Through subsequent analysis and comparison with the author's and other recordings of humpback whale sounds, it has been found that the tentative original identification was in error and that those recordings off Canada were actually the first finding of *humpback* whale songs, even though this species is assumed to have migrated south by this time of the year.

Defensive activities

Scammon (1874) described the evasive manoeuvres of northern right whales as very wild, making the whale difficult to approach and capture. He implied

---

FIG. 12 Sonagrams of sounds from southern right whales. Analysing filter bandwidth, 10 Hz. (A) Two belchlike sounds; (B) one simple and two complex moans; (C) single pulses associated with simple moans; (D) two miscellaneous sounds. (After Cummings *et al.*, 1972.)

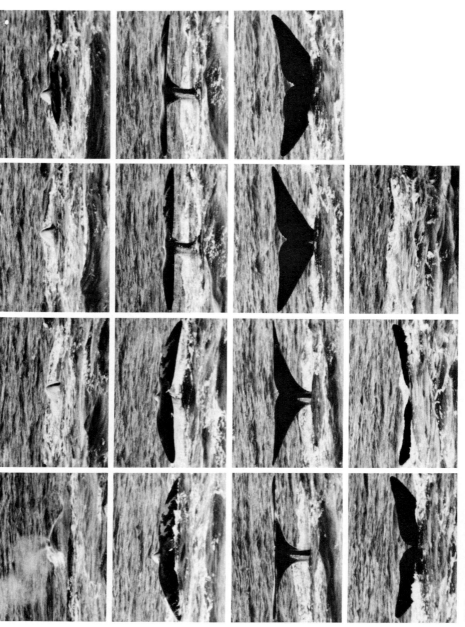

FIG. 13 Series of 15 photos (three per sec) showing movement of flukes preceding a long dive of a southern right whale. (After Cummings *et al.*, 1972.)

this to be a learned behaviour acquired through years of pursuit and manifested by the animal's ability to avoid the catcher boats. When struck by the harpoon, northern right whales fought back by thrashing their flukes from side to side nearly filling the catcher boats with water. Often if the whale could not be dispatched soon enough it would tow the boats so far from the ship that it would have to be cut off. In recent contrast, unmolested groups of mating southern right whales indicated an apparent indifference to our presence, even in close proximity.

In July, 1971, a spectacular encounter took place involving an attack by five killer whales, *Orcinus orca*, on two southern right whales in Golfo San José (Cummings et al., 1971, 1972). The event lasted for 25 min, and although there were no signs of physical damage, such as blood or bits of flesh in the water, it appeared to be a full-fledged attack. After reaching the right whales, the killer whales were in a frenzy, lunging over, between, and under their quarry. Right whale defense consisted of continuous and violent slashing with flukes and flippers and rolling and twisting in tight manoeuvres. After the attack the two right whales moved into very shallow water where they became quiescent, remaining in the same location until nightfall. The fishermen of the area told of similar attacks in the general vicinity of the Valdés Peninsula, many of which resulted in physical harm and even death to the right whales. Underwater recordings made during the above-described attack of 1971 revealed no sounds other than normal water noise from either the right whales or the killer whales. A similar, but less severe, encounter was noted off South Africa (Donnelly, 1967). This event involved three killer whales and two pairs of courting southern right whales. The killer whales swam in tight circles around the large whales, alternating from one pair to the other. One right whale merely swung its tail at the killer whales, but otherwise the encounter apparently caused minimal disturbance.

After successful attempts to induce avoidance in other marine mammals by projecting killer whale sounds (Cummings and Thompson, 1971; Fish and Vania, 1971), the same techniques were tried with southern right whales in Argentina (Cummings et al., 1974). Even when these experiments involved life-sized models of killer whale fins among the *E. glacialis,* the whales surprisingly showed no overt avoidance. Analyses of the killer whale sounds (originally recorded in the northeast Pacific) showed them to be different from those of killer whales of the southern polar region (recorded by G. S. Kooyman, Scripps Institution of Oceanography, California). Although it is only conjecture, perhaps the sounds of northern killer whales were unfamiliar to southern right whales and thus offered no meaningful stimulus. Winn and Winn (1978) and P. O. Thompson, W. C. Cummings, and A. J. Perrone (unpublished report) have described regional dialects among humpback whales, and I have evidence that they occur in blue whales.

Headstanding, breaching, breathing

A common behaviour of southern right whales off Argentina was to assume a vertical position and extend the flukes high into the air, keeping them in this position and often rocking them back and forth for periods up to 2 min (Cummings et al., 1972). Such "headstanding" behaviour was confined to shallow water at Golfo San José, and it was first thought that the whales could have been feeding on benthic organisms. However, headstanding was later noted among whales much farther north in an area too deep for them to reach bottom and surface at the same time. It may have been used as some sort of a sign stimulus. In fact, Donnelly (1967) associated this "up-ending" with courtship by the male partner. Payne (1972) interpreted such behaviour as a resting posture. He also ventured the possibility that the whales accomplish some horizontal movement by sailing in this manner with the aid of the wind, a rather demanding notion from the standpoint of the physics involved.

Jumping, or breaching as it is often called, is common among southern right whales. This spectacular burst and the tumultuous splash upon re-entry may occur 10 times in a series. As I have seen in other species (fin, humpback, grey, and Bryde's) *E. australis* nearly always turns in midair and falls on either the side or back. Although there are many plausible theories connected with this behaviour of large whales, the sudden and repeated appearance of diving Brown-hooded Gulls, *Larus maculipennis*, while right whales jumped led to the belief that the whales were dislodging parasites. This may have been only indirectly associated with jumping, but the birds definitely appeared to be diving and feeding in the wake of the whales' antics. Donnelly (1967) observed that jumping of *E. glacialis* in an area off South Africa was associated with courtship behaviour and suggested that it was a display mechanism for bringing mates together.

Southern right whales also splash on the surface by slapping their huge flukes or flippers, an act which is accompanied by loud underwater sounds that reverberate for considerable periods.

The frequency of blows of southern right whales depends upon their activity. Whales swimming slowly at the surface breathe about once a minute, or once every 2–3 min between "headstands." A more rapidly swimming right whale breathes irregularly, diving for periods that vary between 0.5 and 4.3 min. Another whale followed by a ship may stay down for periods averaging 6 min with a maximum of 8 min. One to three blows are noted between dives. Right whales usually display their flukes just before an extensive dive. However, the last surfacing before a long dive is sometimes accompanied by "false flToken," the flukes being drawn up close to, but not above, the surface.

Feeding

Available data (Klumov, 1962; Omura *et al.*, 1969) indicate that North Pacific right whales feed primarily on at least three species of calanoid copepods (mostly on *Calanus propinquus* and *C. cristatus*) and to a lesser extent on euphausiids (*E. pacifica*). Feeding right whales have been observed at close range in the North Atlantic (Watkins and Schevill, 1976). Plankton tows in the area revealed mostly copepods, *C. finmarchicus*, stage 5, and juvenile euphausiids. According to these authors, two kinds of feeding were observed. On the surface, right whales selectively choose heavy plankton slicks and swim through them with mouth agape. Either the bonnet only is held out of the water during surface feeding, or the baleen may be exposed as the whale rides higher out of the water actually skimming the surface. During surface feeding, the water can be seen flowing through the baleen to the outside and a rattling sound is often heard. Most of the feeding, however, is believed to be accomplished when right whales are totally submerged, presumably working through heavy layers of plankton.

Eubalaena australis from South Georgia has been found to contain *Euphausia superba* (Matthews, 1938) and, off the Patagonia coast, postlarvae of *Munida gregaria*, historically called *Grimothea* by error (Matthews, 1932). *M. gregaria* often occurs in vast swarms which appear to cover the water's surface as do the closely related "red crabs" of California waters, *Pleuroncodes planipes*, the staple of mysticete whales off Baja California. Southern right whales were believed to owe their living to the presence of *M. gregaria*, which the whalers of the mid-nineteenth century called "right whale feed".

Reproduction

Right whales mate and calve in the respective winter and spring seasons of the Northern and Southern hemispheres. Therefore, breeding cycles of *E. glacialis* and *E. australis* are about a half year out of phase. Even if the two species did frequent warm equatorial regions, reproductive isolation would doubtless be enhanced by their respective migrations, also 6 months apart.

In the 1960s, Japanese scientists were granted special scientific permits by the International Whaling Commission to take a total of 13 North Pacific right whales, 5 females and 8 males. The resulting data indicated that males were sexually mature at 15 m, females at 15.5 m, and that males reached full growth at 16.5 m, females at 17.3 m (extrapolated from Omura *et al.*, 1969). As in all large whales, growth is rapid during the first year. *E. glacialis* grows to about 12 m long during the first 18 months, and probably reaches sexual maturity in about 10 years. It is thought that size at first sexual maturity is larger for North Pacific right whales than for those in the North Atlantic.

E. glacialis copulates from December to March, when most of the young are born (Klumov, 1962; Nishiwaki, 1972). Scammon (1874) had reported that southern right whales ventured into protected bays for calving, but he was unaware of any similar habits, or places favored, by mating northern right whales. The gestation period of northern right whales is about 1 year and calves are probably weaned by the age of 6 to 7 months. As with other mysticetes, one calf is born to a sexually mature female, probably every other year. There is usually only one foetus for each pregnancy, but Scammon reported having seen twin northern right whales on rare occasions.

Southern right whales mate and calve mostly in protected bays from June to November, between 20 and 30° S (Chittleborough, 1956; Donnelly, 1967; Best, 1970; Cummings *et al.*, 1971, 1972, 1974).

A large group of southern right whales occurs annually in the vicinity of the Valdés Peninsula, Argentina, as early as June. Thought for many years to have been exterminated, the Patagonian population was rediscovered (Gilmore, 1969) at Valdés. The whales are most numerous in this region in late August and September, and all begin their return to the south by late October and November. Courtship, copulation, and calving occupy much of the whales' time in the protected gulfs of this region.

Just after my second expedition to the area, in 1972, Mr Philippe Cousteau (deceased) and a field party filmed from underwater a spectacular copulation sequence involving one female and three males with penes extruded. We had noted as many as five groups of three southern right whales in a single afternoon, but prior to Cousteau's observations, we could only surmise that polygamous copulation was taking place. Donnelly (1967) observed five pairs of courting southern right whales at Algoa Bay, South Africa, during September. He, too, noted an event of polygamy involving two males and a single female. Mating of Patagonian whales is not confined to the Valdés region, as evidenced by the discovery of an equally large group of 25–30 off Punta Rasa, 250 km father north, in August, 1972 (Cummings *et al.*, 1974).

After much caressing and nuzzling, mating southern right whales roll about randomly exposing flippers, flukes, backs, bellies, and portions of their heads. Donnelly noted that the male would sometimes begin precopulatory behaviour by placing his chin on the exposed hindquarters of the female. Caressing is often accomplished with the aid of the flippers.

If one comes upon a group of right whales suddenly by ship it is often difficult to determine from the surface the number of whales involved in a breeding group. Courting bouts may last for an hour or two, after which participants go their own way. When a female has apparently completed mating she may roll over on her back at the water's surface, a posture which moves her genitalia away from the underlying male. The male may then position himself on his back under the "unwilling" female and grasp her with his flippers. As the female rolls over to breathe, the male is in a favorable coital position and copulation may resume. Donnelly's observations indicated that it is the male which appears at the surface belly up following copulation. I have seen more than one of the breeding group in this position, and it is quite possible that both sexes exhibit the behaviour. Southern right whales appear upside down at the surface even when accompanied by their offspring. Sea birds often land on the bellies (or the backs of upright whales) and walk the length of the exposed surface.

Donnelly (1967) attributed scratches on copulating female southern right whales to barnacles on their male partners, but hastened to point out that all movements were carried out "with utmost tenderness and grace." It should be noted that no animosity is observed among southern right whales between suitors of the same female, an unusual behaviour for mammals. Based upon successive observations of courting whales in the Valdés region of Argentina, sexual activity appears to increase about 1 month after *E. australis* assembles in the gulfs, possibly a manifestation of some requirement for extended pre-coital rituals. Reeves and Brownell (1982) provided a good review of behaviour and the general biology.

To no avail, I and other members of a scientific party cruised the bays and inlets all the way from the Magellan Straits to Valparaiso, Chile, in the austral winter in hopes of finding right whales breeding. Such was not always the case in the southeast Pacific. Captain Day (Scoresby, 1820) noted for this area, "The whales resorting to the latter situations are females, which go into shoal water for the purpose of depositing and rearing their young until nature has given them sufficient strength and powers to follow the older animals in all their meanderings in a deeper element than where they are first brought into existence" (p. 475).

Parasites

Southern right whales are heavily infested with external parasites. Barnard (1932) redescribed three species of cyamid amphipods from southern right whales, *Cyamus ovalis*, *Paracyamus* (= *Cyamus*) *erraticus* and *P. gracilis*. Since that time, Leung (1967) has contributed an excellent key to whale lice. On right whales they appear on excrescences, in the genital groove, around the blowholes, between flippers, and on the tail. Some are nearly completely covered by proliferations of skin. Barnacles also attach to the external surface of southern right whales, and trematodes and tetrabothrids occur in their intestines (Matthews, 1938).

All North Pacific right whales examined by Omura et al. (1969) had heavy infestations of whale lice, *C. erraticus* and *C. ovalis*, on excrescences, in the genital groove, around the eyes, at the base of the flippers, around the blowholes, and on the jaws. Omura et al. (1969) also found a diatom film on the flukes of one right whale and over the entire body surface of two others. The diatom, *Cocconeis ceticola*, is common on baleen plates of both northern and southern right whales. Data on the parasites of North Pacific right whales are not in agreement. Omura et al. (1969) found no internal parasitic worms, two external cyamids, and one external diatom; whereas Klumov (1962) found three species of helminths and only one ectoparasite, the same species of diatom noted by Omura. Best and McCully (1979) found mycosis in a southern right whale.

Conclusion

Beginning in the North Atlantic, and followed in the North Pacific and the southern oceans, man's exploitation of right whales brought these animals to the verge of extinction. Unfortunately these events took place well before the advent of modern scientific inquiry, a fact that explains why so little is known about the biology of right whales compared to our knowledge of some of the other great whales. The southern right whale has been protected since 1935 by enactment of the League of Nations, but this species is still vulnerable to whalers who might want to operate outside the jurisdiction of the International Whaling Commission. Unlike the striking recuperation seen in the number of grey whales, *Eschrichtius robustus*, since their protection at about the same time, right whale populations are still very small and widely scattered, and their existence remains tenuous at best.

Acknowledgement

I thank M. E. Dahlheim for considerable help with the literature survey and typing the earlier versions; J. C. Cummings, T. Rydlinski, and C. L. Gray for assistance in preparing the manuscript; J. S. Leatherwood, W. E. Schevill, and W. A. Watkins, for their helpful comments. Partial support was received from the Office of Naval Research(Contract N00014-78-C-0419), Drs R. C. Tipper and B. Zahuranec, Program Managers.

11

Bowhead Whale
Balaena mysticetus Linnaeus, 1758

Randall R. Reeves and Stephen Leatherwood

Genus and Species

There is uncertainty about the systematic rank used to distinguish the bowhead whale *Balaena mysticetus*, an Arctic species, from the right whale *Eubalaena glacialis*, a widely distributed inhabitant of boreal to subtropical latitudes. Their specific identities were not recognised formally by naturalists until the second half of the nineteenth century (Eschricht and Reinhardt, 1866), even though by that time both had been heavily hunted commercially for several centuries. Now they are assigned to different genera (Allen, 1908), although some authorities continue to view them as congeneric (Rice, 1977). Any decision, based on present evidence, to recognise the genus *Eubalaena* as distinct from *Balaena* is admittedly somewhat arbitrary.

An attempt was made by Bailey and Hendee (1926) to describe two species of bowhead found off Northern Alaska. The *inito* (also called *ingutuk*) was considered by Eskimo whalers to be smaller and lighter in colour than the other kind of bowhead (*ahkalook* or *usingwachaek*), lacking the distinctive bow in the profile of the head, and possessing denser bones and shorter, lighter baleen. American whalers in the Sea of Okhotsk referred to a small whale they called the "poggy," and the so-called "bunchback" was taken in the Arctic Ocean and Sea of Okhotsk (Scammon, 1874). European whalers operating near Spitzbergen believed that there were different "tribes" of bowheads there (Scoresby, 1820) and that "west-ice whales" differed significantly from "south-ice whales" (Zorgdrager, 1720). British whalers in Davis Strait and Baffin Bay sometimes spoke of different "races" of bowheads being found there—designated as "middle-icers," "rock-nosers," or "Pond's Bay fish," according to where they were captured (Brown, 1868).

More recent analyses, using electrophoresis and karyotyping as well as comparison of key characters, indicate that the *ingutuk* is probably a morphological variant within the species *B. mysticetus* (Braham *et al.*, 1980a). Much of the morphological variation reported by early whalers was probably due to age or sex differences. Jarrell (1981) argued that *ingutuks* are yearling (recently weaned) bowheads; however, there is still no satisfactory explanation for the consistent references to *ingutuks* as having dense bone. Fetter and Everitt (1981) confirmed that the bones of *ingutuks* are grossly and microscopically different from those of other bowheads; they suggested that the differences may be due to a congenital defect.

Decisions about how, if at all, to subdivide the genus *Balaena* must await the examination of substantially larger samples from the various populations than are now available. At present, there is no argument for the existence of more than one bowhead species, although the world population of bowheads is considered to include several separate stocks (Tomilin, 1957; Mitchell, 1977; Allen, 1978).

External Characteristics and Morphology

Size

Although we know of no successful attempt to weigh an entire adult bowhead, in whole or in parts, there is little doubt that the weight at a given body length is much greater for bowhead and right whales than it is for any other large baleen whale (Lockyer, 1976). The maximum girth of bowheads can be upwards of seven-tenths of the body length (Eschricht and Reinhardt, 1866; Table 1). The body weight of a bowhead "of typical proportions" was estimated as 75 tonnes by calculating the volume and weight of water displaced by a model made to scale (Gray, 1887). Scoresby (1823) made a similar calculation using the dimensions of an 18.3 m bowhead and obtained an estimate of 114 tonnes. He considered this estimate to be too high but had no doubt that very large individuals could weigh 100 tonnes, and that an "ordinary full grown animal" would weigh 70 tonnes.

TABLE 1 Some measurements of bowhead whales (in cm)[a]

	Specimen number													
	1[b]		2[b]		3[c]		4[d]		5[e]		6[f]		7[f]	
Sex	female		male		(?)		male		male		male		female	
	cm	%	cm	%	cm	%	cm	%	cm	%	cm	%	cm	%
Total body length	1433	100	1372	100	579	100	1554	100	640	100	1234	100	1295	100
Length of flipper	244	17	221	16	—	—	229	15	97	15	183	15	—	—
Maximum width of flipper	122	9	114	8	—	—	152	10	43	7	112	9	—	—
Maximum girth	—	—	853	62	439	76	1067	69	—	—	—	—	—	—
Width of flukes, tip to tip	579	40	488	36	—	—	610	39	182	28	—	—	—	—
Length of blowholes	30	—	34	—	10.2	—	—	—	10	—	—	—	—	—
Maximum blubber thickness	41	—	—	—	—	—	—	—	9[g]	—	—	—	20	—
Average blubber thickness	28	—	23	—	12.7	—	—	—	9[g]	—	—	—	—	—
Length of longest baleen plate	320	22	290	21	30.5	5	320	21	57.2	9	254	21	—	—
Width of longest baleen plate	33	2	30	2	—	—	28	2	10.2	2	25	2	—	—
Distance from tip of snout to eye	—	—	488	36	—	—	—	—	172	27	—	—	427	33
Length of genital slit	56	4	—	—	—	—	—	—	52	8	107	9	46	4
Length of baleen fringes	61	—	—	—	<15.2	—	46	—	—	—	—	—	—	—
Distance from tip of snout to umbilicus	—	—	—	—	—	—	—	—	—	—	671	54	625	48

[a]Note that nineteenth century authors were not adhering to standard measurement techniques. Thus, there is probably substantial error (or inconsistency), especially overestimation due to measurement along the body contour rather than in a straight line. Comparisons should be made with caution. Values have been converted from feet and inches to centimetres.
[b]Whales caught in Bering Sea (Scammon, 1874).
[c]Whale caught east of Greenland (Scoresby, 1823).
[d]"Average" measurements based on "comparison of some two hundred" whales in Spitsbergen stock (Gray, 1887, p. 133–134).
[e]Whale caught off Osaka, Japan, 23 June 1969 (Nishiwaki and Kasuya, 1970).
[f]Whales caught east of Greenland (Gray, 1889).
[g]Including black skin.

A 17.4 m female North Pacific right whale weighed 106.5 tonnes (Klumov, 1962), and a 16.4 m male weighed 78.5 tonnes, with no adjustment made for loss of blood and other body fluids (Omura et al., 1969). Studying a sample of 21 carcasses, Lockyer (1976) developed the formula $W = (.0132)L^{3.06}$ for calculating the body weight of right whales, where W = body weight in tonnes, and L = overall body length in m. This formula indicates that an 18 m right whale should weigh about 92 tonnes. Assuming bowheads grow to be at least as long as right whales and that they are at least as heavy per unit of length, the estimates by Gray and Scoresby appear reasonable.

A female bowhead 16.76 m long was still physically immature (Durham, 1972, as reported in Marquette, 1977), and lengths of 17–18 m are not unusual for females (Nerini et al., 1984). An estimate of length at physical maturity in males from the Bering Sea stock* is 14.02–15.24 m (Marquette, 1977). In general, both sexes are believed to reach physical maturity at 14–18 m (Tomilin, 1957), with some individuals growing to lengths of at least 20 m (Nerini et al., 1984). As is true of other baleen whales (Ralls, 1976), female bowheads grow to a larger size than do males (Johnson et al., 1981).

Description

The bowhead has no dorsal fin. Its flippers are paddlelike or spatulate in shape (Kükenthal, 1922; see photos in Nishiwaki and Kasuya, 1970; Mitchell and Reeves, 1981; Leatherwood et al., 1982; Fig. 1). The flukes, whose combined width can be more than two-fifths of the total body length (Tomilin, 1957), are smooth along the rear margin and divided by a deep notch. The skin of the bowhead is generally free of barnacles and encrustations. The conspicuous callosities found on the head of the right whales are absent on the bowhead. There are no ventral grooves like those found on balaenopterids and on the gray whale, *Eschrichtius robustus*.

Bowheads have proportionately larger heads than other baleen whales. At birth the head can be two-sevenths to one-third of the total body length (Table 1); in adults it can be as much as two-fifths (Eschricht and Reinhardt, 1866). The bowhead's head is disproportionately large not only in length but in vertical thickness. It constitutes more than one-third of the entire bulk of the animal.

The colour of the bowhead is basically black or brown, with limited, well-defined areas of white or light grey. Much of the chin and lower jaw can be white, often with grey or black spotting (Nishiwaki and Kasuya, 1970; McVay, 1973; Fig. 1). Some bowheads have a light grey to whitish band

---

*Many authors in the United States and Canada refer to the Bering Sea stock as the western Arctic stock. It is also known as the Bering–Chukchi–Beaufort seas stock.

FIG. 1 A female bowhead killed by whalers from Kaktovik Village on Barter Island, Alaska, in September, 1983. Note the broad, spatulate flipper (A) and the curved mouthline, long baleen, and white chin patch on the lower jaw (B). (Photographs by Bernd Würsig.)

around the peduncle, usually ending near the origin of the flukes (Ljungblad, 1981, Fig. 16; Würsig *et al.*, 1982). The amount of white present in the fluke or caudal region increases with age (Davis *et al.*, 1983). There is sometimes an irregular white area on the body adjacent to the flippers (Gray, 1887; Nishiwaki and Kasuya, 1970) or a white rear margin on the dorsal side of the flukes (Berzin and Doroshenko, 1981). White patches can be present on other parts of the body as well (Brown, 1868). Large greyish patches are thought to be caused by the sloughing of skin (Würsig *et al.*, 1982); Davis *et al.* (1983, p. 36) attributed mottling on bowheads to "skin sloughing during epidermal molt".

One anomalously coloured bowhead was all white on the top of its head (Bodfish, 1936), and observations of large bowheads, "light to dark brown, with white coloration on the head", were made during aerial surveys in the western Beaufort Sea (Ljungblad, 1981, p. 24). Neonates tend to be lighter in colour than adults, often appearing light grey through the water (Fig. 2). Durham (1980) mentioned an "albino" calf killed by Alaskan Eskimos. Old individuals can be heavily scarred, and scarring together with pigmentation patterns facilitates research in which visual identification of individual whales is required (Davis *et al.*, 1982, 1983; Braham and Rugh, 1983).

A young specimen examined in Japan had a single hair growing from each of the black spots on the chin (Nishiwaki and Kasuya, 1970). There are also

FIG. 2 Bowhead calves are more lightly pigmented than adults, appearing light grey through the water. This mother and small calf were sighted on 13 May 1982 near Point Barrow at 71°34' N, 155°17' W. (Photograph by R. Van Schoik.)

small patches or rows of hairs on the rostrum, on the lips, and near the blowholes (Brown, 1868; Durham, 1980; Haldiman et al., 1981). Even when hairs occur in white or cream-colored areas, the skin immediately surrounding their base is black (Haldiman et al., 1981).

Baleen

The baleen of the bowhead is considerably longer than that of other species; it is nearly twice as long as that of the right whale. It is slender and elastic, with fine fringes lining the inner margin. In texture it resembles most closely that of the sei whale *Balaenoptera borealis* and the right whale (Nemoto, 1970). As in the right whale, the baleen plates increase in length from the front and rear of the jaw, with the longest plates at the middle of the row.

The longest plate measured in the Alaska bowhead fishery since 1973 was 3.13 m, taken from a 16.2 m female (Marquette, 1977). Much longer plates exist. A 3.86 m plate (Bodfish, 1936) and plates up to 5.18 m long (Davis, 1874) were reported during the early commercial fishery. Gray (1940) mentioned a 3.9 m plate taken from a whale in the Greenland Sea, and he gave 3.3–4.3 m as the range noted by his father, a whaling captain. Scoresby (1820) referred to exceptional plates 4.57 m long, but he considered 3.8–4.0 m the usual maximum length. He gave the maximum width as 25–30 cm, the maximum thickness as 1.0–1.3 cm. We are skeptical of the claim that a 5.8 m plate was taken from a bowhead killed by the *Junior* of New Bedford in 1849 (Edwards and Rattray, 1956).

The bowhead's baleen is generally brownish black or bluish black, often with irregular longitudinal white stripes (Scoresby, 1820). The baleen of one anomalously coloured individual reportedly was white except for three streaks of purple and brown along the centre of each plate (Bodfish, 1936). The number of plates on a side usually ranges from 237 to 346 (Marquette, 1977), sometimes up to 360 (Brown, 1868). A total of as many as 780 baleen plates apparently has been counted in a single bowhead (Bodfish, 1936). Bowheads and right whales have no baleen at the front of the palate, in contrast to the balaenopterids (Eschricht and Reinhardt, 1866).

Product yields

Early whalers sought whales for their yield of oil and baleen, and they reported average yields consistently higher for bowheads than for right whales (Scammon, 1874). Individual bowheads yielding more than 275 barrels (32 789 litres) of oil and 3500 pounds (1588 kg) of baleen reportedly were caught in the early years of the Bering, Chukchi, and Beaufort seas fishery. The record yield of oil from the Bering Sea stock reportedly was 327 barrels

(38 986 litres) (Bodfish, 1936). A "rough average" yield for bowheads in the Bering, Chukchi, and Beaufort seas was 100 barrels (11 923 litres) of oil and 1500 pounds (680 kg) of baleen (Bockstoce, 1980). The Scottish whaler *Arctic* caught 28 bowheads in Davis Strait in 1873 (Lubbock, 1937). Fifteen males produced 118.5 tonnes of oil ($\bar{x} = 7.9$ tonnes) and 6.35 tonnes of baleen ($\bar{x} = 0.42$ tonnes); 13 females, 147 tonnes of oil ($\bar{x} = 11.3$ tonnes) and 8.5 tonnes of baleen ($\bar{x} = 0.65$ tonnes).

Internal Anatomical Characteristics

Among the most important descriptions of the bowhead's anatomy are those by van Beneden and Gervais (1880), Eschricht and Reinhardt (1866), Turner (1913), and Tomilin (1957) and the papers contained in Albert (1981). It is from them that most of the following summary is derived.

The mandibles bow outwards. They support massive lips that cover the long rows of baleen rooted in the upper gum. The narrow upper jaw and palate are strongly arched (Fig. 3), accommodating the long baleen plates.

The pink tongue is, as Eschricht and Reinhardt (1866) put it, "the most colossal of all the organs of the Greenland whale" (p. 80). It can be up to 5.5 m long and 3 m wide.

The bowhead has 7 cervical vertebrae, which in adults are fused to form a single unit. In addition, there are 12–13 thoracic, 10–13 lumbar, and 22–24 caudal vertebrae. Most bowheads have 13 pairs of long, curved ribs. Of the 12 pairs found in a 6.4 m male, 10 were two-headed and two were single-headed (Nishiwaki and Kasuya, 1970).

As in other mysticetes, the humerus is short and thick; the radius and ulna, more elongate. The flipper formula is variable. Eschricht and Reinhardt (1866) and van Beneden and Gervais (1880) figured five digits, with digit number I consisting of a single phalanx (Fig. 3). However, Nishiwaki and Kasuya's (1970) young specimen and a foetus examined by Durham (1980) both had only four digits, with digit number I missing. The pisiform cartilage proximal to digit number V is well developed. Rudimentary pelvic and hindlimb bones are usually present (Struthers, 1881).

The epidermis consists of a thin outer cuticle and a lower Malpighian layer, penetrated by long dermal papillae containing small blood vessels (see Albert *et al.,* 1980; Haldiman *et al.,* 1981, 1983). Tomilin (1957) interpreted the thickness of the skin on the head and neck (to 2.5 cm) as an adaptation to frequent, abrasive contact with ice, the thinness of that on the flippers and flukes (to 0.2 cm) as having thermoregulatory significance. According to Scoresby (1823), the epidermis of the calf can be 4.5 cm thick, or twice as thick as that of an adult. This observation was confirmed by Durham (1980), who referred to "two 2 cm thick layers" of pigmented skin on a 4.6 m male neonate. Haldiman et al. (1981) noted that the epidermis of bowheads can be as much as eight times thicker than that of other cetaceans. The relatively thin dermis merges with the underlying hypodermis or blubber layer, which can be 15–70 cm thick. The blubber actually consists of two layers, a fibrous outer layer (technically the hypodermis) and a subhypodermal adipose layer.

The bowhead's karyotype is $2n = 42$ (Jarrell, 1979).

Jones and Tarpley (1981) described the gross anatomy of a partial heart from a sexually mature male bowhead. Because this organ is valued as food in Alaskan whaling communities, specimens have been and probably will remain difficult to obtain. The gross and microscopic structure of the bowhead lung, kidney, brain and skin were described by Haldiman et al. (1981); the reproductive organs, as well as certain endocrine tissues, by Kenney et al. (1981). Tarpley and Stott (1983) examined ovaries from 15 bowheads, only one of which contained a corpus luteum. As many as 26 corpora albicantia were found on the ovaries from one specimen. Of five mature whales, three had more corpora in the left ovary and two had more in the right.

Home (1812) described and illustrated the gross anatomy of the hearing apparatus of a juvenile bowhead. Norris and Leatherwood (1981) described and illustrated the microscopic morphology of a small sample of bowhead cochleas. Dubielzig and Agiurre (1981) examined four bowhead eyes.

The mean brain weight of six bowheads, of body lengths ranging from 8.7 m to 10.8 m ($\bar{x} = 9.53$ m), was 2738 g. The bowhead has a smaller brain and a proportionately larger cerebellum than the fin whale *Balaenoptera physalus*, humpback whale *Megaptera novaeangliae*, and sei whale. The bowhead's brain, like those of other mysticetes, is less convoluted than those of odontocetes (Ridgway, 1981; see also Haldiman et al., 1981, 1983).

The bowhead has a four-chambered stomach, consisting of a nonglandular forestomach and three glandular compartments (Tarpley and Stott, 1983). A surface erosion in the anal canal of one bowhead was identified as an ulcer (Sis and Tarpley, 1981).

Meek (1918) described the urogenital organs of a female foetus. The renicules of the bowhead kidney have overall dimensions and a basic structure similar to those of other cetaceans (Abdelbaki et al., 1984). Medway (1980, 1981, 1983) analysed the urine and blood from a small series of specimens but was unable to reach any major interpretive conclusions.

The respiratory system appears to be similar to that of other cetaceans (Henry et al., 1983; Haldiman et al., 1984). The nares, located up to 4.8 m posterior to the tip of the snout and at the dorsalmost area of the skull in an area known to the whalers as the crown, are 15–20 cm long and 3 cm apart anteriorly, diverging to 18 cm posteriorly (McVay, 1973; Haldiman et al.,

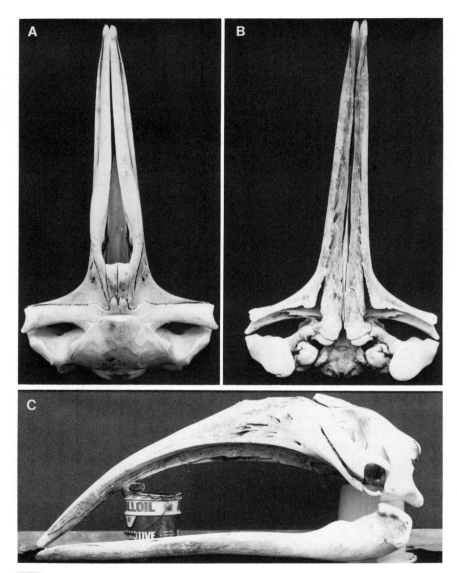

FIG. 3 Dorsal (A), ventral (B), and lateral (C) views of skull of a bowhead captured in Osaka Bay, Japan, 23 June 1969 (from Nishiwaki and Kasuya, 1970), and (D) lateral view of skull and skeleton of a bowhead (drawn by B. S. Irvine after Eschricht and Reinhardt, 1866). B, Left parietal bone; C, cervical vertebrae; F, frontal bone; H, humerus; I, intermaxillary bone; K, right occipital condyle; L, left lacrimal bone; M, maxilla bone; O, occipital bone; P, pelvis; R, radius; S, scapula; T, left temporal bone; U, ulna; Z, zygomatic arch.

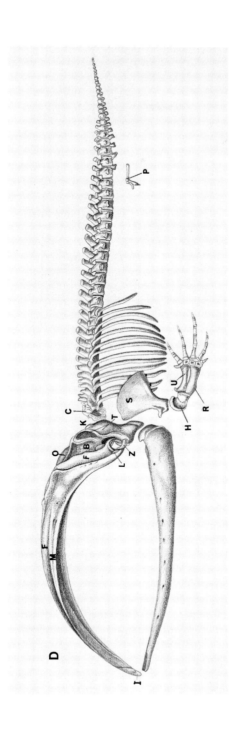

1981). They produce the V-shaped blow characteristic of the species. The rostral protuberance of the mobile larynx is blunter than that of most cetaceans (Haldiman *et al.*, 1981).

Physiology

Using many assumptions, Brodie (1981) made some theoretical calculations concerning bowhead energetics. He estimated daily heat loss for an adult bowhead to be 95×10^3 Kcal, exclusive of demands from reproduction, growth, and feeding activity. A 13.72 m long bowhead would need to ingest, according to Brodie's assumptions, about 4000 kg of lipids during a feeding season to sustain it throughout 1 year. Lowry and Frost (1984) used results of examinations of the gastrointestinal tracts of bowheads with information on the fat and caloric content of Arctic zooplankton to make inferences about bowhead energetics. They concluded that the whales could meet their annual energy requirements during a 130-day feeding season in the Beaufort Sea. It is important, however, to recognise the theoretical nature of these calculations and the need for more empirical work, particularly to test the validity of Brodie's assumptions.

Distribution

For management, at least four geographic stocks of bowheads are recognised (Mitchell, 1977; Allen, 1978). These stocks are partially separated from one another by land masses and broad expanses of ice (Fig. 4). There is anecdotal evidence from the early whaling period suggesting at least a limited degree of interchange (Scoresby, 1820; Tomilin, 1957; Clark, 1887), but much of this evidence is inconclusive (Southwell, 1898; Reeves *et al.*, 1983).

The Spitzbergen stock is centered in the Greenland Sea, with some movement into the Norwegian and Barents seas (Southwell, 1898; Reeves, 1980). Historically, individuals occurred at least as far east as Novaya Zemlya (Tomilin, 1957), and there have been a few sightings and strandings near there in recent years (Jonsgård, 1964, 1981, 1982; Benjaminsen *et al.*, 1976; Vasilchuk and Yablokov, 1981; A. Yablokov, personal communication, 1981). The northern limit is north of 80°N, the southern limit at about 62 to 63°N (Eschricht and Reinhardt, 1866).

The Davis Strait stock is found between Labrador and Smith Sound (Eschricht and Reinhardt, 1866; Brown, 1868; Southwell, 1898; Davis and Koski, 1980; Reeves *et al.*, 1983). Lancaster Sound, Pond Inlet, Admiralty Inlet, Prince Regent Inlet, and the east coast of Baffin Island are important

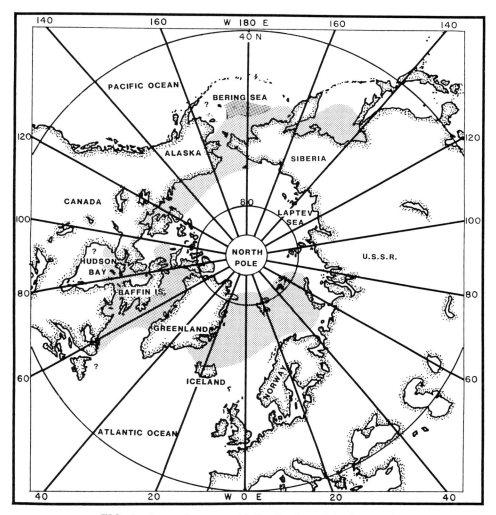

FIG. 4 Approximate world distribution of the bowhead.

summering areas for this stock; the wintering grounds appear to be centered along the west coast of Greenland (Klinowska and Gerslund, 1983), off southeast Baffin Island, at the mouth of Hudson Strait, and off northeast Labrador.

A putative third stock—the Hudson Bay stock—inhabits northern Hudson Bay and Foxe Basin (Mansfield, 1971; Ross, 1974, 1975; Reeves *et al.*, 1983). The presence of whales in or near Fury and Hecla Strait and Hudson

Strait casts some doubt on the Hudson Bay stock's discreteness from the Davis Strait stock (e.g., Southwell, 1898; Finley *et al.*, 1982).

A skull found on the south shore of the St Lawrence estuary in 1949, buried above the high tide line, was identified as a bowhead (Cameron, 1951). If this identification were to be confirmed, it would indicate the former presence of bowheads well inside the Gulf of St Lawrence. Some of the early records of a Basque whale fishery in the Gulf, apparently centered in the Strait of Belle Isle but extending along the north shore to the Saguenay River, pertain to right whales (Tuck and Grenier, 1981; Thurston, 1983) and possibly humpback whales rather than bowheads. However, the "Grand Bay whale" hunted by the Basques and early American whalers in the Strait of Belle Isle apparently was the bowhead (Eschricht and Reinhardt, 1866; True, 1904).

The Bering Sea stock follows a well-documented migration between principally the northern half of the central and western Bering Sea (winter) and the eastern Beaufort Sea and Amundsen Gulf (summer) [Townsend, 1935; Johnson *et al.*, 1966; Braham and Krogman, 1977; Braham *et al.*, 1980b,c (original data republished by Brueggeman, 1982); Fraker and Bockstoce, 1980; Hazard and Cubbage, 1982; Braham *et al.*, 1984; Ljungblad *et al.*, 1984]. Although their failure to find appreciable numbers of bowheads during four research cruises led Dahlheim *et al.* (1980) to conclude that "few, if any, bowheads remain[ed] in the area south of the ice front in the Bering and Chukchi Seas during the summer months" (p. 56), some bowheads reportedly have been seen during summer near the Chukotsk Peninsula in the western Chukchi Sea and the northwest Bering Sea (Bogoslovskaya *et al.*, 1982). The question of whether there is a separate stock of bowheads that regularly remains in the Bering and Chukchi seas during summer was unresolved at the time of this writing (see Miller *et al.*, 1983); animals seen along the northeast Siberian coast in autumn "are considered to be early returns from the Beaufort Sea, and not to indicate a western Chukchi Sea substock" (Bannister, 1984, p. 131). The autumn migration route is thought to take some whales from the Beaufort Sea to as far west as Herald and Wrangel islands and into the eastern part of the East Siberian Sea before they move south through Bering Strait (Braham *et al.*, 1984).

The Okhotsk Sea stock appears to consist of year-round residents of the Sea of Okhotsk (Berzin and Kuz'min, 1975; Berzin and Doroshenko, 1981). A 6.4 m bowhead was killed by fishermen in Osaka Bay, Japan, in late June, 1969 (Nishiwaki and Kasuya, 1970). At 33°28′ N, this is the southernmost confirmed record of the species' occurrence anywhere.

The distributional boundary between bowheads and right whales is poorly defined. Gray (1930, 1937a) believed right whales rather than bowheads were the whales hunted in summer in the ice-free waters of Spitzbergen bays during the early years of that fishery, but so far this question has not been resolved. Similarly, the southern limit of Davis Strait bowheads and the northern limit of northwest Atlantic right whales appear to converge near the Strait of Belle Isle and along the Labrador coast, where large numbers of balaenids were captured by sixteenth-century Basque (Barkham, 1978; Tuck and Grenier, 1981; Thurston, 1983) and eighteenth-century Yankee whalers (Reeves and Mitchell, in press). It is difficult to ascertain whether certain historical whaling records in Denmark Strait and lower Davis Strait pertain to right whales or bowheads. Townsend's (1935) charts of the Bering Sea suggest some overlap in the two species' range there. However, Nasu's (1960) alleged sighting of two right whales in the Chukchi Sea requires better documentation. In general, bowheads and right whales seem allopatric, the former often associating with ice and the latter generally avoiding it. The two species may occupy, or at least may have occupied during historic times, some of the same areas, but at different seasons.

Abundance and History of Exploitation

The world population of bowheads at the beginning of the seventeenth century, soon after which intensive commercial whaling began near Spitzbergen, was probably at least 56 000. Mitchell (1977) attempted to reconstruct catches and to estimate preexploitation size for each of five stocks (Allen, 1978). He estimated 25 000 for the Spitsbergen stock in 1679, 6000 for the Davis Strait stock, 700 for the Hudson Bay stock, 18 000 for the Bering Sea stock in 1850, and 6500 for the Okhotsk Sea stock. Mitchell considered these to be conservative, crude approximations. The estimate for the Davis Strait stock was revised upwards to 11 000 for 1825 (Mitchell and Reeves, 1981). Additional detailed work on the historic catch in the Bering, Chukchi, and Beaufort seas (e.g., Breiwick et al., 1980; Breiwick and Mitchell, 1983; Bockstoce and Botkin, 1983) has left Mitchell's estimate for the Bering Sea stock essentially unchanged (see Best, 1981).

The total world population of bowheads today is probably < 10% of what it was prior to commercial whaling.

The European bowhead fishery centering around Spitzbergen began in the early 1600s, prompted by reports of explorers who found these northern waters to be well stocked with whales (Zorgdrager, 1720; Scoresby, 1820; Conway, 1906; Jenkins, 1921, 1948; Lubbock, 1937; de Jong, 1983). It was dominated by the Dutch until the mid-eighteenth century, after which the British assumed an increasingly significant role. By the late nineteenth century the Spitzbergen fishery had become unprofitable due to severe depletion of the regional bowhead stock (Gray, 1929; Southwell, 1898), which remains at a very low level (Reeves, 1980). Jonsgård (1981, 1982) expressed the belief that some of the few reported sightings between Eurasia and east Greenland in recent years may be of Bering Sea or Davis Strait "strays" rather than of survivors belonging to the Spitzbergen stock. We believe Spitzbergen bowheads still survive but that their number is a small fraction of what it was at the time of their discovery.

The Davis Strait fishery began about 1719 and lasted until the early twentieth century (Ross, 1979). There were two 10-year peaks in pelagic whaling, one involving primarily Dutch vessels from 1729 to 1738, the other primarily Scottish vessels from 1825 to 1834 (Ross, 1979; Mitchell and Reeves, 1981). In addition, the Danish colonial administration of west Greenland carried on a shore-based bowhead fishery from the first half of the eighteenth century to the second half of the nineteenth century (Eschricht and Reinhardt, 1866; Klinowska and Gerslund, 1983). The Davis Strait stock had been severely depleted when commercial whaling for bowheads in Davis Strait and Baffin Bay ended about 1915. The small remaining population has continued to be hunted opportunistically by northern residents (Mitchell and Reeves, 1982). Little stock recovery has been documented. The population today is thought to number no more than a few hundred (Davis and Koski, 1980).

The Hudson Bay commercial fishery took place mainly between 1860 and 1915, during which time an estimated 688 bowheads were caught by American and Scottish whalers (Ross, 1974, 1975). Hunting by Eskimo whalers in Foxe Basin and Hudson Bay has continued at a low level since 1915 (Mitchell and Reeves, 1981, 1982), and no stock recovery has been shown to have occurred.

The Bering–Chukchi–Beaufort seas fishery for bowheads, primarily an American enterprise, did not begin until 1848, and it lasted until about 1914 (Scammon, 1874; Cook, 1926; Bodfish, 1936; Bockstoce, 1980; Bockstoce and Botkin, 1983). Between 1880 and 1910, steamships based in San Francisco dominated the fishery (Bockstoce, 1977), concentrating more on whalebone (baleen) as oil decreased in value. An estimate of the total kill by pelagic whaling vessels in the Bering, Chukchi, and Beaufort seas between 1848 and 1915 is 18 650 bowheads (Bockstoce and Botkin, 1983). After 1914, shore-based bowhead whaling continued in Alaska, and less extensively in northeast Siberia (Tomilin, 1957; Marquette and Bockstoce, 1980; Bogoslovskaya et al., 1982) and Canada (Reeves and Mitchell, 1985).

The controversial Alaskan aboriginal fishery has been viewed by some as one of the most serious whale conservation problems in the world (Mitchell and Reeves, 1980; Tillman, 1980; IWC, 1982; Gambell, 1983). The documented landed catch between 1973 and 1977, inclusive, totaled 149, with at least an additional 177 whales reported as struck-but-lost (Marquette, 1979; Braham et al., 1979). There is disagreement about the current size of the

Bering Sea stock. The Scientific Committee of the International Whaling Commission (IWC) agreed at its 1983 annual meeting on a best available estimate of 3871 ± SE 254 animals in the Bering–Chukchi–Beaufort seas stock (Tillman, 1984; see Zeh *et al.*, 1983). Efforts by the IWC to set a catch quota during the late 1970s resulted in formation of the Alaska Eskimo Whaling Commission, which is now responsible for management of the hunt in cooperation with the U.S. National Marine Fisheries Service.

The Yankee-dominated bowhead fishery in the Sea of Okhotsk has not been well documented (see Scammon, 1874; Mitchell, 1977), but a detailed historical study is underway (Henderson, in prep.). Like the Bering, Chukchi, and Beaufort seas bowhead fishery, it was an extension of the North Pacific right whale fishery. It occurred principally from about 1846 to the 1880s. Catch records of Okhotsk Sea bowheads are often confounded by the fact that oil and baleen returns are lumped with those referring to right whales and Bering Sea bowheads. There is little evidence of further exploitation of bowheads in the Sea of Okhotsk after about 1880, but this population is still thought to be severely depleted (Berzin and Kuz'min, 1975; Berzin and Doroshenko, 1981).

Behaviour

Migration

The bowhead is migratory; its appearance in all areas is seasonal. The timing and routes of migration are influenced by the distribution of ice cover. Bowheads are often found in leads and along the edges of pack ice (Eschricht and Reinhardt, 1866; Fig. 5). As Scoresby (1820, Vol. 2, p. 219) put it:

> They seem to have a preference for close packs and patches of ice, and for fields under certain circumstances; for deep bays or bights, and sometimes for clear water situations; occasionally for detached streams of drift ice; and most generally, for extensive sheets of bay ice. Bay ice is a very favourite retreat of the whales, so long as it continues sufficiently tender, to be conveniently broken, for the purpose of respiration. In such situations, whales may frequently be seen in amazing numbers, elevating and breaking the ice with their crowns, where they are observed to remain much longer at rest than when seen in open water, or in the clear interstices of ice, or indeed in almost any other situation.

Results of quantitative studies of distribution in the Bering Sea during March and April suggest that bowheads inhabit areas with 3–4 octa ice (three-eighths to one-half covered) near St. Lawrence and St. Matthew islands (Braham *et al.*, 1980c; Brueggeman, 1982). Even in Amundsen Gulf and the eastern Beaufort Sea during summer, "sea ice may be a habitat characteristic of bowheads" (Hazard and Cubbage, 1982, p. 523).

FIG. 5 Bowheads are often found close to pack ice. This individual, photographed at the floe edge in Lancaster Sound in late June or July, 1983, is swimming almost directly towards the photographer (A) and past the photographer (B). (Photographs by Gregory Silber.)

Bowheads can break new ice up to at least 22 cm thick, and they can push under young ice to make hummocks for breathing space (Carroll and Smithhisler, 1980). The significance of their association with ice is not clearly known. However, it may improve the whales' foraging capabilities or discourage predation (Dunbar, 1981; Mitchell and Reeves, 1982).

Several sites have been used in recent years for observing bowheads during their migration. Greendale and Brousseau-Greendale (1976) censused bowheads and other marine mammals swimming westward in Lancaster Sound, past Cape Hay on Bylot Island, during the period 21 June to 28 July 1976. Davis and Koski (1980) used a 200-m-high coastal cliff at Cape Adair on northeast Baffin Island for monitoring the southward fall migration of bowheads from 13 September to 7 October 1978. Finley et al. (1982), during a study of white whales, *Delphinapterus leucas*, censused bowheads from Cape Hopes Advance on the south shore of Hudson Strait. In the Chukchi Sea, ice camps along the lead system off north Alaska and a land station at Cape Lisburne have been used for censusing and studying bowheads during their spring eastward migration (Braham et al., 1979, 1980b; Krogman, 1980; Rugh and Cubbage, 1980; Marquette et al., 1982; Zeh et al., 1983). While bowheads were still plentiful in the Greenland Sea, they could be seen migrating along the coast of Jan Mayen Island in late March and April (Southwell, 1898).

Summer distribution of bowheads in all areas is relatively well known, as whalers took advantage of the open-water season to hunt them. Attempts were made to radio-tag and track bowheads in the Beaufort Sea during summer, 1980 (Hobbs and Goebel, 1982), and aerial surveys from July through October have been used since 1979 to document the species' summer and fall distribution in the Beaufort and Chukchi seas (Ljungblad et al., 1983, and contained references; Davis et al., 1982).

The wintering grounds of bowheads are less well documented but are probably centered in polynyas and at the edges of pack ice zones. Of the proposed geographic stocks mentioned above, the winter whereabouts of the Davis Strait stock and the Bering Sea stock are the best known. A long series of observations at Holsteinsborg, west Greenland, indicate that between 1780 and 1822 bowheads were present only from 22 November to 9 April (Eschricht and Reinhardt, 1866). At Godhavn, west Greenland, between 1780 and 1837, the earliest date of the bowhead's arrival was 12 November; the latest date of departure, 25 June. Davis Strait bowheads also winter off Labrador and at the mouth of Hudson Strait (Southwell, 1898; Reeves et al., 1983). The wintering grounds of the Bering Sea stock are in the pack ice from St Lawrence Island south to St Matthew Island (Braham et al., 1980b,c; Brueggeman, 1982; Leatherwood et al., 1983) and in large polynyas in the western Bering Sea (Bogoslovskaya et al., 1982). In heavy-ice years they can be found as far southeast as the Pribilof Islands (Braham et al., 1980b).

Sociability

Because of the depleted status of bowhead stocks, their social organization as observed in most areas today may not represent what it was before intensive whaling began. Some early accounts suggest that bowheads are not strongly gregarious. According to Scoresby (1820, Vol. 1, p. 473)

> [Bowhead] whales, though often found in great numbers together, can scarcely be said to be gregarious; for they are found most generally solitary, or in pairs, except when drawn to the same spot, by the attraction of an abundance of palatable food, or of a choice situation of the ice.

Brown (1868) had a different view, describing the bowhead as "a gregarious animal . . . generally found in small 'schools' of three or four, but when travelling from one part of the ocean to another they will sometimes collect in large parties" (p. 540).

Only individuals and infrequently pairs are seen in the Spitzbergen stock area today (Jonsgård, 1981; A. V. Yablokov, personal communication). Groups of up to 20 reportedly have been seen along the east coast of Baffin Island (Mansfield, 1971), but singles and groups of two or three individuals are now more common there as well (Davis and Koski, 1980; Reeves et al., 1983). Larger aggregations are seen in the Beaufort Sea, where "herds" of 50–60 have been reported along the Plover Islands just east of Point Barrow (Durham, 1979). In the eastern Beaufort Sea during summer, up to several hundred bowheads can be found within a radius of 20 to 50 km, socialising or feeding in different ways but with most animals engaged in similar activities at the same time (Würsig et al., 1982).

During migration, interacting groups of 2 to 7 whales, often apparently involved in courtship, have been observed (McVay, 1973; Everitt and Krogman, 1979; Braham et al., 1980c; Ljungblad, 1981; Fig. 6). However, during spring migration most bowheads appear to be traveling alone or in pairs (Carroll and Smithhisler, 1980). Much socialising occurs in summer, although sexual activity near the surface is usually less boisterous then than it is during migration (Würsig et al., 1982).

The female–calf bond is strong (Scoresby, 1820). Whalers sometimes harpooned a calf in order to gain easier access to its mother, which was reluctant to abandon her wounded offspring (Brown, 1868).

So-called "runs," "pulses," or "waves" of bowheads have been noted during migration (Scoresby, 1820; Southwell, 1898; Marquette, 1977; Durham, 1979; Carroll and Smithhisler, 1980; Braham et al., 1980b). There is some evidence that the waves of bowheads passing through a narrow migration corridor represent different age classes. Durham (1979) believed that the first run off Barrow consists mainly of young animals and that the fourth (and usually last) run involves a high proportion of females accompanied by calves. Other investigators have found that calves are present during all phases of the spring migration off northern Alaska (Braham et al., 1980b), but at least in 1982 a high rate of calf sightings occurred during the last few days of the spring migration off Barrow (Dronenburg et al., 1983).

FIG. 6 Four bowheads interacting in Bering Strait during the northward spring migration, 24 April 1978. (Photograph by S. Leatherwood.)

There is evidence from whalers (summarised by Southwell, 1898) that bowheads east of Greenland were partitioned on the whaling grounds according to age and sex. Older whales, particularly males, were found in open water more often than were females with their young. The latter appeared to prefer the vicinity of fast and solid ice. Scoresby (1823, pp. 292–293) claimed that he "almost invariably found an indiscriminate mixture of males and females" except in summer, when he guessed that females with young "retire into the interior of bays and sounds". Gray (1931) found a similar "separation of the sexes" during summer. The whalers also believed there to be a marked degree of segregation in the Davis Strait stock, with females and calves penetrating deep into Lancaster Sound and Prince Regent Inlet, possibly circumnavigating Baffin Island, while adult males remained along the east coast of Baffin Island where they were frequently caught during autumn (Reeves et al., 1983). Observations made during aerial surveys in recent years in the eastern Beaufort Sea have tended to corroborate the belief that, on their summer feeding grounds at least, bowheads are segregated by age and sex (Davis et al., 1982, 1983).

Swimming and diving

Estimates of the percentage of time spent at the surface by bowheads range from < 5% (spring migration; Carroll and Smithhisler, 1980) to 16% (autumn migration; Davis and Koski, 1980) or 25% (summer; Würsig et al., 1982). Würsig et al. (1984) estimated the mean proportions of the time bowheads were visible from the air in the eastern Beaufort Sea during summer as 0.38 for noncalves and 0.29 for calves. Among the more conspicuous activities that can be seen are breaching, spyhopping, lobtailing or tail slapping, flipper slapping, and lunging. Occasionally, migrating bowheads perform long series of breaches: one whale breached 57 times during a 96-min period (Carroll and Smithhisler, 1980); another, 39 times in an unbroken series (Rugh and Cubbage, 1980). Breaching seems to be more frequent during the spring migration than it is during summer and autumn (Würsig et al., 1982). Whales seen rolling and turning close together, slapping each other with their flippers or flukes, and breaching near one another are often believed to be involved in courtship.

Estimates of mean swimming speed of bowheads during the different phases of their migration vary from about 1.5 to 6 km/hr. In comparison with the large whales in the genus *Balaenoptera*, bowheads and right whales are slow swimmers. A bowhead calf, closely observed from shore and tracked with a theodolite (surveyor's transit), swam at a maximum rate of 22.7 km/hr (Thomas, 1982), with a mean rate during a 1 hr, 49 min, observation period of $8.9 \pm SD\ 5.57$ km/hr (Würsig et al., 1982). Such speeds are probably exceptional. Individual bowheads on their summer feeding grounds have been known to travel as far as 154 km in 121 hr (Davis et al., 1983).

According to Scoresby (1820), harpooned bowheads generally dive for periods of 30 min; the longest dive timed by Scoresby himself lasted 56 min. Scammon (1874) related a second-hand account of an 80-min dive by a harpooned bowhead, and D. C. Foote (cited in Carroll and Smithhisler, 1980) claimed that bowheads under duress could remain submerged for up to 75 min. These reports of exceptionally long dives are difficult to evaluate because of the possibility that the whales used air sources under the ice. Scoresby (1820) stated that periods at the surface for undisturbed bowheads seldom last longer than 2 min, and regular dives last 5 to 15 min. Bodfish (1936) agreed generally, noting that an average-sized bowhead surfaces about three times every hour, at intervals of 18 to 22 min. Recent observations are consistent with those of Scoresby and Bodfish. Timed dives of more than 28 min (Braham et al., 1979), 31 min (Würsig et al., 1984) and 20 min (Davis and Koski, 1980) have been reported. Of 63 timed dives by migrating bowheads off Alaska, the estimated mean duration was 15.6 min (Carroll and Smithhisler, 1980).

Observations of harpooned bowheads led Gray (1937b) to assert that adults dive to depths of 1463 m; juveniles, 732–1097 m. He considered blubber thickness a key factor in determining a whale's diving capabilities. Scoresby's (1820) crude estimate of rate of descent for a harpooned bowhead was 13–16 km/hr. He judged that some individuals died near the bottom in water he estimated to be 1100–1200 m deep.

Bowheads sometimes rest for long periods at the surface and possibly beneath the surface as well (Gray, 1933). Scoresby (1823) observed a group of bowheads remaining motionless at the surface for periods of 5 to 30 min. Bowheads have been seen to rest at the surface for more than an hour during the migration off Alaska (Carroll and Smithhisler, 1980). They seem to have somewhat longer surface times during certain phases of their migration than while on their summer feeding grounds (Würsig et al., 1982). Munn (n.d.) described the way several bowheads sought refuge from killer whales, *Orcinus orca*, in close pack ice, wedging themselves into tight crevices and not moving for a long period, seemingly indifferent to airborne noise and human activity.

Sound production

Like other mysticetes, the bowhead produces intense underwater sounds at relatively low frequencies (Ljungblad et al., 1980). Many of these sounds can be described as either simple moans with a general tonal quality or complex moans with a more pulsive character (Ljungblad et al., 1982). The fundamental frequencies of the former often range from a median low of 90 Hz to a median upper limit of 158–179 Hz; the latter, from 52–70 Hz to 500–900 Hz. The average duration of both types of sound is just over 1 sec, and their total frequency range is 20 to 2000 Hz. Occasional "outbursts of harsh, strident, pulsive, and broadband calls" have been heard during the spring migration (Clark and Johnson, 1984, p. 286). The duration of these complex calls ranges between 0.3 and 7.2 sec, and they contain energy between 100 and 3500 Hz.

Source levels have been estimated as high as 175–185 dB re 1 μPa at 1 m (Würsig et al., 1982). The measured intensity level of the sounds of one whale, recorded while the animal was 100–150 m from the receiver, was approximately 156 dB re 1 μPa at 1 m (Clark and Johnson, 1984). Spring recordings have produced some "unique songlike sequences of two- and three-note syllables." Certain of the bowhead's higher-frequency sounds, with energy content up to 5 kHz, have been described as "elephant trumpeting" or "screeching" (Ljungblad *et al.*, 1982).

Feeding

Bowheads on their northern feeding grounds are known to feed by "skimming" (Nemoto, 1970; Würsig *et al.*, 1982, 1984). A description of this behaviour was given by an experienced Scottish whaler, David Gray (Goode, 1884, p. 23):

> When the food is near the surface they usually choose a space between two pieces of ice, from three to four hundred yards apart, which we term their beat, and swim backwards and forwards, until they are satisfied that the supply of their food is exhausted. They often go with the point of their nose so near the surface that we can see the water running over it just as it does over a stone in a shallow stream; they turn round before coming to the surface to blow, and lie for a short time to lick the food off their bone before going away for another mouthful. They often continue feeding in this way for hours . . . .

Bowheads feeding in this manner often do so in groups of 2 to 10 individuals, frequently swimming in echelon formation (Würsig *et al.*, 1982). Observations of bowheads in open water, diving for long periods and releasing "clouds of red-orange" faeces, were judged to be feeding in the water column (Würsig *et al.*, 1982).

Stomach contents indicate that small- to medium-sized zooplankton are the principal prey (Lowry and Burns, 1980). Euphausiids (Lowry *et al.*, 1978) and copepods (Marquette, 1979) are especially important, the former constituting approximately 65% and the latter 30% of the diet in the Beaufort Sea during autumn (Frost and Lowry, 1981; Marquette *et al.*, 1982). The presence of benthic gammarid amphipods and pebbles in stomach contents, together with direct aerial observations of mud streaming from the whales' mouths, indicate that foraging sometimes occurs at or near the bottom in shallow areas (Lowry *et al.*, 1978; Lowry and Burns, 1980; Würsig *et al.*, 1982). A whale killed in the Beaufort Sea in autumn had approximately 45 litres of stomach contents (Lowry and Burns, 1980).

The stomachs of whales killed off Alaska during their spring migration are often empty (Marquette, 1977; Braham *et al.*, 1980c; Marquette *et al.*, 1982), although one whale killed near St. Lawrence Island, Bering Strait, on 1 May had 20–40 litres of food, mainly gammarid amphipods, in its stomach (Hazard and Lowry, 1984). Lack of sampling in winter means that there is no information on winter feeding. All bowhead populations appear to undertake a seasonal migration, motivated at least in part by the availability of extensive feeding banks at high latitudes in summer (Gray, 1931; Fraker and Bockstoce, 1980). It has been suggested that bowheads migrate to the Beaufort Sea not only to take advantage of the abundant, high-energy food available there, but also because competition for food is less there (Lowry and Frost, 1984).

The detection of volatile fatty acids and significant levels of bacteria in the forestomachs of bowheads has been taken to suggest that microbial forestomach fermentation occurs in this species (Herwig *et al.*, 1984). However, it is unclear what substance serves as the substrate for such a fermentation process.

Reproduction

Scoresby (1820, Vol. 1, p. 470) claimed to have observed mating most often "about the latter end of summer". This statement was contradicted by Eschricht and Reinhardt (1866), who considered most mating off west Greenland to take place in January and February. Sexual activity has been documented during the spring migration off northern Alaska in mid-March (Braham *et al.*, 1980c) and in May (Everitt and Krogman, 1979; D. C. Foote, as reported in Marquette, 1977; Carroll and Smithhisler, 1980). Such behaviour may occur at various seasons, but the apparent spring peak in natality (see below) suggests a corresponding seasonal peak in fertility.

Scoresby (1820) believed that most births in the Greenland Sea occur in February or March, although he knew of a newborn taken in late April. Neonates have been examined in west Greenland at late as 6 May (Eschricht and Reinhardt, 1866) and in Alaska as late as 6 June (Marquette, 1977, Table 25). Most calving appears, then, to take place during a protracted spring season, possibly peaking in April–May (Nerini *et al.*, 1984). A calf estimated to be 4.1 m long was observed on 12 August in the eastern Beaufort Sea (Davis *et al.*, 1983); thus, some calves are probably born as late as July or August.

The length of gestation is uncertain. Some indirect evidence supports the hypothesis that it is about 12 months (Marquette, 1977) or 13–14 months (Nerini *et al.*, 1984).

Length at birth has been stated by whalers to be about 3.0 to 4.3 m (Scoresby, 1820; Bodfish, 1936), although smaller calves (1.52–1.83 m, Bodfish, 1936) and larger foetuses (5.18 m, Richards, 1949) have been reported (Fig. 7). The mean is about 4.0–4.5 m (Nerini *et al.*, 1984).

FIG. 7 A foetus, 128 cm long, taken from a bowhead killed near Barrow, Alaska, on 3 September 1976. This specimen was incorrectly attributed to Wainwright, Alaska, May 1976, by Leatherwood et al. (1982, fig. 2). (Photograph by Jeff Cushman, courtesy of Naval Arctic Research Laboratory, Barrow, Alaska.)

Growth rate and duration of lactation are not known with certainty. Scoresby (1820) considered calf dependency to last for a year or more. However, there is no record of lactating females killed in autumn, which has been taken to suggest that weaning is complete in less than a year, perhaps by the time the calf is 5 to 6 months old (Marquette, 1977), or, alternatively, that the caught sample is unrepresentative of the whale population (Nerini et al., 1984). Davis et al. (1983) interpreted the length distribution of calves photographed in the eastern Beaufort Sea during August–September (4.1–7.6 m; mean of 6.45 m ± 0.695 SD; median of 6.4 m) to mean some calves could be weaned and others still be nursing by the time they migrate past Point Barrow in October. A female with a young calf may occasionally be accompanied by a larger juvenile, possibly an offspring more than 1 year old (Bodfish, 1936; Foote, 1964, as cited in Carroll and Smithhisler, 1980; but see Davis et al., 1983).

Estimates of length at sexual maturity for the Bering Sea stock are 11.58 m for males (Durham, 1972, as reported in Marquette, 1977) and 14.0 to 14.5 m for females (Nerini et al., 1984, contra Johnson et al., 1981). Using estimates of length from aerial photographs, Davis et al. (1983) determined that a 12.2 m female was accompanied by a calf, although the majority of females with calves were judged to be more than 13.5 m long.

Age Determination

A major problem in life-history studies of bowheads is that no adequate method of age determination has been devised (Nerini, 1983). The waxy earplugs used for age determination of balaenopterids are present in balaenids, but their soft consistency makes them difficult to collect and fix (Omura et al., 1969; Braham et al., 1980c; Nerini, 1983). Also, the laminae either are not present or are not easily detected and counted in the earplugs of bowheads and right whales. Scoresby (1820) used ridges on the surface of the baleen as a crude index of age in bowheads. He guessed that a whale's age could be estimated by doubling the length in feet of the longest blade of baleen, a method he considered useful only for individuals not yet physically mature. The use of ridges on baleen to calculate absolute age is confounded by the wearing away of the plates at their distal end. Nishiwaki and Kasuya (1970) estimated the age of one bowhead as 16 months, assuming the growth rate of baleen to be constant and each ridge to represent 1 year.

Growth layers have been detected in the periosteal bone of the involucrum of the auditory bulla of a bowhead (Mitchell, 1984). Such layering may prove useful in attempts to estimate ages of dead bowheads.

An aspartic acid racemisation method of age determination has been applied to the ocular lens nuclei from a small sample of killed bowheads (Marquette et al., 1982; Nerini, 1983). At present, the method is still unproven and requires further testing.

In the absence of a reliable method of assessing an individual bowhead's age, it is difficult to know how long these whales can live. Most records of harpoons found imbedded in killed whales leave some doubt as to the origin and time of implantation of the harpoon (see Reeves et al., 1983).

Predation and Mortality

The bowhead probably has only one natural enemy: the killer whale (Mitchell and Reeves, 1982). There is adequate anecdotal evidence of killer whale predation on bowheads to establish that it occurs. However, the frequency of successful attacks is a matter of speculation. Entrapment by ice and forced stranding due to ice accumulation and movement contribute to bowhead mortality as well (Tomilin, 1957; Mitchell and Reeves, 1982).

In the absence of any data with which to estimate the natural mortality rate for the Bering Sea stock of bowheads, Chapman (1984) used an interspecific relationship based on maximum length to calculate adult natural mortality rates as in the range of 0.05–0.06. In an age-structured model of the Bering Sea stock, Breiwick et al. (1984) considered 0.03–0.07 as "a realistic range of adult bowhead instantaneous natural mortality rates" (p. 487).

Parasites found associated with bowhead tissue include at least three nematodes, two trematodes, one acanthocephalan, and two protozoans in the digestive system; one cestode in the blubber; and four diatoms and an amphipod (whale louse) on the skin (Scammon, 1874; Dailey and Brownell, 1972; Haldiman et al., 1981; Heckmann, 1981; Nerini et al., 1984). Because balaenids apparently do not regularly consume fish, large crustaceans, and molluscs – common intermediate hosts for parasitic helminths – their parasite load may be less diverse than those of most other marine mammals (Klumov, 1965).

Tissues, swab samples, and serums from four bowheads were tested for specific antibodies and processed to grow cell cultures and isolate viruses (Smith et al., 1981). Leptospire antibodies were absent from this sample, but antibodies to three caliciviruses known to infect other marine mammals in the Bering Sea were present in these whales' serum. Also, two adenoviruses were isolated from colon samples. A study of the microflora in nine sampled bowheads was inconclusive because of the uncertain origin of the isolated bacteria (Johnston and Shum, 1981).

Acknowledgements

We thank E. D. Mitchell, H. W. Braham, B. Würsig, and D. Croll for critically reading the manuscript, and W. M. Marquette for checking certain data on our behalf.

References

Abdelbaki, Y. Z., Henk, W. G., Haldiman, J. T. Albert, T. F., Henry, R. W., and Duffield, D. W. (1984). Macroanatomy of the renicule of the bowhead whale (*Balaena mysticetus*). *Anat. Rec.* **208**, 481–490.

Albert, T. (Ed.) (1981). Tissue structural studies and other investigations on the biology of endangered whales in the Beaufort Sea. Final Report for the period April 1, 1980 through June 30, 1981. Prepared for U.S. Dept. Interior, Bureau of Land Management, Alaska OCS Office, Anchorage.

Albert, T. F., Migaki, G., Casey, H. W., and Philo, L. M. (1980). Healed penetrating injury of a bowhead whale. *Mar. Fish. Rev.* **42**(9–10), 92–96.

Allen, J. A. (1908). The North Atlantic right whale and its near allies. *Bull. Am. Mus. Nat. Hist.* **24**(18), 277–329.

Allen, K. R., Chairman. (1978). Report of the Scientific Committee. *Rep. int. Whal. Commn* **28,** 38–89.
Bailey, A. M., and Hendee, R. W. (1926). Notes on the mammals of northwestern Alaska. *J. Mammal.* **7,** 9–28.
Bannister, J. (Convenor). (1984). Annex. G. Report of the Sub-Committee on Other Protected Species and Aboriginal/Subsistence Whaling. *Rep. int. Whal. Commn* **34,** 130–143.
Barkham, S. (1978). The Basques: filling a gap in our history between Jacques Cartier and Champlain. *Can. Geogr. J.* **96**(1), 8–19.
van Beneden, P.-J., and Gervais, P. (1880). "Ostéographie des Cétacés Vivants et Fossiles, comprenant la description et l'iconographie du squelette et du système dentaire de ces animaux; ainsi que des documents relatifs à leur histoire naturelle". A. Bertrand, Paris.
Benjaminsen, T., Berlund, J., Christensen, D., Christensen, I., Huse, I., and Sandnes, O. (1976). Merking, observasjoner og adferdsstudier av hval i Barentshavet og ved Svalbard i 1974 og 1975. (Marking, sightings and behavior studies of whales in the Barents Sea and at Svalbard in 1974 and 1975). *Fisken Hav.* **76**(2), 9–23. (In Norwegian; English abstract.)
Berzin, A. A., and Doroshenko, N. V. (1981). Right whales of the Okhotsk Sea. *Rep. int. Whal. Commn* **31** 451–455.
Berzin, A. A., and Kuz'min, A. A. (1975). Serye i gladkie kity okhotskogo moria. (Gray and right whales of the Okhotsk Sea.) *In* "Morskie Mlekopitauishchie. Chast' 1. Materialy VI vsesoyuznogo soveshchaniya (Kiev, Oktyabr' 1975 g.)" (Eds G. B. Agarkov *et al.*), pp. 30–32. Izdatel'stvo "Naukova Dumka," Kiev.
Best, P. B. (Convenor). (1981). Report of the sub-committee on other protected species and aboriginal whaling. *Rep. int. Whal. Commn* **31,** 133–139.
Bockstoce, J. R. (1977). "Steam Whaling in the Western Arctic". New Bedford Whaling Museum, Old Dartmouth Historical Society, New Bedford, Massachusetts.
Bockstoce, J. (1980). A preliminary estimate of the reduction of the Western Arctic bowhead whale population by the pelagic whaling industry: 1848–1915. *Mar. Fish. Rev.* **42**(9–10), 20–27.
Bockstoce, J. R., and Botkin, D. B. (1983). The historical status and reduction of the Western Arctic bowhead whale (*Balaena mysticetus*) population by the pelagic whaling industry, 1848–1914. *Rep. int. Whal. Commn (Special Issue 5)*, 107–141.
Bodfish, H. H. (1936). "Chasing the Bowhead. As told by Captain Hartson H. Bodfish and recorded for him by Joseph R. Allen". Harvard Univ. Press, Cambridge, Massachusetts.
Bogoslovskaya, L. S., Votrogov, L. M., and Krupnik, I. I. (1982). The bowhead whale off Chukotka: migrations and aboriginal whaling. *Rep. int. Whal. Commn* **32,** 391–399.
Braham, H. W., and Krogman, B. D. (1977). Population biology of the bowhead (*Balaena mysticetus*) and beluga (*Delphinapterus leucas*) whale in the Bering, Chukchi and Beaufort Seas. Northwest and Alaska Fisheries Center Processed Report. U.S. Dept. of Commerce, NOAA, NMFS, Seattle, Washington.

Braham, H. W., and Rugh, D. J. (1983). A photographic identification system for bowhead whales. Northwest and Alaska Fisheries Center, National Marine Fisheries Service, U.S. Department of Commerce, NWAFC Processed Report 83-20.

Braham, H., Krogman, B., Leatherwood, S., Marquette, W., Rugh, D., Tillman, M., Johnson, J., and Carroll, G. (1979). Preliminary report of the 1978 spring bowhead whale research program results. *Rep. int. Whal. Commn* **29**, 291–306.

Braham, H. W., Durham, F. E., Jarrell, G. H., and Leatherwood, S. (1980a). Ingutuk: a morphological variant of the bowhead whale, *Balaena mysticetus*. *Mar. Fish. Rev.* **42**(9–10), 70–73.

Braham, H. W., Fraker, M. A., and Krogman, B. D. (1980b). Spring migration of the Western Arctic population of bowhead whales. *Mar. Fish. Rev.* **42**(9–10), 36–46.

Braham, H., Krogman, B., Johnson, J., Marquette, W., Rugh, D., Nerini, M., Sonntag, R., Bray, T., Brueggeman, J., Dahlheim, M., Savage, S., and Goebel, C. (1980c). Population studies of the bowhead whale (*Balaena mysticetus*); results of the 1979 spring research season. *Rep. int. Whal. Commn* **30**, 391–413.

Braham, H. W., Krogman, B. D., and Carroll, G. M. (1984). Bowhead and white whale migration, distribution, and abundance in the Bering, Chukchi, and Beaufort seas, 1975–1978. *NOAA Tech. Rep. NMFS* SSRF-778.

Breiwick, J. M., and Mitchell, E. D. (1983). Estimated initial population size of the Bering Sea stock of bowhead whales (*Balaena mysticetus*) from logbook and other catch data. *Rep. int. Whal. Commn (Spec. Issue 5)*, 147–151.

Breiwick, J. M., Mitchell, E. D., and Chapman, D. G. (1980). Estimated initial population size of the Bering Sea stock of bowhead whale, *Balaena mysticetus:* an iterative method. *Fish. Bull.* **78**, 843–853.

Breiwick, J. M., Eberhardt, L. L., and Braham, H. W. (1984). Population dynamics of Western Arctic bowhead whales (*Balaena mysticetus*). *Can. J. Fish. Aquat. Sci.* **41**(3), 484–496.

Brodie, P. F. (1981). A preliminary investigation of the energetics of the bowhead whale (*Balaena mysticetus*). *Rep. int. Whal. Commn* **31**, 501–502.

Brown, R. (1868). Notes on the history and geographical relations of the Cetacea frequenting Davis Strait and Baffin's Bay. *Proc. Zool. Soc. London 1868*, 533–556.

Brueggeman, J. J. (1982). Early spring distribution of bowhead whales in the Bering Sea. *J. Wildl. Manage.* **46**, 1036–1044.

Cameron, A. W. (1951). Greenland right whale recorded in Gaspe County, Quebec. Annual Report of the National Museum of Canada for the fiscal year 1949–1950, Bull. No. 123, pp. 116–119.

Carroll, G. M., and Smithhisler, J. R. (1980). Observations of bowhead whales during spring migration. *Mar. Fish. Rev.* **42**(9–10), 80–85.

Chapman, D. G. (1984). Estimates of net recruitment of Alaska bowhead whales and of risk associated with various levels of kill. *Rep. int. Whal. Commn* **34**, 469–471.

Clark, A. H. (1887). The whale-fishery. 1.-History and present condition of the fishery. *In* "The Fisheries and Fishery Industries of the United States. Section V. History and methods of the fisheries" (Ed. G. B. Goode), Vol. II, Part XV,

pp. 3–218. U.S. Commission of Fish and Fisheries, Government Printing Office, Washington, D.C.
Clark, C. W., and Johnson, J. H. (1984). The sounds of the bowhead whale, *Balaena mysticetus*, during the spring migrations of 1979 and 1980. *Can. J. Zool.* **62,** 1436–1441.
Conway, M. (1906). "No Man's Land. A history of Spitsbergen from its discovery in 1596 to the beginning of the scientific exploration of the country". Cambridge Univ. Press, Cambridge.
Cook, J. A. (1926). "Pursuing the whale. A Quarter-century of Whaling in the Arctic". Murray, London.
Dahlheim, M., Bray, T., and Braham, H. (1980). Vessel survey for bowhead whales in the Bering and Chukchi seas, June–July 1978. *Mar. Fish. Rev.* **42**(9–10), 51–57.
Dailey, M. D., and Brownell, R. L. Jr., (1972). A checklist of marine mammal parasites. *In* "Mammals of the Sea. Biology and Medicine" (Ed. S. H. Ridgway), pp. 528–589. Thomas, Springfield, Illinois.
Davis, R. A., and Koski, W. R. (1980). Recent observations of the bowhead whale in the eastern Canadian high Arctic. *Rep. int. Whal. Commn* **30,** 439–444.
Davis, R. A., Koski, W. R., Richardson, W. J., Evans, C. R., and Alliston, W. G. (1982). Distribution, numbers and productivity of the Western Arctic stock of bowhead whales (*Balaena mysticetus*) in the eastern Beaufort Sea and Amundsen Gulf, summer 1981. SC/34/DOC PS 20. International Whaling Commission, Cambridge.
Davis, R. A., Koski, W. R., and Miller, G. W. (1983). Preliminary assessment of the length–frequency distribution and gross annual reproductive rate of the Western Arctic bowhead whale as determined with low-level aerial photography, with comments on life history. Unpublished report prepared by LGL Ltd., environmental research associates, Toronto, Ontario, and Anchorage, Alaska, for National Marine Mammal Laboratory, National Marine Fisheries Service, National Oceanic and Atmospheric Administration, Seattle, Washington. i-ix + 1–91.
Davis, W. M. (1874). "Nimrod of the sea; or, the American whaleman". Harper, New York.
Dronenburg, R. B., Carroll, G. M., Rugh, D. J., and Marquette, W. M. (1983). Report of the 1982 spring bowhead whale census and harvest monitoring including 1981 fall harvest results. *Rep. int. Whal. Commn* **33,** 525–537.
Dubielzig, R., and Agiurre, G. (1981). Morphological studies of the visual apparatus of the bowhead whale, *Balaena mysticetus*. *In* "Tissue structural studies and other investigations on the biology of endangered whales in the Beaufort Sea", (Ed. T. Albert), pp. 157–171. Final Report for the period April 1, 1980 through June 30, 1981. Prepared for U.S. Dept. Interior, Bureau of Land Management, Alaska OCS Office, Anchorage.
Dunbar, M. J. (1981). Physical causes and biological significance of polynyas and other open water in sea ice. *In* "Polynyas in the Canadian Arctic" (Eds. I.

Stirling and H. Cleator), pp. 29–43. Occasional Paper No. 45, Canadian Wildlife Service.

Durham, F. E. (1979). The catch of bowhead whales (*Balaena mysticetus*) by Eskimos, with emphasis on the Western Arctic. Los Angeles Co. Mus. Contrib. Sci. **314,** 1–14.

Durham, F. E. (1980). External morphology of bowhead fetuses and calves. *Mar. Fish. Rev.* **42**(9–10), 74–80.

Edwards, E. J., and Rattray, J. E. (1956). " 'Whale Off!' The story of American shore whaling". Coward–McCann, New York.

Eschricht, D. F., and Reinhardt, J. (1866). On the Greenland right-whale. (*Balaena mysticetus*, Linn.), with especial reference to its geographical distribution and migrations in times past and present, and to its external and internal characteristics. (Transl. from Danish publ. of 1861). *In* "Recent Memoirs on the Cetacea by Professors Eschricht, Reinhardt and Lilljeborg" (Ed. W. H. Flower), pp 1–150. The Ray Society, London.

Everitt, R. D., and Krogman, B. D. (1979). Sexual behavior of bowhead whales observed off the north coast of Alaska. *Arctic* **32,** 277–280.

Fetter, A. W., and Everitt, J. I. (1981). Determination of the gross and microscopic structures of selected tissues and organs of the bowhead whale, *Balaena mysticetus*, with emphasis on bone, blubber and the lymphoimmune and cardiovascular systems. *In* "Tissue structural studies and other investigations on the biology of endangered whales in the Beaufort Sea", (Ed. T. Albert), pp. 51–88. Final Report for the period April 1, 1980 through June 30, 1981. Prepared for U.S. Dept. Interior, Bureau of Land Management, Alaska OCS Office, Anchorage.

Finley, K. J., Miller, G. W., Allard, M., Davis, R. A., and Evans, C. R. (1982). The belugas (*Delphinapterus leucas*) of northern Quebec: distribution, abundance, stock identity, catch history and management. *Can. Tech. Rep. Fish. Aquat. Sci.* **1123,** 1–57.

Fraker, M. A., and Bockstoce, J. R. (1980). Summer distribution of bowhead whales in the eastern Beaufort Sea. *Mar. Fish. Rev.* **42**(9–10), 57–64.

Frost, K., and Lowry, L. (1981). Feeding and trophic relationships of bowhead whales and other vertebrate consumers in the Beaufort Sea. Final report to Natl. Mar. Mammal Lab., Natl. Mar. Fish. Serv., NOAA, Seattle, Washington.

Gambell, R. (1983). Bowhead whales and Alaskan Eskimos: a problem of survival. *Polar Rec.* **21**(134), 467–473.

Goode, G. B. (1884). A.—The whales and porpoises. *In* "The Fisheries and Fishery Industries of the United States. Section I. Natural History of Useful Aquatic Animals" (Ed. G. B. Goode), pp. 7–32. United States Commission of Fish and Fisheries, Washington, D.C.

Gray, R. (1887). Notes on a voyage to the Greenland Seas in 1886. *Zoologist, Third Series* **11**(122), 48–57, (123), 94–100, (124), 121–136.

Gray, R. (1889). Notes on a voyage to the Greenland Sea in 1888. *Zoologist, Third Series,* **13**(145), 1–9, (146), 41–51, (147), 95–104.

Gray, R. W. (1929). The extermination of whales. *Nature (London)* **123,** 314–315.
Gray, R. W. (1930). Spitsbergen whale fishery of the seventeenth century. *Nature (London)* **126,** 204.
Gray, R. W. (1931). The colour of the Greenland Sea and the migrations of the Greenland whale and narwhal. *Geog. J. (London)* **78,** 284–290.
Gray, R. W. (1933). The sleeping habits of whales. *Naturalist,* pp. 257–260.
Gray, R. W. (1937a). The Atlantic or Biscay whale, *Balaena glacialis* and the Spitsbergen whale fishery of the seventeenth century. *Naturalist,* pp. 153–156.
Gray, R. W. (1937b). The blubber of whales. *Naturalist,* pp. 55–57.
Gray, R. W. (1940). *Balaena mysticetus*—the Greenland, Arctic or bow-head whale. *Naturalist,* pp. 193–199.
Greendale, R. G., and Brousseau-Greendale, C. (1976). Observations of marine mammals at Cape Hay, Bylot Island during the summer of 1976. *Can. Fish. Mar. Serv. Tech. Rep.* **680,** 1–25.
Haldiman, J. T., Abdelbaki, Y. Z., Al-Bagdadi, F. K., Duffield, D. W., Henk, W. G., and Henry, R. W. (1981). Determination of the gross and microscopic structure of the lung, kidney, brain and skin of the bowhead whale, *Balaena mysticetus.* In "Tissue structural studies and other investigations on the biology of endangered whales in the Beaufort Sea", (Ed. T. Albert), pp. 305–662. Final Report for the period April 1, 1980 through June 30, 1981. Prepared for U.S. Dept. Interior, Bureau of Land Management, Alaska OCS Office, Anchorage.
Haldiman, J. T., Abdelbaki, Y. Z., Duffield, D. W., Henk, W. G., and Henry, R. W. (1983). Observations on the morphology of the skin, respiratory system, urinary system, and brain of the bowhead whale, *Balaena mysticetus.* Doc. SC/35/PS16, pp. 1–20. Document presented to Scientific Committee, International Whaling Commission, Cambridge.
Haldiman, J. T., Henk, W. G., Henry, R. W., Albert, T. F., Abdelbaki, Y. Z., and Duffield, D. W. (1984). Microanatomy of the major airway mucosa of the bowhead whale, *Balaena mysticetus. Anat. Rec.* **209,** 219–230.
Hazard, K. W., and Cubbage, J. C. (1982). Bowhead whale distribution in the southeastern Beaufort Sea and Amundsen Gulf, summer 1979. *Arctic* **35,** 519–523.
Hazard, K. W., and Lowry, L. F. (1984). Benthic prey in a bowhead whale from the northern Bering Sea. *Arctic* **37**(2), 166–168.
Heckmann, R. A. (1981). Parasitological study of the bowhead whale, *Balaena mysticetus.* In "Tissue structural studies and other investigations on the biology of endangered whales in the Beaufort Sea", (Ed. T. Albert), pp. 275–304. Final Report for the period April 1, 1980 through June 30, 1981. Prepared for U.S. Dept. Interior, Bureau of Land Management, Alaska OCS Office, Anchorage.
Henderson, D. (In preparation). Whaling in the Okhotsk Sea, 1845–early 20th century [provisional title]. Old Dartmouth Historical Society, New Bedford Whaling Museum, New Bedford, Massachusetts.
Henry, R. W., Haldiman, J. T., Albert, T. F., Henk, W. G., Abdelbaki, Y. Z., and Duffield, D. W. (1983). Gross anatomy of the respiratory system of the bowhead whale, *Balaena mysticetus. Anat. Rec.* **207,** 435–449.

Herwig, R. P., Staley, J. T., Nerini, M. K., and Braham, H. W. (1984). Baleen whales: preliminary evidence for forestomach microbial fermentation. *Appl. Environ. Microbiol.* **47**(2), 421–423.

Hobbs, L. J., and Goebel, M. E. (1982). Bowhead whale radio tagging feasibility study and review of large cetacean tagging. NOAA Technical Memorandum NMFS F/NWC-21, Available through National Technical Information Service, U.S. Dept. of Commerce, Springfield, Virginia.

Home, E. (1812). An account of some peculiarities in the structure of the organ of hearing in the *Balaena mysticetus* of Linnaeus. *Philos. Trans. R. Soc. London, Pt. 1,* 83–89.

IWC. (1982). Aboriginal/subsistence whaling (with special reference to the Alaska and Greenland fisheries). *Rep. int. Whal. Commn (Spec. Issue 4),* 1–86.

Jarrell, G. (1979). Karyotype of the bowhead whale (*Balaena mysticetus*). *J. Mammal.* **60,** 607–610.

Jarrell, G. (1981). Cytogenetic and morphological investigation of variability in the bowhead whale, *Balaena mysticetus*. In "Tissue structural studies and other investigations on the biology of endangered whales in the Beaufort Sea", (Ed. T. Albert), pp. 213–231. Final Report for the period April 1, 1980 through June 30, 1981. Prepared for U.S. Dept. Interior, Bureau of Land Management, Alaska OCS Office, Anchorage.

Jenkins, J. T. (1921). "A History of the Whale Fisheries from the Basque Fisheries of the Tenth Century to the Hunting of the Finner Whale at the Present Date". Witherby, London.

Jenkins, J. T. (1948). Bibliography of whaling. *J. Soc. Bibliogr. Nat. Hist.* **2**(4), 71–166.

Johnson, M. L., Fiscus, C. H., Ostenson, B. T., and Barbour, M. L. (1966). Marine mammals. In "Environment of the Cape Thompson Region, Alaska" (Ed. N. J. Wilimovsky), pp. 877–924. U.S. Atomic Energy Commission, Washington, D.C.

Johnson, J. H., Braham, H. W., Krogman, B. D., Marquette, W. M., Sonntag, R. M., and Rugh, D. J. (1981). Bowhead whale research: June 1979 to June 1980. *Rep. int. Whal. Commn* **31,** 461–475.

Johnston, D. G., and Shum, A. C. (1981). Bacteriological study of the bowhead whale, *Balaena mysticetus*. In "Tissue structural studies and other investigations on the biology of endangered whales in the Beaufort Sea", (Ed. T. Albert), pp. 255–274. Final Report for the period April 1, 1980 through June 30, 1981. Prepared for U.S. Dept. Interior, Bureau of Land Management, Alaska OCS Office, Anchorage.

Jones, C. L., and Tarpley, R. J. (1981). Observations of the heart of the bowhead whale. In "Tissue structural studies and other investigations on the biology of endangered whales in the Beaufort Sea", (Ed. T. Albert), pp. 805–828. Final Report for the period April 1, 1980 through June 30, 1981. Prepared for U.S. Dept. Interior, Bureau of Land Management, Alaska OCS Office, Anchorage.

de Jong, C. (1983). The hunt of the Greenland whale: a short history and statistical sources. *Rep. int. Whal. Commn (Spec. Issue 5),* 83–106.

Jonsgård, Å. (1964). A right whale (*Balaena* sp.), in all probability a Greenland right whale (*Balaena mysticetus*) observed in the Barents Sea. *Nor. Hvalfangst-Tid.* **53**, 311–313.

Jonsgård, Å. (1981). Bowhead whales, *Balaena mysticetus*, observed in Arctic waters of the eastern North Atlantic after the Second World War. *Rep. int. Whal. Commn* **31**, 511.

Jonsgård, Å. 1982. Bowhead (*Balaena mysticetus*) surveys in Arctic Northeast Atlantic waters in 1980. *Rep. int. Whal. Commn* **32**, 355–356.

Kenney, R. M., Garcia, M. C., and Everitt, J. I. (1981). The biology of the reproductive and endocrine systems of the bowhead whale, *Balaena mysticetus*, as determined by evaluation of tissues and fluids from subsistence harvested whales. In "Tissue structural studies and other investigations on the biology of endangered whales in the Beaufort Sea", (Ed. T. Albert), pp. 89–155. Final Report for the period April 1, 1980 through June 30, 1981. Prepared for U.S. Dept. Interior, Bureau of Land Management, Alaska OCS Office, Anchorage.

Klinowska, M., and Gerslund, E. (1983). The Cetacea of West Greenland—Progress report on stage 1 and proposal for stage 2. Document submitted to the Scientific Committee, International Whaling Commission, Cambridge.

Klumov, S. K. (1962). Gladkie (yaponskie) kity tikhogo okeana. *Tr. Inst. Okeanol.* **58**, 202–297. (In Russian, with English abstract.)

Klumov, S. K. (1965). Food and helminth fauna of whalebone whales (Mystacoceti) in the main whaling regions of the world ocean. Fish. Res. Board of Canada Translation Series No. 589. (Transl. of Pitanie i gelmintofauna usatykh kitov (Mystacoceti) v osnovnykh promyslovykh raionakh mirovogo okeana. *Tr. Inst. Okeanol.* **71**, 94–194, 1963.)

Krogman, B. D. (1980). Sampling strategy for enumerating the Western Arctic population of the bowhead whale. *Mar. Fish. Rev.* **42**(9–10), 30–36.

Kükenthal, W. (1922). Die brustflosse des Grönlandswales (*Balaena mysticetus* L.) *Bijdr. Dierk.*, pp. 59–63.

Leatherwood, S., Reeves, R. R., Perrin, W. F., and Evans, W. E. (1982). Whales, dolphins, and porpoises of the eastern North Pacific and adjacent Arctic waters. With Appendix A on tagging by Larry Hobbs. *NOAA Tech. Rep. NMFS Circ.* **444**, 1–245.

Leatherwood, S., Bowles, A. E., and Reeves, R. R. (1983). Endangered whales of the eastern Bering Sea and Shelikof Strait, Alaska; results of aerial surveys, April 1982 through April 1983 with notes on other marine mammals seen. HSWRI Tech. Rep. 83–159. Final Report to Office of Marine Pollution Assessment, National Oceanic and Atmospheric Administration, Juneau, Alaska. Contract No. NA82RAC00039.

Ljungblad, D. K. (1981). Aerial surveys of endangered whales in the Beaufort Sea, Chukchi Sea and northern Bering Sea. Technical Document 44, pp. 1–49. Naval Ocean Systems Center, San Diego, California.

Ljungblad, D. K., Leatherwood, S., and Dahlheim, M. E. (1980). Sounds recorded in the presence of an adult and calf bowhead whale. *Mar. Fish. Rev.* **42**(9–10), 86–87.

Ljungblad, D. K., Thompson, P. O., and Moore, S. E. (1982). Underwater sounds recorded from migrating bowhead whales, *Balaena mysticetus,* in 1979. *J. Acoust. Soc. Am.* **71,** 477–482.

Ljungblad, D. K., Moore, S. E., and Van Schoik, D. R. (1983). Aerial surveys of endangered whales in the Beaufort, Eastern Chukchi and Northern Bering seas, 1982. Technical Document 605. Naval Ocean Systems Center, San Diego.

Ljungblad, D. K., Moore, S. E., and Van Schoik, D. R. (1984). Aerial surveys of endangered whales in the northern Bering, eastern Chukchi and Alaskan Beaufort seas, 1983: with a five year review, 1979–1983. Technical Report 955. Naval Ocean Systems Center, San Diego.

Lockyer, C. (1976). Body weights of some species of large whales. *J. Cons. Cons. int. Explor. Mer* **36**(3), 259–273.

Lowry, L., and Burns, J. (1980). Foods utilized by bowhead whales near Barter Island, Alaska, autumn 1979. *Mar. Fish. Rev.* **42**(9–10), 88–91.

Lowry, L., and Frost, K. (1984). Foods and feeding of bowhead whales in western and northern Alaska. *Sci. Rep. Whales Res. Inst.* **35,** 1–16.

Lowry, L. F., Frost, K. J., and Burns, J. J. (1978). Food of ringed seals and bowhead whales near Point Barrow, Alaska. *Can. Field Nat.* **92,** 67–70.

Lubbock, B. (1937). "The Arctic Whalers". Brown, Son & Ferguson, Glasgow.

Mansfield, A. W. (1971). Occurrences of the bowhead or Greenland right whale (*Balaena mysticetus*) in Canadian Arctic waters. *J. Fish. Res. Board Can.* **28,** 1873–1875.

Marquette, W. M. (1977). The 1976 catch of bowhead whales (*Balaena mysticetus*) by Alaskan Eskimos, with a review of the fishery, 1973–1976, and a biological summary of the species. Processed Report, U.S. Dept. of Commerce, Natl. Oceanic Atmos. Admin., Natl. Mar. Fish. Serv., Northwest and Alaska Fisheries Center, Seattle, Washington.

Marquette, W. M. (1979). The 1977 catch of bowhead whales (*Balaena mysticetus*) by Alaskan Eskimos. *Rep. int. Whal. Commn* **29,** 281–289.

Marquette, W. M., and Bockstoce, J. R. (1980). Historical shore-based catch of bowhead whales in the Bering, Chukchi, and Beaufort Seas. *Mar. Fish. Rev.* **42**(9–10), 5–19.

Marquette, W. M., Braham, H. W., Nerini, M. K., and Miller, R. V. (1982). Bowhead whale studies, autumn 1980–spring 1981: harvest, biology and distribution. *Rep. int. Whal. Commn* **32,** 357–370.

McVay, S. (1973). Stalking the Arctic whale. *Am. Sci.* **61**(1), 24–37.

Medway, W. (1980). Some observations on urine from a bowhead whale. *Mar. Fish. Rev.* **42**(9–10), 91–92.

Medway, W. (1981). The cytological and clinical evaluation of blood and urine of the bowhead whale, *Balaena mysticetus. In* "Tissue structural studies and other investigations on the biology of endangered whales in the Beaufort Sea", (Ed. T. Albert), pp. 201–212. Final Report for the period April 1, 1980 through June 30, 1981. Prepared for U.S. Dept. Interior, Bureau of Land Management, Alaska OCS Office, Anchorage.

Medway, W. (1983). Examination of blood and urine from Eskimo killed bowhead whales (*Balaena mysticetus*). *Aquat. Mamm.* **10**(1), 1–8.

Meek, A. (1918). The reproductive organs of Cetacea. *J. Anat.* **52,** 186–210.
Miller, R. V., Johnson, J. H., and Rugh, D. J. (1983). Notes on the distribution of bowhead whales, *Balaena mysticetus*, in the western Chukchi Sea, 1979 to 1982. Doc. SC/35/PS9. Document submitted to the Scientific Committee, International Whaling Commission, Cambridge.
Mitchell, E. (1977). Initial population size of bowhead whale (*Balaena mysticetus*) stocks: cumulative catch estimates. SC/29/Doc 33. International Whaling Commission, Cambridge.
Mitchell, E. (1984). Ecology of North Atlantic boreal and arctic monodontid and mysticete whales. *In* "Arctic Whaling: Proceedings of the International Symposium, February 1983". Works of the Arctic Centre No. 8, Univ. of Groningen, the Netherlands.
Mitchell, E., and Reeves, R. R. (1980). The Alaska bowhead problem: a commentary. *Arctic* **33,** 686–723. (Also see errata in **34,** 100).
Mitchell, E., and Reeves, R. R. (1981). Catch history and cumulative catch estimates of initial population size of cetaceans in the Eastern Canadian Arctic. *Rep. int. Whal. Commn* **31,** 645–682.
Mitchell, E., and Reeves, R. R. (1982). Factors affecting abundance of bowhead whales *Balaena mysticetus* in the eastern Arctic of North America, 1915–1980. *Biol. Conserv.* **22,** 59–78.
Munn, H. T. (n.d.). "Tales of the Eskimo. Being impressions of a strenuous, indomitable, and cheerful little people" (with photographs by the author). Chambers, London.
Nasu, K. (1960). Oceanographic investigation in the Chukchi Sea during the summer of 1958. *Sci. Rep. Whales Res. Inst.* **15,** 143–157.
Nemoto, T. (1970). Feeding pattern of baleen whales in the ocean. *In* "Marine Food Chains". (Ed. J. H. Steele), pp. 241–252. Univ. of California Press, Berkeley.
Nerini, M. K. (1983). Age determination techniques applied to mysticete whales. Unpublished M.S. thesis, Univ. of Washington, Seattle.
Nerini, M. K., Braham, H. W., Marquette, W. M., and Rugh, D. J. (1984). Life history of the bowhead whale, *Balaena mysticetus*. *J. Zool.* **204,** 443–468.
Nishiwaki, M., and Kasuya, T. (1970). A Greenland right whale caught at Osaka Bay. *Sci. Rep. Whales Res. Inst.* **22,** 45–62.
Norris, J. C., and Leatherwood, S. (1981). Hearing in the bowhead whale, *Balaena mysticetus*, as estimated by cochlear morphology. *In* "Tissue structural studies and other investigations on the biology of endangered whales in the Beaufort Sea", (Ed. T. Albert), pp. 745–787. Final Report for the period April 1, 1980 through June 30, 1981. Prepared for U.S. Dept. Interior, Bureau of Land Management, Alaska OCS Office, Anchorage.
Omura, H., Ohsumi, S., Nemoto, T., Nasu, K., and Kasuya, T. (1969). Black right whales in the North Pacific. *Sci. Rep. Whales Res. Inst.* **21,** 1–78.
Ralls, K. (1976). Mammals in which females are larger than males. *Q. Rev. Biol.* **51,** 245–276.
Reeves, R. R. (1980). Spitsbergen bowhead stock: a short review. *Mar. Fish. Rev.* **42**(9–10), 65–69.

Reeves, R. R., and Mitchell, E. (1985). Shore-based bowhead whaling in the eastern Beaufort Sea and Amundsen Gulf. *Rep. int. Whal. Commn* **35,** in press.

Reeves, R. R., and Mitchell, E. (in press). American pelagic whaling for right whales in the North Atlantic. *Rep. int. Whal. Commn (Special Issue)*.

Reeves, R., Mitchell, E., Mansfield, A., and McLaughlin, M. (1983). Distribution and migration of the bowhead whale, *Balaena mysticetus*, in the eastern North American Arctic. *Arctic* **36**(1), 5–64.

Rice, D. W. (1977). A list of marine mammals of the world (third edition). *NOAA Tech. Rep. NMFS SSRF* **711.** 1–15 p.

Richards, E. A. (1949). "Arctic Mood, a Narrative of Arctic Adventures". Caxton Printers, Caldwell, Idaho.

Ridgway, S. H. (1981). Some brain morphometrics of the bowhead whale. *In* "Tissue structural studies and other investigations on the biology of endangered whales in the Beaufort Sea", (Ed. T. Albert), pp. 837–844. Final Report for the period April 1, 1980 through June 30, 1981. Prepared for U.S. Dept. Interior, Bureau of Land Management, Alaska OCS Office, Anchorage.

Ross, W. G. (1974). Distribution, migration, and depletion of bowhead whales in Hudson Bay, 1860 to 1915. *Arct. Alp. Res.* **6,** 85–98.

Ross, W. G. (1975). Whaling and Eskimos: Hudson Bay 1860–1915. *Natl. Mus. Can. Natl. Mus. Man Publ. Ethnol.* **10,** 1–164.

Ross, W. G. (1979). The annual catch of Greenland (bowhead) whales in waters north of Canada 1719–1915: a preliminary compilation. *Arctic* **32,** 91–121.

Rugh, D. J., and Cubbage, J. C. (1980). Migrations of bowhead whales past Cape Lisburne, Alaska. *Mar. Fish. Rev.* **42**(9–10), 46–51.

Scammon, C. M. (1874). "The Marine Mammals of the North-western Coast of North America, described and illustrated: together with an account of the American whale-fishery". Carmany & Co., San Francisco.

Scoresby, W. (1820). "An Account of the Arctic Regions, with a history and description of the northern whale-fishery", two volumes. Archibald Constable and Co., Edinburgh, and Hurst, Robinson and Co., London.

Scoresby, W., Jr. (1823). "Journal of a Voyage to the Northern Whale-Fishery; including researches and discoveries on the eastern coast of West Greenland, made in the summer of 1822, in the ship Baffin of Liverpool". Archibald Constable and Co., Edinburgh, and Hurst, Robinson and Co., London.

Sis, R. F., and Tarpley, R. J. (1981). Structural studies of the stomach and small intestine of the bowhead whale *Balaena mysticetus*. *In* "Tissue structural studies and other investigations on the biology of endangered whales in the Beaufort Sea", (Ed. T. Albert), pp. 663–743. Final Report for the period April 1, 1980 through June 30, 1981. Prepared for U.S. Dept. Interior, Bureau of Land Management, Alaska OCS Office, Anchorage.

Smith, A. W., Skilling, D. E., and Benirschke, K. (1981). Investigations of the serum antibodies and viruses of the bowhead whale, *Balaena mysticetus*. *In* "Tissue structural studies and other investigations on the biology of endangered whales in the Beaufort Sea", (Ed. T. Albert), pp. 233–254. Final Report for the period

April 1, 1980 through June 30, 1981. Prepared for U.S. Dept. Interior, Bureau of Land Management, Alaska OCS Office, Anchorage.

Southwell, T. (1898). The migration of the right whale (*Balaena mysticetus*). *Nat. Sci.* **12,** 397–414.

Struthers, J. (1881). On the bones, articulations, and muscles of the rudimentary hind-limb of the Greenland right-whale (*Balaena mysticetus*). *J. Anat. Physiol.* **15,** 141–176.

Tarpley, R. J., and Stott, G. G. (1983). Preliminary observations on the anatomy of the reproductive and alimentary systems of the bowhead whale (*Balaena mysticetus*). Doc. SC/35/PS15. Document presented to the Scientific Committee, International Whaling Commission, Cambridge, July 1983. 1–35.

Thomas, P. O. (1982). Calf breaching. *In* "Behavior, Disturbance Responses and Feeding of Bowhead Whales *Balaena mysticetus* in the Beaufort Sea, 1980–81" (Ed. W. J. Richardson), pp. 126–130. Chapter by New York Zoological Society in unpublished report from LGL Ecol. Res. Assoc., Inc., Bryan, Texas, for the U.S. Bureau of Land Management, Washington, D.C.

Thurston, H. (1983). The Basque connection. *Equinox* **12,** 46–59.

Tillman, M. F. (1980). Introduction: a scientific perspective of the bowhead whale problem. *Mar. Fish. Rev.* **42**(9–10), 2–5.

Tillman, M. F. (chairman). (1984). Report of the Scientific Committee. *Rep. int. Whal. Commn* **34,** 35–181.

Tillman, M. F., Breiwick, J. M., and Chapman, D. G. (1983). Reanalysis of historical whaling data for the Western Arctic bowhead whale population. *Rep. int. Whal. Commn (Special Issue 5)*, 143–146.

Tomilin, A. G. (1957). "Mammals of the U.S.S.R. and Adjacent Countries. Vol. IX: Cetacea" (Ed. V. G. Heptner), Nauk S.S.S.R., Moscow. (English Translation, 1967, Israel Program for Scientific Translations, Jerusalem).

Townsend, C. H. (1935). The distribution of certain whales as shown by logbook records of American whaleships. *Zoologica* **19**(1), 1–50.

True, F. W. (1904). The whalebone whales of the western North Atlantic compared with those occurring in European waters with some observations on the species of the North Pacific. *Smithson. Contrib. Knowl.* **33,** 1–332.

Tuck, J. A., and Grenier, R. (1981). A 16th-century Basque whaling station in Labrador. *Sci. Am.* **245**(5), 180–189.

Turner, W. (1913). The right whale of the North Atlantic, *Balaena biscayensis:* its skeleton described and compared with that of the Greenland right whale, *Balaena mysticetus*. *Trans. R. Soc. Edinburgh,* **48,** Part IV, (No. 33), 889–922.

Vasilchuk, Y. K., and Yablokov, A. V. (1981). Grenlandskiy kit v Obskoy gube. (Greenland whale in Obskaya Inlet.) *Priroda* (Moscow) **181**(3), 116.

Würsig, B., Clark, C. W., Dorsey, E. M., Fraker, M. A., and Payne, R. S. (1982). Normal behavior of bowheads. *In* "Behavior, Disturbance Responses and Feeding of Bowhead Whales *Balaena mysticetus* in the Beaufort Sea, 1980–81". (Ed. W. J. Richardson), pp. 33–143. Chapter by New York Zoological Society in unpublished report from LGL Ecol. Res. Assoc., Inc., Bryan, Texas, for the U.S. Bureau of Land Management, Washington, D.C.

Würsig, B., Dorsey, E. M., Fraker, M. A., Payne, R. S., Richardson, W. J., and Wells, R. S. (1984). Behavior of bowhead whales, *Balaena mysticetus*, summering in the Beaufort Sea: surfacing, respiration, and dive characteristics. *Can. J. Zool.* **62,** 1910–1921.

Zeh, J. E., Ko, D., Krogman, B. D., and Sonntag, R. (1983). Minimum population estimates of the Western Arctic stock of the bowhead whale from ice-based census results 1978–82. Doc. SC/35/PS12. Document submitted to the Scientific Committee, International Whaling Commission, Cambridge.

Zorgdrager, G. F. (1720). "Bloeyende Opkomst der Aloude en Hedendaagsche Groenlandsche Visschery". ("Development of the old and contemporary Greenland fishery".) Oosterwyk, Amsterdam.

12

Pygmy Right Whale
Caperea marginata (Gray, 1846)

Alan N. Baker

Genus and Species

A single species exists in this Southern Hemisphere genus. The common name infers (correctly) a relationship to the right whale, *Eubalaena glacialis*, which resembles *Caperea* in its strongly arched jaw.

External Characteristics and Morphology

The common name also implies that this is a small whale; certainly at a maximum length of 645 cm (female), the species is much smaller than the right whale (18 m). There is apparently sexual dimorphism in size, as the largest male recorded is 609 cm (McManus *et al.*, 1984).

The body is streamlined much like that of a rorqual (Figs. 1 and 2). The dorsal surface is weakly convex except forward of the blowholes, where the narrow rostrum curves progressively downwards. The snout itself is rather

FIG. 1 The pygmy right whale, *Caperea marginata*. A composite drawing based on photographs of several stranded animals.

FIG. 2 Two views of a 553-cm female pygmy right whale stranded at Port Underwood, New Zealand, 22 January 1966. Note white baleen gum, dark, oval body scars, and prominent falcate dorsal fin.

bluntly rounded, owing to the thick tip of the lower jaw. The mouth is strongly arched, and its downward curving line extends posteriorly below the eye. It contains 213–230 baleen plates on each side. The plates are narrow (five to seven times as long as wide), have a fringe of fine soft hair, and are coloured shiny whitish yellow, with a narrow, dark brown marginal band on the external edge. The shape and colour pattern of these baleen plates is diagnostic for the species—the name *marginata* refers to the dark marginal band.

The ventral surface of the pygmy right whale is shallowly convex, but has the additional feature of a distinct swelling on the posterior area of the throat, bearing a pair of shallow, longitudinal grooves (Ross *et al.*, 1975). The anteriorly directed, V-shaped blowholes are situated in a slight depression and a small median ridge runs from near the blowhole to the snout tip.

A dorsal fin with a distinctly concave posterior margin is situated at about two-thirds to three-fourths of the body length from the snout, and the flippers are narrow and rounded at the tip. The tail flukes are broad, with a distinct median notch.

The body colour is black or grey dorsally, merging to light grey on the lower flanks and whitish ventrally. An albino specimen has been recorded (Ross *et al.*, 1975). The rostrum is dark, and the baleen gum is a strongly contrasting white, which shows when the mouth is slightly open. The lower jaws are dark or light grey, merging to a pale throat area. The outer surface of the flipper is dark and the inner surface is light grey, although the tip and sometimes the leading edge are almost black. The tail flukes are black above and light grey below, with a dark marginal band along the posteroventral edge. It seems likely that young pygmy right whales are paler than adults; however, as data on colour patterns have been obtained mostly from stranded animals, it is possible that postmortem darkening is a complicating factor.

Skeleton

The pygmy right whale's skeleton is unusual in several respects. The skull has a long brain case—almost as long as the rostrum—and, viewed from above and below (Fig. 3A,B), is broadest across the squamosals, with a relatively narrow, sharply pointed rostrum. There is a strong median longitudinal ridge on the massive supraoccipitals (Fig. 4). Viewed from the side (Fig. 3C), the skull has a somewhat inverted shallow V-shape, with the apex being the skull's vertex. The upper profile is straight from condyles to vertex, and slightly curved from vertex to rostrum tip. The lower profile of the skull ahead of the orbits is curved, the outer edge of the maxillaries forming an

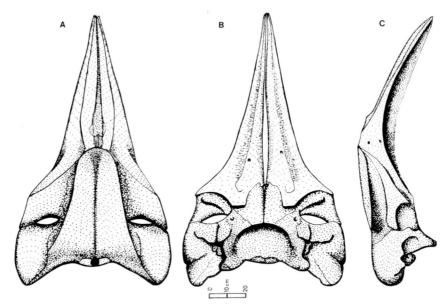

FIG. 3 (A) Dorsal, (B) ventral, and (C) lateral views of a 145-cm CBL skull of *Caperea marginata*. Rule indicates 20 cm.

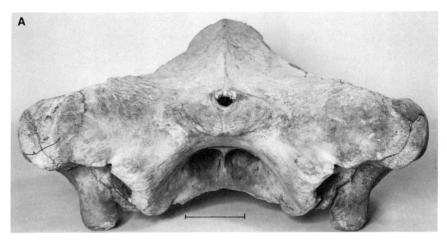

FIG. 4 (A) Posterior view of skull; (B) anterior view of skull; (C) mandible. Rules, 10 cm. (Photo courtesy of U.S. National Museum of Natural History.)

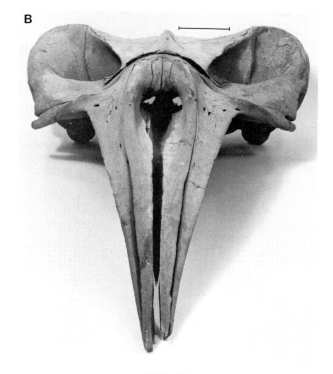

FIG. 4B

FIG. 4C

even, sweeping curve towards the orbits, with the prominent median palatal ridge being slightly flatter.

The cervical vertebrae are fused, and the thoracic and lumbar vertebrae have very broad and flat transverse processes. The ribs are also mostly very broad and flat, and the scapulae are flat and smooth. The number of vertebrae (based on seven specimens) ranges as follows: cervical, 7; thoracic, 17–18; lumbar, 1–4; caudal, 14–16; total, 40–44.

TABLE 1 Data from two pygmy right whales from the South Atlantic[a]

Organs	Female	Male
Length of body	621 cm	547 cm
Weight		
body	3 200 kg	2 850 kg
heart	7 800 g	8 600 g
lungs	9 000 g	5 500 g
	7 700 g	6 000 g
kidneys	4 700 g	4 200 g
	5 100 g	4 200 g
liver	42 000 g	34 850 g
spleen	—	400 g
pancreas	2 700 g	2 050 g
thyroid	—	370 g
right ovary[b]	260 g	
left ovary[c]	160 g	
right testis		900 g
left testis		970 g

[a]From Ivashin et al. (1972).
[b]Six corpora albicantia.
[c]Three corpora albicantia.

Body and Organ Weights

Two pygmy right whales from the South Atlantic were weighed and dissected by Ivashin et al. (1972). Data from these specimens are given in Table 1.

Distribution

The pygmy right whale lives in the Southern Hemisphere between about 31° and 52°S. Most records of this species are from strandings in South Africa, southern Australia, and New Zealand (Ross et al., 1975; Aitken, 1971; Davies and Guiler, 1957; Gaskin, 1968; Guiler, 1978; McManus et al., 1984), and these, together with a few sea sightings and catch records (Ivashin et al., 1972; Budylenko et al., 1973), indicate a circumpolar distribution approximately within the sea surface isotherms, 5–20°C (Fig. 5).

Strandings have occurred in all months throughout the species' range, but in the South African, South Australian, and Tasmanian regions, most have been reported in spring and summer—28 in the period September to February, 11 between March and August. This could indicate an inshore movement of pygmy right whales in spring, but the number of dated occurrences is too small to indicate a definite migration pattern. Besides, spring and summer are times when coastal observation is at its maximum. That pygmy right whales have been sighted far offshore at least indicates that they are not confined to coastal waters of the temperate and sub-Antarctic Southern Hemisphere.

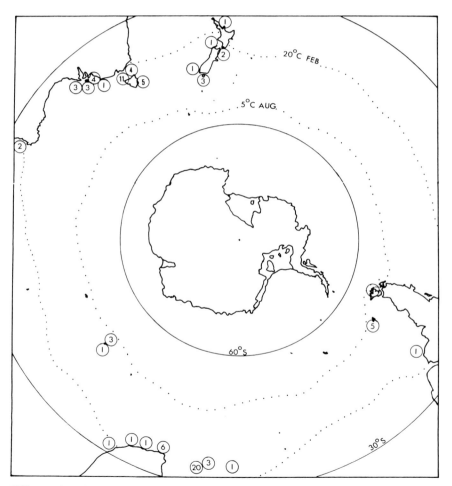

FIG. 5   Southern Hemisphere distribution of *Caperea marginata*. Numbers refer to records of individuals sighted or stranded at approximate localities indicated.

Reproduction

Only minimal information is available on reproduction in this species of whale, but it is thought that it may have an extended breeding season. Pregnant females 600–645 cm long have stranded on the Tasmanian coast in June and September, and have been captured in the South Atlantic ocean in November and December. The foetuses from these females ranged from 24 to 60 cm in length and 230 to 600 g in weight (Ivashin et al., 1972; Munday et al., 1982). The size of pygmy right whales at first maturity is not known, but a 621 cm female specimen bore the functioning corpus luteum of pregnancy and up to six corpora albicantia in one ovary (Ivashin et al., 1972).

The length of pygmy right whales at birth is also unknown, but the 2.7–3.5 m length juveniles so far recognised have not been newborn, and calculations based on data from other baleen whales indicates lengths of 1.6–2.2 m at birth, and 1.9–3.5 m at weaning (Ross et al., 1975). A 2.0 m fetus, regarded as "full term," was reported from a 6.0 m female by McManus et al. (1984). The occurrence of juveniles or subadults in coastal situations in spring and summer has been interpreted as a possible inshore influx of young whales after weaning.

Feeding

The stomach contents of two pygmy right whales caught in the South Atlantic consisted of copepods of the genus *Calanus*, a food item consistent with the fine nature of the baleen plate fringes, an adaptation for filtering small planktonic organisms.

Behaviour

The pygmy right whale has rarely been observed at sea. One suggested reason for this is that the whales spend long periods submerged, even resting on the sea floor (hence the wide, flat, viscera-supporting ribs). Observations of these whales at sea and of one in captivity have shown, however, that they are not prolonged divers. Diving times of between 40 sec and 4 min have been recorded. Nevertheless, they apparently do not spend long periods exposed on the surface, and when they do emerge, the dorsal fin is often not shown, the blow is small and hardly visible, and the animals do not indulge in spectacular surface behaviour as has been well documented for right whales and some balaenopterids. This inconspicuous surfacing behaviour and the possibility that sightings of pygmy right whales and minke whales

(*Balaenoptera acutorostrata*) could be confused at sea may account for the rarity of observations.

Pygmy right whales have been seen swimming alone or in pairs, in gams of five to eight animals over an area of 2 to 3 square miles (Ivashin *et al.*, 1972). They are slow swimmers; in two independent observations the whales did not exceed 5 knots. The swimming behaviour of one individual was filmed underwater and the animal showed extensive flexing of the entire body, with waves of motion passing from head to tail, increasing in amplitude along the body (Ross *et al.*, 1975).

These whales have been observed swimming with or near other species of cetaceans—white-sided dolphins, *Lagenorhynchus* sp.; pilot whales, *Globicephala* sp.; and a sei whale, *Balaenoptera borealis*, with calf.

Diseases and Predators

Crescentic and oval scars, sores, and wounds, attributed to small squaloid sharks of the genus *Isistius*, have been recorded from the flanks of pygmy right whales. The specimen filmed underwater had strange, rakelike, and radiating scars of unknown origin on its flanks and belly.

Acknowledgements

P. F. Aitken, R. H. Green, and R. M. Warneke supplied unpublished data on strandings in southern Australia, and N. F. Herron photographed the Port Underwood, New Zealand, pygmy right whale.

References

Aitken, P. F. (1971). Whales from the coast of South Australia. *Trans. R. Soc. South Aust.* **95**(2), 95–103.
Budylenko, G. A., Pantilov, B. G., Pakhomova, A. A., and Sazhinov, E. G. (1973). New data on pygmy right whales *Neobalaena marginata* (Gray, 1848). *Tr. Atl. Nauchno-Issled. Inst. Rybn. Khoz. Okeanogr.* **51**, 122–132.
Davies, J. L., and Guiler, E. R. (1957). A note on the pygmy right whale, *Caperea marginata* Gray. *Proc. Zool. Soc. London* **129**(4), 579–589.
Gaskin, D. E. (1968). The New Zealand Cetacea. *Fish. Res. Bull. N.Z.* **1**, 1–92.
Guiler, E. R. (1961). A pregnant female pygmy right whale. *Aust. J. Sci.* **24**, 297–298.
Guiler, E. R. (1978). Whale strandings in Tasmania since 1945 with notes on some seal reports. *Pap. Proc. R. Soc. Tasmania* **112**, 189–213.

Ivashin, M. V., Shevchenko, V. I., and Yuchov, V. L. (1972). Karlikovyi gladkii kit *Caperea marginata* (Cetacea). *Zool. Zh.* **51,** 1715–1723.

McManus, T. J., Wapstra, J. E., Guiler, E. R., Munday, B. L., and Obendorf, D. L. (1984). Cetacean strandings in Tasmania from February 1978 to May 1983. *Pap. Proc. R. Soc. Tasmania* **118,** 117–135.

Munday, B. L., Green, R. H., and Obendorf, D. L. (1982). A pygmy right whale *Caperea marginata* (Gray, 1846) stranded at Stanley, Tasmania. *Pap. Proc. R. Soc. Tasmania* **116,** 1–3.

Ross, G. J. B., Best, P. B., and Donnelly, B. G. (1975). New records of the pygmy right whale (*Caperea marginata*) from South Africa, with comments on distribution, migration, appearance and behaviour. *J. Fish. Res. Board Can.* **32,** 1005–1017.

Index

Abundance, *see under* species
Age determination, 112, 219, 331
Aggression, 259
Albinism, 347
Anal tonsil, 70
Anatomy; *see under* species for skull, skeleton, and under organ
Aphrodisiac, dugong tears, 23

Balaena mysticetus, 305–354
  abundance, 319–321
  age, 331
  anatomy, 312–316
  baleen, 311, 331
  behaviour, 321–326
    feeding, 328
    social, 323–326
  birth, 329
  blood, 313
  brain, 313
  calves, 310, 324, 326, 329
  coloration, 308–310
  dimensions, 306–308
  distribution, 316–319
  diving, 326–327
  external characteristics, 306–311
  feeding and food, 316, 328
  genus, 305–306

growth, 330
lactation and milk, 330
mating, 329
maturity, 308, 330
migration, 318–319, 321–323, 329
morphology, 306–311
morphometrics, 306
parasites, 332
population studies, 319–320
predators, 331
products, 311–312
reproduction, 329–330
skeleton, 312, 315
skull, 312–314
sound production, 327–328
species, 305–306
swimming, 326–327
weights, 306–308
Balaenoptera acutorostrata, 91–136
  abundance, 103–111
  anatomy, 116, 120
  baleen, 94–95, 120
  behaviour, 116–119
    feeding, 120
    reproductive, 112–115
    social, 117–119
  birth, 112–113
  blood, 120

B. acutorostrata—cont.
  calves, 112–113
  captivity, 124
  coloration, 94–96, 98–99
  dimensions, 93–94, 114–115
  distribution, 94–103
  diving, 119
  exploitation, 103–109
  external characteristics, 93–94
  feeding and food, 109–112, 120
  genus, 91–93
  growth, 112–115
  lactation, 113
  life history, 103–115
  mating, 112–113
  maturity, 112–115
  migration, 94–103
  morphology, 93–94
  parasites, 123–124
  population studies, 109–111
  predators, 124
  reproduction, 112–115
  skeleton, 120
  skull, 116
  sound production, 123
  species, 91–93
  swimming, 118–119
  taxonomy, 92–93
  throat grooves, 94–95
  twins, 113
  weights, 93
Balaenoptera borealis, 155–170
  abundance, 160–161
  anatomy, 162–165
  baleen, 156, 165–166
  behaviour, 167
  birth, 167
  calves, 161, 167
  coloration, 157
  dimensions, 156
  distribution, 157–159
  diving, 167
  external characteristics, 156–157
  feeding and food, 162
  genus, 155
  lactation, 161, 167
  life history, 161

  longevity, 161
  maturity, 167–168
  migration, 158–161
  morphology, 156–157
  parasites, 168
  population studies, 159, 161
  reproduction, 167–168
  skeleton, 162–163
  skull, 162
  species, 155
  throat grooves, 157
  weights, 157, 165
Balaenoptera edeni, 137–154
  abundance, 146–148
  anatomy, 142–143
  baleen, 139–143
  behaviour, 148–149
  birth, 151
  calves, 151
  coloration, 138
  dimensions, 138
  distribution, 143–146
  diving, 148
  external characteristics, 138–142
  eye, 139
  feeding and food, 150
  genus, 137–138
  lateral ridges, 141–142
  mating, 151
  maturity, 151
  migration, 143–146
  morphology, 138, 143
  population studies, 148, 152
  reproduction, 151
  skeleton, 142–143
  skull, 142
  sound production, 149–150
  species, 137–138
  throat grooves, 139
Balaenoptera musculus, 193–240
  abundance, 208–215
  age determination, 219
  anatomy, 219–222
  baleen, 197–199, 219, 222
  behaviour, 223
    feeding, 215–217
    social, 223

birth, 217
blood, 222
blubber, 222
brain, 222
calves, 217
coloration, 196–197, 215
dimensions, 195–198
diseases, 224–225
distribution, 201–207
diving, 223
exploitation, 208–210
external characteristics, 195–201
feeding and food, 215
genus, 193–195
growth, 217–219
life history, 208–215
longevity, 219
marking, 214–215
mating, 217
migration, 201–207
milk, 218
morphology, 195–201
morphometrics, 197–198
mortality, 217
parasites, 224–225
pollutants, 225
population studies, 210–214
pygmy blue whale, 194–197, 210, 218
reproduction, 222
skeleton, 222
skull, 199, 219
sound production, 224
species, 193–195
swimming, 223
throat grooves, 199
twinning, 217
weights, 195–198
Balaenoptera physalus, 171–192
  abundance, 179
  anatomy, 186
  baleen, 174, 186
  behaviour, 186
  birth, 180
  brain, 186
  calves, 180
  coloration, 173
  dimensions, 172, 175

distribution, 175
diving, 187
exploitation, 179
external characteristics, 173–174
feeding and food, 180–181, 187
genus, 171
lactation, 188
life history, 180–181
longevity, 181
mating, 180, 188
maturity, 188
migration, 175–176
parasites, 188
population studies, 176–180
reproduction, 186–187
skeleton, 185
skull, 181–184
sound production, 187
species, 171
stranding, 189
swimming, 187
throat grooves, 172–173
weights, 174, 185
Baleen, *see under* species
Behaviour, feeding, reproductive, and social, *see under* species
  bubble, 259–261
  epimeletic, 81
  friendly, 81
  headstanding, 296
  spyhopping, 78
Birth, *see under* species
Biscayan right whale, 276
Blood, 50–51, 120, 222, 313
Blubber, 222, 248
Blue whale, 193–240, *see also Balaenoptera musculus*
Boltering, 78
Bonnet, 278–279
Bowhead, 305–354, *see also Balaena mysticetus*
Bradycardia, 50
Brain, 15, 51–52, 69, 186, 222, 249, 287, 313
Bread eating, by manatee, 54
Bryde's whale, 137–154, *see also Balaenoptera edeni*

358

Calves, *see under* species
Caperea marginata, 345–354
  anatomy, 347–350
  baleen, 347
  behaviour, 352–352
  birth, 352
  calves, 352
  captivity, 352
  coloration, 347
  dimensions, 345–347
  distribution, 350–351
  diving, 352
  external characteristics, 345–347
  feeding and food, 352
  genus, 345
  growth, 352
  maturity, 352
  migration, 351
  morphology, 345–347
  reproduction, 352
  skeleton, 347–349
  skull, 347–349
  species, 345
  stranding, 350
  swimming, 353
  weights, 350
  wounds, 353
Captivity, 21, 61–63, 85, 124, 352
Cardiovascular system, 17, 50–51, 186, 222, 313
Chevron bones, 143
Chromosomes, 120, 165, 186, 313
Coloration, *see under* species
Common names; *see* Whales and under species
Common rorqual, 171–192
Conservation, 23–25, 42
Copulation, *see* mating under species
Curious whales, 81

Dentition, 13–14, 45–46
Digestive system, 18–19, 47–50, 222
Dimensions, *see under* species
Diseases, 22–23, 61–63, 83, 224–225, 267, 332, 353

Distribution, *see under* species
Diving, *see under* species
Dugong dugon, 1–31
  abundance, 6–9
  anatomy, 12–19
  behaviour, 19–21
    feeding, 19–20
    reproductive, 22
    social, 19
  birth, 22
  brain, 15
  calves, 22
  captivity, 21
  coloration, 2
  dentition, 13–14
  dimensions, 6
  diseases, 22–23
  distribution, 6–9
  diving, 20–21
  exploitation, 23–25
  external characteristics, 2–6
  food, 19–20
  genus, 1–2
  growth, 6
  kidney, 17
  lactation, 21–22
  life history, 21–23
  locomotion, 20
  longevity, 21–22
  mating, 22
  morphology, 2–6
  morphometrics, 6
  muzzle, 2–3
  parasites, 23
  population studies, 6–11
  protection, 23–25
  reproduction, 21–23
  size, 6
  skeleton, 14–15
  skull, 12–13
  social behaviour, 19
  sound production, 19
  species, 1–2
  stomach, 18–20
  swimming, 20
  teeth, 13–14

tusks, 14
weights, 6
Dugongidae, 1–2
Duodenum, 18, 50

Ear, 15, 222, 313
Endocrine organs, 18, 51, 120, 186, 222
Epiglottis, 50
Epimeletic behaviour, 81
Eschrichtius robustus, 67–90
  abundance, 74
  anatomy, 69–71
  baleen, 68, 80
  behaviour, 77–82
    feeding, 78–80
    friendly, 81
    reproductive, 81
    social, 78, 81
  birth, 82
  brain, 69
  calves, 69, 80
  captivity, 85
  coloration, 68
  dimensions, 69
  diseases, 83
  distribution, 70
  diving, 77–78
  external characteristics, 68–69
  feeding and food, 78–80
  genus, 67–68
  growth, 69
  lactation, 82
  life history, 82–83
  locomotion, 77
  mating, 81–83
  migration, 73, 76–77
  morphometrics, 69
  parasites, 83
  population studies, 73–77
  reproduction, 82–83
  scent gland, 69
  skeleton, 69
  skull, 69
  social behaviour, 78, 81
  sound production, 81–82
  species, 67–68
  stomach, 80
  swimming, 77
  weights, 69
Eubalaena glacialis and *E. australis*, 275–304
  abundance, 288–289
  anatomy, 285–287
  baleen, 279–280, 285–286
  behaviour, 291–297
    headstanding, 296
  birth, 297
  blowhole, 282
  brain, 287
  calves, 281
  coloration, 276–278
  dimensions, 280–282
  distribution, 289–290
  external characteristics, 276–283
  feeding and food, 297
  genus, 275–276
  gonads, 286
  kidneys, 287
  lactation and milk, 298
  mating, 297–299
  maturity, 297
  migration, 289
  morphology, 276–283
  parasites, 299–300
  population studies, 288–289
  predators, 295
  reproduction, 297–299
  skeleton, 283–284
  skull, 283
  sound production, 291–293
  species, 275–276
  weights, 279–281
Excrescences, 278–281
Exploitation, 23, 42–43, 103–109, 160, 179, 208–210, 255–257
Eye, 52–53, 139, 222

Feeding and food; *see under* species
Fin whale, 171–192; *see also Balaenoptera physalus*
Friendly behaviour, 81

Gestation periods, 22, 59, 112, 151, 167, 188, 217, 265, 298, 329
Gonads, 286, see also Ovary, Testis
Gray whale, 69–90, see also *Eschrichtius robustus*
Greenland right whale, 395–354, see also *Balaena mysticetus*
Growth, see under species

Headstanding, 296
Hearing, 15, 52, 313
Heart, 17, 50, 165, 185, 313, 350
Humpback whale, 241–273; see also *Megaptera novaeangliae*

Integument, 17, 37, 186, 287, 312–313
Intestine, 18–19, 47–50

Karyotypes, 120, 165, 186, 313
Kidney, 17, 51, 165, 186, 287, 350

Lactation, see under species
Larynx, 50
Lateral ridges, 141–142
Lens, 331
Lesser rorqual, 91–136
Life history, see under species
Locomotion, 20, 55, 77, see also swimming
Longevity, 21–22, 161, 181, 219, 242
Lungs, 15–16, 50, 186, 222

Manatee, 33–66; see also *Trichechus manatus*
Marking (tracking), 102, 113, 204, 214–215, 250–252
Mating, 22, 57–58, 81–83, 112–115, 151, 167–168, 217, 257, 266, 297, 329
Maturity, sexual, 22, 58, 82, 112–115, 151, 167–168, 188, 217–219, 242, 247, 266, 297, 308, 330, 352
Megaptera novaeangliae, 241–273
  abundance, 255–257
  anatomy, 245–248
  baleen, 243
  behaviour, 257–263
  feeding, 257–263
  reproductive, 257, 266
  social, 258, 265
  spyhopping, 257–259
  birth, 265–266
  brain, 249
  blubber, 248
  calves, 265–266
  coloration, 243–244
  dimensions, 242
  diseases, 267
  distribution, 249–255
  diving, 263
  exploitation, 255–257
  external characteristics, 242–245
  feeding and food, 257–263
  genus, 241
  growth, 265
  lactation and milk, 265
  life history, 255–265
  longevity, 242
  marking (tracking), 250–252
  mating, 257, 266
  maturity, 242, 247, 266
  migration, 249–252
  morphology, 242–245
  parasites, 267
  pollutants, 267
  population studies, 252–257
  predators, 267
  reproduction, 249, 265
  skeleton, 247–248
  skull, 245–247
  sound production, 264
  species, 241
  swimming, 263
  ventral grooves, 243
  ventral pouch, 248
  weights, 242–243, 250
Migration, see under species
Milk, 218, 265
Minke whale, 91–136; see also *Balaenoptera acutorostrata*
Morphology, see under species
Mortality, 217, see also Diseases

INDEX 361

Musculature, 15, 46
Muzzle, 2–3, 17

Nares, 156, 313; *see also* external characteristics under species
Nervous system, 15, 51–52, 69, 120, 165, 186, 222, 249, 287, 313
North Atlantic right whale, 276
North Pacific right whale, 276

Organs, *see* heart, stomach, etc.
Ovary, 18, 51, 168, 250, 266, 286, 352

Parasites, 23, 62–63, 83, 123–124, 168, 188, 244, 267
Penis, 18, 83, 139, 266, 283
Placenta, 18
Pollution, 42, 225, 267
Population studies, *see under* species
Predators, 43, 124, 267, 295, 331
Prostate, 18, 51
Protection, *see under* species, abundance, exploitation
Pygmy blue whale, 194–195, 196–199, 207, 210, 218
Pygmy right whale, 345–354

Razorback, 172
Reproduction and reproductive behavior, *see under* species
Respiratory system, 15–16, 50, 186, 222, 313
Retia, 17, 50–51, 186
Right whales, 275–304, 345–354
  Biscayan, 276
  black, 275
  Greenland, 305–354
  North Atlantic, 276
  North Pacific, 276
  pygmy, 345–354
  southern, 276
Rorquals
  common, 171–192
  lesser, 91–136
  Sibbald's, 193–240

Scent gland, 69
Sea cow, 1, 34
Sei whale, 155–170, *see also Balaenoptera borealis*
Serology, *see* blood *under* species
Sex ratio, 180
Sibbald's rorqual, 193–240
Sinus hairs, 17, 165
Skeleton and skull, *see under* species
Smell, 53
Social behaviour, *see under* species
Sound production, 19, 55–56, 81–82, 149–150, 123, 187, 224, 264, 327
Southern right whale, 276
Spyhopping, 78, 257–259
Stomach, 18–20, 47–49, 80, 313
Stranding, 189, 350
Sulphur bottom, 193–240
Swimming, 20, 54–55, 77, 118, 176, 187, 223, 263, 326, 353

Tags, *see* Marking
Taste, 53
Tears, dugong, 23
Teeth
  Dugong, 13–14
  Trichechus, 45–46
Testis, 17, 51, 83, 250, 286
Thermoregulation, 17
Throat grooves, 68, 92, 94–95, 139, 157, 172–173, 199, 241, 243
Tongue, 47, 199, 286, 312
Trachea, 50
Trichechidae, 33–34
Trichechus inunguis, 33–66
Trichechus manatus, 33–66
  abundance, 41
  anatomy, 45–53
  behaviour, 53–58
    reproductive, 57–59
    social, 56–58
  birth, 58–59
  calves, 59
  coloration, 36
  dentition, 45–46
  dimensions, 37

T. manatus—cont.
  diseases, 61–63
  distribution, 38–45
  diving, 50, 54
  external characteristics, 35–38
  feeding and food, 44, 53–54
  genus, 33–34
  kidney, 51
  lactation, 59
  life history, 41–58
  locomotion, 55
  mating, 57–58
  maturity, 58
  migration, 43–45
  morphology, 35–38
  parasites, 62–63
  population studies, 42–44
  protection, 42–43
  reproduction, 58–61
  skeleton and skull, 45
  sound production, 55–56
  species, 33–34
  stomach, 48
  swimming, 54–55
  teeth, 45–46
  weights, 37–38
Trichechus senegalensis, 33–66
Tusks, 14
Twinning, 113, 217

Urine, 45
Uterus, 18, 51

Vertebrae, 46, 69, 120, 142, 162, 185, 222, 247, 283, 312
Ventral grooves, 68, 92, 94–95, 139, 157, 172–173, 199, 241, 243
Ventral pouch, 165, 248
Viruses, 85, 332
Vision, 15, 53

Weights, *see under* species
Whales
  common names, *see also* under species
    Biscayan right whale, 276
    black right whale, 275
    blue whale, 193–240
    bowhead, 305–354
    Bryde's whale, 137–154
    California gray whale, 67–90
    common rorqual, 171–192
    fin whale, 171–192
    gray whale, 67–90
    Greenland right whale, 305–354
    humpback whale, 241–273
    lesser rorqual, 91–136
    little piked whale, 91–136
    minke whale, 91–136
    North Atlantic right whale, 276
    North Pacific right whale, 276
    pygmy blue whale, 194–199, 210, 218
    pygmy right whale, 345–354
    razorback, 172
    right whales, 275–304
    rorquals, 91, 171, 193
    sei whale, 155–170
    Sibbald's rorqual, 193–240
    southern right whale, 276
    sulphur bottom, 194
  scientific names
    Balaena mysticetus, 305–354
    Balaenoptera acutorostrata, 91–136
    B. borealis, 155–170
    B. edeni, 137–154
    B. musculus, 193–240
    B. physalus, 171–192
    Caperea marginata, 345–354
    Eschrichtius robustus, 67–90
    Eubalaena australis, 275–304
    E. glacialis, 275–304
    Megaptera novaeangliae, 241–273